SOCIETY FOR EXPERIMENTAL BIOLOGY
SEMINAR SERIES: 51

OXYGEN TRANSPORT IN BIOLOGICAL SYSTEMS

SOCIETY FOR EXPERIMENTAL BIOLOGY SEMINAR SERIES

A series of multi-author volumes developed from seminars held by the Society for Experimental Biology. Each volume serves not only as an introductory review of a specific topic, but also introduces the reader to experimental evidence to support the theories and principles discussed, and points the way to new research.

2. Effects of pollutants on aquatic organisms. *Edited by A.P.M. Lockwood*
6. Neurones without impulses: their significance for vertebrate and invertebrate systems. *Edited by A. Roberts and B.M.H. Bush*
8. Stomatal physiology. *Edited by P.G. Jarvis and T.A. Mansfield*
10. The cell cycle. *Edited by P.C.L. John*
11. Effects of disease on the physiology of the growing plant. *Edited by P.G. Ayres*
12. Biology of the chemotactic response. *Edited by J.M. Lackie and P.C. Williamson*
14. Biological timekeeping. *Edited by J. Brady*
15. The nucleolus. *Edited by E.G. Jordan and C.A. Cullis*
16. Gills. *Edited by D.F. Houlihan, J.C. Rankin and T.J. Shuttleworth*
17. Cellular acclimatisation to environmental change. *Edited by A.R. Cossins and P. Sheterline*
19. Storage carbohydrates in vascular plants. *Edited by D.H. Lewis*
20. The physiology and biochemistry of plant respiration. *Edited by J.M. Palmer*
21. Chloroplast biogenesis. *Edited by R.J. Ellis*
23. The biosynthesis and metabolism of plant hormones. *Edited by A. Crozier and J.R. Hillman*
24. Coordination of motor behaviour. *Edited by B.M.H. Bush and F. Clarac*
25. Cell ageing and cell death. *Edited by I. Davies and D.C. Sigee*
26. The cell division cycle in plants. *Edited by J.A. Bryant and D. Francis*
27. Control of leaf growth. *Edited by N.R. Baker, W.J. Davies and C. Ong*
28. Biochemistry of plant cell walls. *Edited by C.T. Brett and J.R. Hillman*
29. Immunology in plant science. *Edited by T.L. Wang*
30. Root development and function. *Edited by P.J. Gregory, J.V. Lake and D.A. Rose*
31. Plant canopies: their growth, form and function. *Edited by G. Russell, B. Marshall and P.G. Jarvis*
32. Developmental mutants in higher plants. *Edited by H. Thomas and D. Grierson*
33. Neurohormones in invertebrates. *Edited by M. Thorndyke and G. Goldsworthy*
34. Acid toxicity and aquatic animals. *Edited by R. Morris, E.W. Taylor, D.J.A. Brown and J.A. Brown*
35. Division and segregation of organelles. *Edited by S.A. Boffey and D. Lloyd*
36. Biomechanics in evolution. *Edited by J.M.V. Rayner and R.J. Wootton*
37. Techniques in comparative respiratory physiology: an experimental approach. *Edited by C.R. Bridges and P.J. Butler*
38. Herbicides and plant metabolism. *Edited by A.D. Dodge*
39. Plants under stress. *Edited by H.G. Jones, T.J. Flowers and M.B. Jones*
40. *In situ* hybridisation: application to developmental biology and medicine. *Edited by N. Harris and D.G. Wilkinson*
41. Physiological strategies for gas exchange and metabolism. *Edited by A.J. Woakes, M.K. Grieshaber and C.R. Bridges*
42. Compartmentation of plant metabolism in non-photosynthetic tissues. *Edited by M.J. Emes*
43. Plant growth: interactions with nutrition and environment. *Edited by J.R. Porter and D.W. Lawlor*
44. Feeding and the texture of foods. *Edited by J.F.V. Vincent and P.J. Lillford*
45. Endocytosis, excytosis and vesicle traffic in plants. *Edited by G.R. Hawes, J.O.D. Coleman and D.E. Evans*
46. Calcium, oxygen radicals and cellular damage. *Edited by C.J. Duncan*
47. Fruit and seed production: aspects of development, environmental physiology and ecology. *Edited by C. Marshall and J. Grace*
48. Perspectives in plant cell recognition. *Edited by J.A. Callow and J.R. Green*
49. Inducible plant proteins: their biochemistry and molecular biology. *Edited by J.L. Wray*
50. Plant organelles: compartmentation of metabolism in photosynthetic cells. *Edited by A.K. Tobin*

The study of oxygen transfer in animals and plants represents a large and diverse field of research, covering not only different levels of organisation ranging from the whole organism through to subcellular organelles, and different mechanisms such as convection and simple or facilitated diffusion, but also, of necessity, invoking diverse methodological approaches. This book offers a range of perspectives on the modelling of oxygen transport, covering the broadest possible spectrum of disciplines within the field and emphasising the different methodological approaches used, so encouraging the communication of ideas from individual, often very specialised fields, to a wider audience. The volume presents an authoritative and stimulating review of theoretical and methodological developments in this exciting field of study and concludes with a critical evaluation of the most recent data, providing a unique synthesis from which the research can move forward.

OXYGEN TRANSPORT IN BIOLOGICAL SYSTEMS: Modelling of pathways from environment to cell

Edited by

S. Egginton and **H. F. Ross**

Department of Physiology, University of Birmingham

Published by the Press Syndicate of the University of Cambridge
The Pitt Building, Trumpington Street, Cambridge CB2 1RP
40 West 20th Street, New York, NY 10011–4211, USA
10 Stamford Road, Oakleigh, Victoria 3166, Australia

First published 1992

Printed in Great Britain by Redwood Press Limited, Melksham, Wiltshire

A catalogue record of this book is available from the British Library

Library of Congress cataloguing in publication data

Oxygen transport in biological systems: modelling of pathways from
environment to cell / edited by S. Egginton and H.F. Ross.
p. cm.–(Seminar series / Society for experimental biology; 51)
Includes index.
1. Oxygen–Physiological transport–Models. 2. Respiration–Models. I.
Egginton, S. II. Ross, H. F. III. Series: Seminar series (Society for
Experimental biology (Great Britain)); 51.
QP121.O87 1992
574.1'2—dc20 92–15898 CIP

ISBN 0 521 41488 1 hardback

WV

Contents

Contributors

ALEXANDER, R.McN.
Department of Pure and Applied Biology, University of Leeds, Leeds LS2 9JT, UK.
ARMSTRONG, W.
Department of Applied Biology, University of Hull, Hull HU6 7RX, UK.
BECKETT, P.M.
Department of Applied Mathematics, University of Hull, Hull HU6 7RX, UK.
BELLELLI, A.
CNR Centre of Molecular Biology and Department of Biochemical Sciences, University 'La Sapienza', Piazza A. Moro 5, 00185 Roma, Italy.
DI PRISCO, G.
Institute di Biochimica delle Proteine ed Enzimologia, 80125 Napoli, Italy.
EGGINTON, S.
Department of Physiology, University of Birmingham Medical School, Birmingham B15 2TT, UK.
GROEBE, K.
Institut für Physiologie und Pathophysiologie, Johannes Gutenberg–Universität Mainz, Saarstr. 21, 6500-Mainz, Germany.
HOOFD, L.
Department of Physiology, Faculty of Medicine, Catholic University of Nijmegen, 6500 HB Nijmegen, The Netherlands.
MAYHEW, T.M.
Department of Human Morphology, Queen's Medical Centre, University of Nottingham, Nottingham NG7 2UH, UK.
PERRY, S.F.
Respiratory Physiology Research Group, University of Calgary, 3330 Hospital Drive NW, Calgary, Alberta T2N 4N1, Canada.

ROSS, H.F.
Department of Physiology, University of Birmingham Medical School, Birmingham B15 2TT, UK.
SHELTON, G.
School of Biological Sciences, University of East Anglia, Norwich NR4 7TJ, UK.
VAN BEEK, J.H.G.M.
Laboratory of Physiology, Free University, van der Boechorststraat 7, 1081 BT Amsterdam, The Netherlands.
YOUNG, I.S.
Department of Pure and Applied Biology, University of Leeds, Leeds LS2 9JT, UK.

Preface

This volume presents a sampler of the types of models which have been employed to describe various aspects of the transfer of oxygen from the environment to a cell or organelle. This field is very large and quite diverse, so no model or group of workers can cover its full extent, and hence contributors of the 10 chapters approach the problem from a range of perspectives and utilise a variety of methodologies.

The first contribution (Shelton) offers an overview of model application to the respiratory system in general, and describes the contrast between steady-state and non-steady-state situations. The second (Alexander & Young) shows how simple mechanical analogies may be used to gain insight into the limiting components of the respiratory system. The following two contributions (Perry, Mayhew) show how structural limits to gas transfer may be analysed; from a global consideration of the respiratory system, to a more local example of how morphometric analysis may help explain the relevance of physiological adaptations which are not amenable to direct experimentation. Next, Bellelli & di Prisco take the fundamental problem one stage further, by providing an analysis of how oxygen transfer is accomplished at the molecular level by the use of respiratory pigments.

We then move on to contributions where direct analogies are less evident, and purely theoretical analyses play a larger part. Van Beek gives an example of how an abstract mathematical concept has been used to explore whether such a framework may be useful in explaining physiological phenomena. Next, the need for a realistic anatomical basis for models of intramuscular diffusion is addressed by use of geometric constructs that lend themselves to unbiased sampling and various elaborations (Egginton & Ross). From this prerequisite for implementing mathematical models, we see its application in a review of how structural and functional inhomogeneities affect oxygen diffusion at a cellular level (Hoofd), and an examination of which factors may ultimately play a limiting role in this process (Groebe). Finally, we have a comprehensive overview of gas transfer in plants, which emphasises many parallels in the fundamental processes common to oxygen transfer wherever it occurs (Beckett & Armstrong).

This volume arose out of a symposium organised on behalf of the Respiration Group of the Society for Experimental Biology, held at the University of Birmingham in April 1991.

S. Egginton
H. F. Ross
Birmingham

G. SHELTON

Model applications in respiratory physiology

Introduction

Models give structure to complex relationships by isolating and simplifying important features of a system and, at their least creative level, act as a useful framework for existing knowledge. It is clearly possible for conceptual models to have properties or components that cannot actually be physically realised. Analogues can be used in conceptual models to yield useful relationships or propositions though, as Dainty (1960) wrote, 'no one should use an electrical analogue who is not very familiar with electricity'. This warning applies across a much wider range of modelling possibilities than the electrical one. On the whole, biologists are unhappy with purely imaginary concepts and prefer their models to have some recognisable parallels with structural and physical reality. Confusion between purely analytical concepts and physical reality can be a major problem in modelling (Pringle, 1960), particularly when the model is in an unfamiliar idiom. Imaginary concepts can be endowed with a physical reality not intended in the original model. Despite these problems there is no doubt that models are of enormous value in understanding complex relationships (Riggs, 1963). The most valuable models are predictive. It is difficult to see any worthwhile distinction between such models and hypotheses.

Physical models

The most straightforward models are those based directly in the same frame of reference as the system being modelled. Thus, mechanical systems can most simply be modelled mechanically albeit with considerable simplification. Models that represent reality closely and are, at the same time, simple enough to allow actual construction are not common in biology, except perhaps in studies of animal locomotion where they have sometimes proved to be valuable. Modelling of this type is often costly and complicated. In addition, the problems of scaling and design ensure

Society for Experimental Biology Seminar Series 51: *Oxygen Transport in Biological Systems*, ed. S. Egginton & H.F. Ross.

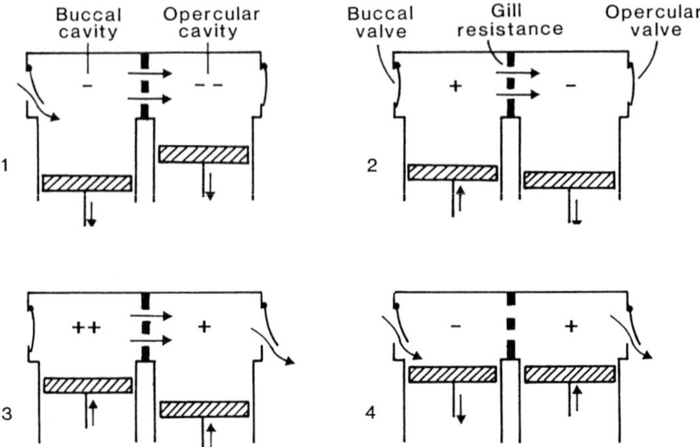

Fig. 1. Model of the double pumping mechanism for ventilating the gills of fishes. The two major phases of the breathing cycle are (1) in which the opercular pump dominates and water is drawn through the gills, and (3) in which the buccal pump dominates and forces water through the gills. During the transition phases, (2) and (4), water flow through the gills slows briefly.

that it is rarely a productive or a feasible option. For these reasons, the total transport system for oxygen from environment to cell, though often considered in the modelling idiom (e.g. Weibel, 1984), has never actually been modelled in this way. However, components within the total system, especially those related to the mechanics of ventilation, do lend themselves to physical modelling and, ultimately, to actual construction. Ventilation pumps typically consist of a source of energy, some passive elements that enable the energy source to move the respiratory medium and, finally, the medium itself which is a fluid contained within a system of channels.

Mechanics of ventilation

An example of such a model is the two-pump system of gill ventilation in teleost fish (Fig. 1) proposed by Hughes & Shelton (1958, 1962). This was constructed and produced continuous water flow over the gills by means of buccal and opercular pumps operating with a slight phase difference. The evidence for its design came largely from pressure measurements made in the buccal and opercular cavities of fish. In the animal the two pumps were clearly not independent because of skeletal linkages but equally were not rigidly linked. The full extent of their interrelationships, affecting parameters such as the volume and timing of the pumping strokes, the relative volumes of the chambers, and the hydrodynamic properties of the chambers and gills, were not well established and have still not been rigorously determined for any species of fish.

The fish model illustrates many of the techniques and the consequences of modelling. It is clearly one of the 'artificial contraptions' discussed by Weibel (1984). As many structural complications as possible are eliminated, either because they are not understood or because the physiological problems are thus simplified, amenable to manipulation and even to construction. The inevitable problem such simplification creates is how far the model can now reflect reality and produce useful insights into the breathing process in the real animal. The fish model clearly has some didactic value and illuminates the basic mechanism. The concepts it illustrates prompted analyses of the work done in breathing (Alexander, 1967). However, the simplifications inherent in its design meant that the value of the model in studying pressure, volume and flow relationships as they might be in fish was quite limited.

Similar rather simplified mechanical representations of ventilation mechanisms have been proposed for a wide variety of animals, though the models have usually remained in a diagrammatic rather than a more fully analysed physical form. In invertebrates, the paddle-like oscillations of the scaphognathite of decapod crustaceans (Hughes *et al.*, 1969; Wilkens & McMahon, 1972), and the rhythmic volume changes of the mantle cavity in cephalopod molluscs (Wells & Wells, 1982; Wells & Smith, 1985) are understood and could be modelled at this level. In amphibians, the intermittent lung inflations caused by a buccal pump linked to active nostril and glottal valves has been extensively described (De Jongh & Gans, 1969; West & Jones, 1975; Brett & Shelton, 1979; Vitalis & Shelton, 1990). A variety of mechanisms exists in reptiles, from movements of the girdles and limbs in chelonians (Gans & Hughes, 1967), through the pumping action of the visceral mass and liver moved by abdominal and diaphragmatic muscles in crocodiles (Gans & Clark, 1976), to flank movements produced by the intercostal musculature in lizards and snakes (Templeton & Dawson, 1963; Rosenberg, 1973). The mechanics of bird lung ventilation, using distensible air sacs to ventilate the rigid lungs, are also clearly established (Bretz & Schmidt-Nielsen, 1972; Brackenbury, 1973).

Though all of these systems are well enough understood for simple mechanical models to be developed, the fact that none has actually been built suggests that the returns for such modelling do little to justify the effort involved. They can be more fruitfully studied as mechanical abstractions to which the principles of statics and dynamics apply.

Even in mammals, in which the reciprocating bellows action of the chest wall has been extensively studied, the modelling process has stopped short of actual construction. The resistive, elastic and inertial forces opposing motion in the respiratory musculature have been quantified by measuring volumes and pressures in the lungs and chest wall. The system can be modelled in simple mechanical assemblies of masses, springs and dashpots. Some applications of this type of model appear in the chapter by Alexander & Young (this volume).

Flow in bird lungs

A final example of physical modelling, in which reality and simplification characteristic of the modelling idiom have been combined in a unique manner, is the study of unidirectional flow in bird lungs carried out by Scheid *et al.* (1972). Isolated pieces of the bronchial regions of the lungs were chemically fixed and connected to a pump producing sinusoidal pressures and flows. The model exhibited the sort of unidirectional flow, especially in the dorsobronchi, that was thought to occur in the bronchi and parabronchi of the living animal. Since the bronchial tubes in the model were inelastic and totally unreactive, flow rectification must have been a result of inactive valving caused by the aerodynamic properties of the tubes and their connections.

Conceptual models

This category of model is in much wider use. As the previous section shows, there is considerable overlap between conceptual and physical models. The advantage of conceptual models is that all the stages in a complex system do not need to be known, even grosser simplifications can be made and analogues can be more widely used. Logic becomes the main justification for these more abstract models though ultimately they must be tested against the living system. They tend to become increasingly mathematical and the availability of computing facilities has led to an enormous expansion in this area. Examples occur in several chapters of this book.

Steady-state models of oxygen transport

The basic model of oxygen transport from environment to cells in vertebrates (and applicable, with modifications, to many invertebrates) is shown in Fig. 2. The model has been developed in various forms either in whole (Taylor & Weibel, 1981; Piiper, 1982; Weibel, 1984) or in parts,

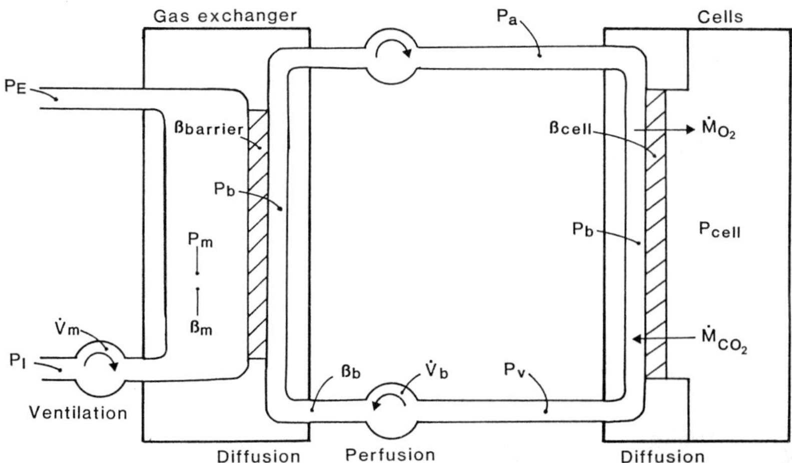

Fig. 2. Model of oxygen transport system from environment to cells comprising two convection (ventilation and perfusion) components and two diffusion (in gas exchanger and cells) components. The symbols are those used in the equations in the text and refer to the flow rate ($\dot{M}$) of O_2 or CO_2, the flow rate ($\dot{V}$) of external medium (m) or blood (b), and the partial pressure (P) and capacitance coefficient (β) for O_2 in various parts of the system. The model is based on a mammal or bird with two circulation pumps and complete separation of pulmonary and systemic circuits; deletion of the second pump makes it applicable to fish.

particularly those parts concerned with gas exchange (Piiper & Scheid, 1972, 1977) and tissue transfer (Krogh, 1922; Piiper, 1982; chapters by Hoofd and Groebe, this volume). The whole model consists of an alternating sequence of two convection and two diffusion components leading from the environment through the gas exchanger and circulatory system to the cells. Simple equations describing oxygen movement through this sequential system are as follows:

(a) Oxygen flow rate ($\dot{M}_{O_2}$) from environment into gas exchanger by convection is the product of flow rate in the medium ($\dot{V}$m) and concentration difference of oxygen between inspired (CI) and expired medium (CE). Concentration differences can be expressed as the product of the capacitance coefficient for oxygen in the medium (βm) and the difference in P_{O_2} in inspired (PI) and expired (PE) medium:

$$\dot{M}_{O_2} = \dot{V}\mathrm{m}(C\mathrm{I} - C\mathrm{E}) = \dot{V}\mathrm{m}\beta\mathrm{m}(P\mathrm{I} - P\mathrm{E}) \quad (1)$$

where β, the capacitance coefficient, is the increment of concentration per increment of partial pressure. In the gas phase, β for all gases is equal to

$1/RT$ (R = gas constant, T = absolute temperature). In water, β for a non-reactive gas such as oxygen is equal to its solubility coefficient.

(b) Oxygen flow rate through the gas exchanger from medium to blood by diffusion is proportional to the area of the diffusion barrier (A_1) and the oxygen concentration difference across it from medium to blood ($C\mathrm{m} - C\mathrm{b}$) and inversely proportional to the barrier thickness (x_1). The constant of proportionality is the diffusion coefficient (d_1) for oxygen in the tissues of the gill barrier. It is convenient to express concentration difference as the product of partial pressure difference and capacitance coefficient for O_2 in the tissues of the gill barrier ($\beta\mathrm{barrier}$). The capacitance and diffusion coefficients can be combined into a single constant, referred to as the Krogh diffusion constant (K_1):

$$\begin{aligned}\dot{M}_{O_2} &= \frac{A_1}{x_1} d_1(C\mathrm{m} - C\mathrm{b}) \\ &= \frac{A_1}{x_1} d_1 \beta\mathrm{barrier}(P\mathrm{m} - P\mathrm{b}) \\ &= \frac{A_1}{x_1} K_1(P\mathrm{m} - P\mathrm{b})\end{aligned} \tag{2}$$

(c) Oxygen flow rate from gas exchanger to capillaries in the tissues by convection in the circulating blood is the product of blood flow rate ($\dot{V}\mathrm{b}$) and oxygen concentration difference in arterial ($C\mathrm{a}$) and venous ($C\mathrm{v}$) blood. As before, the concentration difference can be expressed as the product of capacitance coefficient for oxygen in blood ($\beta\mathrm{b}$) and oxygen partial pressure difference between arterial and venous blood ($P\mathrm{a} - P\mathrm{v}$). The non-linear relationship between O_2 partial pressure and concentration, as seen in the S-shaped dissociation curve, makes this expansion more complicated than in the equations relating to the medium. An additional complication is that regulator molecules such as carbon dioxide, adenosine triphosphate and 2,3-diphosphoglycerate can modify the relationships. An assumption commonly made is that the relationship is linear and that $\beta\mathrm{b}$ is the slope of a straight line between the measured arterial and venous points on a partial pressure–concentration plot. Closer approximations such as hyperbolic curves or actual data on blood curves (Malte & Weber, 1985; Shelton & Croghan, 1988) have been used but most models contain some simplifications:

$$\dot{M}_{O_2} = \dot{V}\mathrm{b}(C\mathrm{a} - C\mathrm{v}) = \dot{V}\mathrm{b}\beta\mathrm{b}(P\mathrm{a} - P\mathrm{v}) \tag{3}$$

(d) Oxygen flow rate from capillaries to cells and mitochondria by diffusion is proportional to O_2 concentration difference and the dimensions of the tissue and cell diffusion barrier (A_2, x_2), as in the case of the gill barrier in the gas exchanger:

$$\dot{M}_{O_2} = \frac{A_2}{x_2} d_2(C\text{b} - C\text{cell})$$

$$= \frac{A_2}{x_2} d_2 \beta\text{cell}(P\text{b} - P\text{cell})$$

$$= \frac{A_2}{x_2} K_2(P\text{b} - P\text{cell}) \tag{4}$$

In a steady-state system O_2 will flow at the same rates in all four parts of the transport system. At each step the flow rate is the product of a P_{O_2} difference and a parameter which was called conductance (G) by Piiper & Scheid (1972) (Fig. 3). In the convective parts of the system, where medium and blood are both actively pumped, conductance (Gvent and Gperf, respectively) is the product of capacitance coefficient and flow rate. From Eqns 1 and 3 it follows that conductance can also be determined by dividing oxygen flow by the P_{O_2} difference in medium or blood:

$$G\text{vent} = \dot{V}\text{m}\beta\text{m} = \frac{\dot{M}_{O_2}}{(P\text{I} - P\text{E})} \tag{5}$$

$$G\text{perf} = \dot{V}\text{b}\beta\text{b} = \frac{\dot{M}_{O_2}}{(P\text{a} - P\text{v})} \tag{6}$$

The concept of conductance (originally called capacity-rate factor by Hughes & Shelton, 1962) is fundamental to the study and modelling of gas transport. Use of the term in convective transport does not, however, imply a strict electrical analogy. The determinant of current flow, equivalent to O_2 flow rate ($\dot{M}_{O_2}$/time, is potential difference in an electrical conductance, and the analogue in gas transport is P_{O_2} difference. Important determinants of O_2 movement in a convective system are the rates of flow and capacitance coefficients of the conducting medium. The electrical analogue has no equivalent for the conducting medium, such as air, water or blood, in which O_2 is transported. In many ways a better analogy than the electrical one would be one using heat exchange and transport, as will appear later.

In the diffusion regions of the model, the cross-sectional area (A_1, A_2) and the thickness (x_1, x_2) of the diffusion pathway determines the conductance of the gas exchanger (Gdiff, frequently referred to as diffusing capacity, D) and the cells (Gcell). Both the diffusion coefficients (x_1 and x_2) and the Krogh constants (K_1 and K_2) may differ slightly in the exchanger and the cells since there are small variations in tissues of different type. Again, it is possible to determine the diffusion conduc-

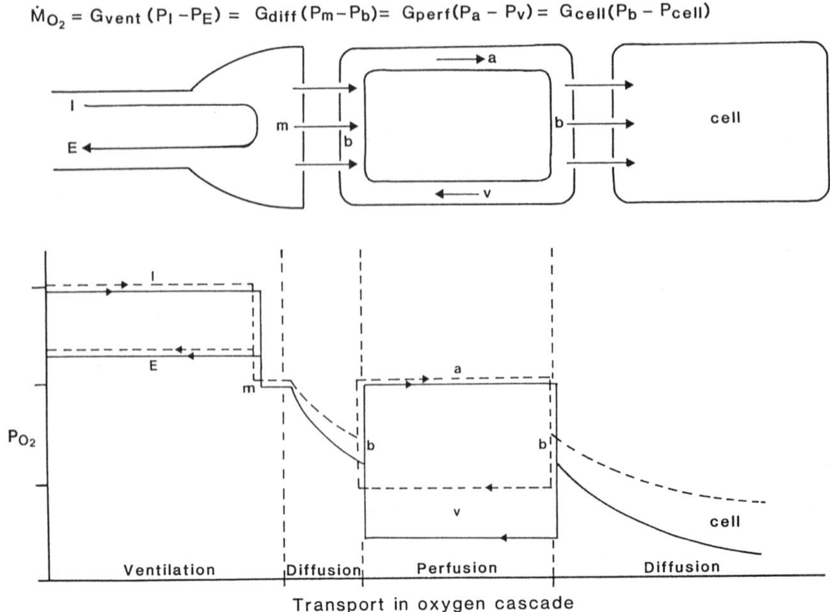

Fig. 3. Model of O_2 cascade from environment to cell showing the fall in P_{O_2} through the system and the dependence of O_2 flow rate on the product of conductance (G) and P_{O_2} difference at each stage of the cascade. The change in P_{O_2} gradient from rest (dashed line) to heavy exercise (solid line) shows that there is potential for a limited increase during activity. Since O_2 can increase approximately 10-fold most of the adjustment in the system must rely on conductance changes in all parts of the cascade (after Weibel, 1984).

tances as the quotient of O_2 consumption and mean P_{O_2} difference between medium and blood or between blood and cells, though the relationship between the two types of derivation may be complicated, as will emerge later:

$$G\text{diff} = D = \frac{A_1}{x_1} d_1 \beta\text{barrier} = \frac{A_1}{x_1} K_1 = \frac{\dot{M}_{O_2}}{(P\text{m} - P\text{b})} \tag{7}$$

$$G\text{cell} = \frac{A_2}{x_2} d_2 \beta\text{cell} = \frac{A_2}{x_2} K_2 = \frac{\dot{M}_{O_2}}{(P\text{b} - P\text{cell})} \tag{8}$$

The analogy between diffusive and electrical conductance is much more precise than in convective transport because the containing medium does not move. The only driving force in diffusive transfer is the P_{O_2} gradient.

The partial pressure differences at each stage of the O_2 transport sequence will depend on the conductance of that stage. Clearly, if conductance is high, partial pressure difference will be low and vice versa (Fig. 3). The total partial pressure difference available for O_2 transfer is that from the value in inspired medium to the value, presumably approaching zero, as O_2 acts as the terminal H^+ acceptor in the respiratory chain of the mitochondria. The way in which that total range, over which the animal has little or no active control, is disposed to the four stages of the transport sequence is of major importance in understanding the total O_2 transport process (Weibel, 1984). The P_{O_2} levels at which each stage operates are subject to regulatory processes as the activity of the organism varies (Fig. 3) but the overall limits of operation of each stage must be determined by evolutionary processes. They are also subject to adaptational change during the individual's lifetime, depending on the need for O_2 flow imposed by the levels of activity or by environmental restrictions such as low P_{O_2}.

Two rather opposing but nevertheless extremely useful concepts have emerged from such considerations. The first is that of the limiting stage (Piiper *et al.*, 1976; Piiper & Scheid, 1977; Piiper, 1982). In conditions in which demands on the transport system increase, as in exercise or hypoxia for example, relationships between stages almost certainly change because each has a different capacity to respond. It is argued that one or other of the stages will ultimately become a major limiting process though not, of course, to the exclusion of lesser limitations in other stages. In particular, diffusion limitation may increase during exercise (Piiper, 1982) or hypoxic stress. A limitation index can be calculated for each stage by comparing actual gas transport with that theoretically possible if the stage concerned had an infinitely high conductance. The second concept is what Taylor & Weibel (1981) have called the principle of symmorphosis. In a multi-stage system such as that for O_2 transport, structural design at each stage is optimised by the processes of natural selection so that the stages are all matched and adjusted precisely to the functional needs of the organism. Taylor and Weibel suggest that this matching will occur when the organism is working at its maximal rate of O_2 consumption. The implication is that all stages become equally limiting when working at the maximum activity levels of the animal because their structures have evolved to sustain that level of O_2 flow.

Interactions between the four stages of the transport sequence in both the model and the organism make optima difficult to determine, as

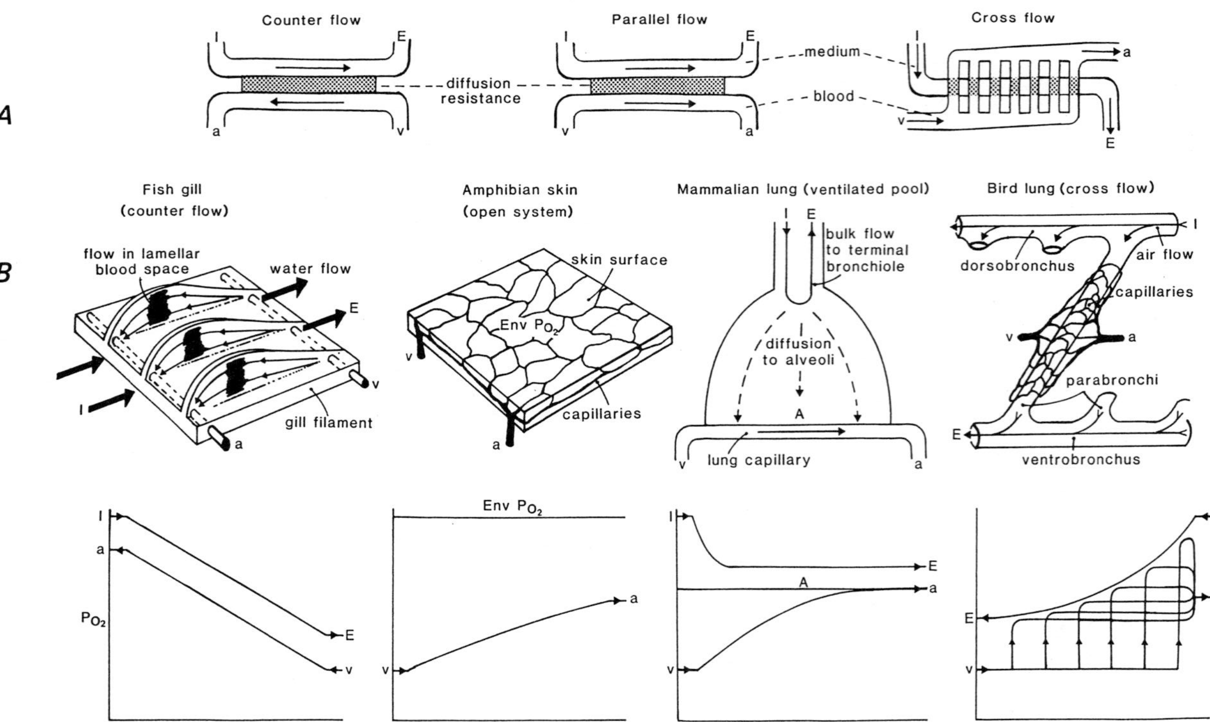

Fig. 4. Design of gas-exchange organs. *A*, Three basic types of gas exchanger as described in the text, showing relationships between flow in medium and blood. *B*, Four variations on basic gas-exchange models as found in vertebrates, showing general relationships between convective flow and anatomy (above) and partial pressure profiles in gas exchanger (below). Note that in both counter- and cross-flow arrangements arterial P_{O_2} can exceed expired P_{O_2}.

Weibel (1984) points out. For example, it is difficult to compare the design of diffusive and convective systems in these terms. In the former, area and thickness of diffusion pathway are important and dictate the overall size of the transfer region that has to be accommodated in, and maintained by, the body (see Mayhew, this volume). In the latter, flow rate is the most important variable, affecting design of both pump and system of channels. Capacitance coefficient of the blood (but not of the medium) is also an important variable and is related to red cell number and haemoglobin properties. The model cannot be used to predict the optimum balance between such diverse components. The principle of symmorphosis can be tested only by looking at the results of evolution in the organism itself. Taylor & Weibel (1981) suggest that symmorphosis can best be seen in the effects that the size of organisms or their activity levels have on the design and performance of the four stages of the total transport sequence. There should be just enough structure at each stage to cope with the maximum rate of O_2 transport. That structure should scale in a predictable way with size or with the capacity for activity (see Perry, this volume).

Design of gas-exchange organs

Although there are good reasons for looking at the O_2 transport process as a whole, it is more often considered in two major categories: those of the gas exchanger and the cells. Three basic types of gas exchanger, combining one diffusion and two convection stages, can be recognised and modelled (Fig. 4*A*). These are counter-flow, parallel-flow and cross-flow systems. They differ in the relative arrangement of blood and medium flow on either side of the diffusion barrier.

It is interesting that evolution has produced all three types in vertebrate animals, though inevitably with some modifications. Piiper & Scheid (1972, 1975, 1977, 1989) have suggested that four variations on these basic themes can be usefully identified (Fig. 4*B*). A typical counter-flow system is seen in teleost gills and, in a modified form, in elasmobranch gills. The cross-flow pattern occurs in bird lungs. The ventilated pool, based on the mammalian lung, is fundamentally a parallel-flow system modified by the addition of dead space. The latter is the result of tidal rather than unidirectional ventilation in a lung with a long conducting zone of several orders of dividing bronchi. Finally, the infinite pool, based on amphibian skin, shows no ventilation on the medium side and consists only of diffusion and perfusion conductances between environment and blood. It represents an extreme case of all three of the basic systems with their ventilated compartment completely open to the environment so that ventilation conductance is infinitely high.

Models of aquatic gas exchange

A concept that has been used, along with symmorphosis and limitation, to assess performance and design in the different types of gas exchanger, is that of efficiency, often with an underlying sense that natural selection will favour high efficiency. There are two difficulties. The first is that, in a sequential system, realistic estimates of efficiency of components must ultimately be seen in the context of the whole; a chain is as strong as its weakest link, which returns us to the principle of symmorphosis. The second problem, made more significant by the colloquial nature of the term, is one of defining appropriate criteria of efficiency so that they can be applied with some precision to the total transport system.

The concept of efficiency has been particularly important in the development of models of aquatic gas exchange. Because water is a high-viscosity, high-density medium in which both capacitance coefficient and diffusion coefficient for oxygen are low, breathing in water has always seemed likely to be energetically costly when compared to breathing in air (Hughes & Shelton, 1962; Dejours, 1981; Piiper, 1982). Adaptations, such as continuous, unidirectional water flow, counter-current design of the gas exchanger and short diffusion pathways between ventilated water and gill epithelium, help overcome these restrictions and probably increase efficiency however it may be defined. Van Dam (1938) suggested that the extent to which O_2 was removed from the ventilated water (the utilisation) was a good indicator:

$$\text{Utilisation } \% = \frac{(P_I - P_E)}{P_I} \times 100 \tag{9}$$

Hughes & Shelton (1962) were doubtful that efficiency could be measured in precisely this way but, using the term effectiveness, again looked for some measure of O_2 extraction and transfer. By limiting the maximum range of extraction to the difference between partial pressures in inspired medium and venous blood, they claimed that a more realistic evaluation of the gas-transfer process could be achieved:

$$\text{Effectiveness } \% = \frac{(P_I - P_E)}{(P_I - P_v)} \times 100 \tag{10}$$

Using an analogy with heat exchanger function, they proposed that effectiveness depended on gill area, flow arrangement and capacity-rate ratio (C_w/C_b; equivalent to the conductance ratio G_{vent}/G_{perf}) between the two pumped media, water and blood (Fig. 5). High values of effectiveness of O_2 extraction from water could be achieved if the conductance ratio was low, e.g. if perfusion of the gills with blood took place at very much

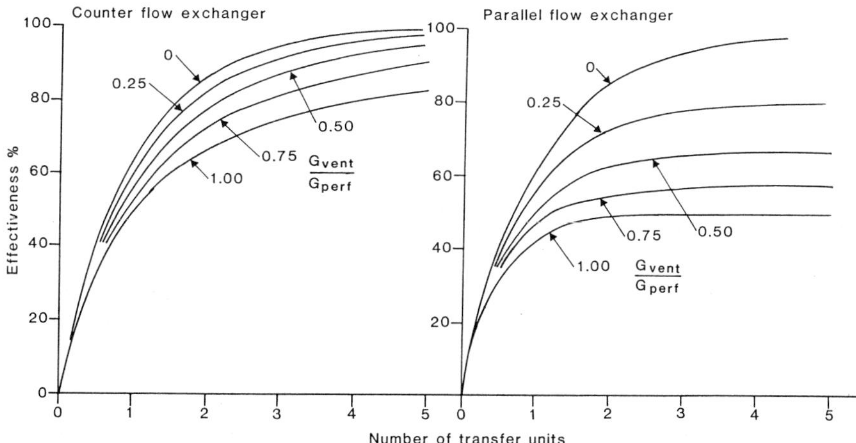

Fig. 5. An early model to show the important factors affecting gas exchange, using a heat exchanger analogue. Effectiveness is the ratio of O_2 actually removed from the medium to the maximum that could be removed. Transfer units in the original model are equivalent to the G_{diff}/G_{vent} ratio. The model shows the significance of conductance ratios and the conditions under which counter-flow performs better than parallel-flow.

higher rates than their ventilation with water. Under such conditions, however, effectiveness of O_2 transfer into blood would be extremely low and only small quantities of O_2 would be transferred. The best compromise in terms of O_2 extracted from water and transferred to blood would be found when the conductance ratio was unity (Hughes & Shelton, 1961). Under these conditions, counter-flow is a much better arrangement than parallel-flow, as the analogy with heat-exchange systems showed (Fig. 5).

Both utilisation and effectiveness relate to either the medium or the blood side of the gas-exchange organ. To overcome this restriction, Piiper & Scheid (1975, 1977) suggested that the overall efficiency (sometimes called efficacy) of the gas exchanger could best be assessed in terms of its total conductance (G_{tot}), the transfer rate of O_2 per unit of overall partial pressure difference in the exchanger:

$$G_{tot} = \frac{\dot{M}_{O_2}}{(P_I - P_v)} \tag{11}$$

Applied to the different models of gas exchanger described above, this measure shows that efficiency declines from counter-flow, through cross-

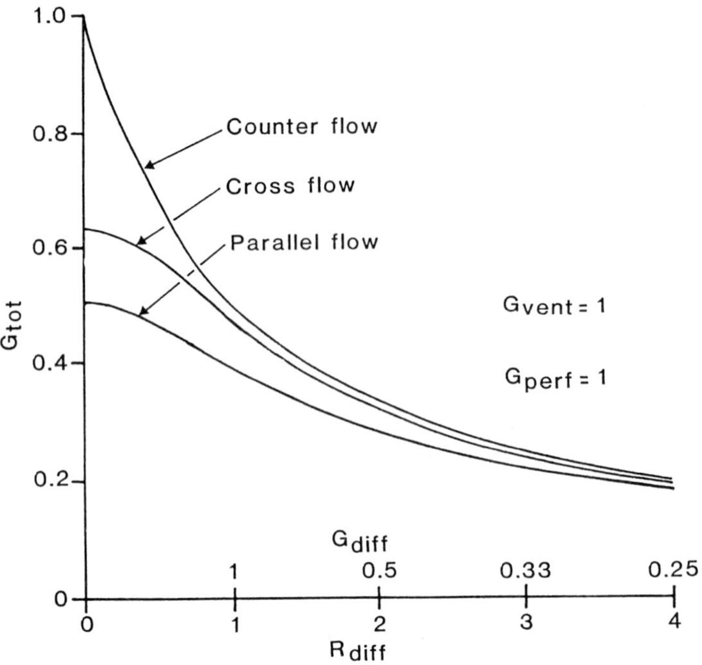

Fig. 6. Comparison of the total conductance (*G*tot: transfer of O_2 per unit of overall P_{O_2} difference) in the three basic types of gas exchanger with constant ventilation and perfusion conductances. Note that the difference between the three types diminishes as the diffusion conductances of the exchangers go down (after Piiper, 1982).

flow to ventilated pool models (Fig. 6), provided that diffusion conductance is reasonably high. Applied to real animals, however, this proposal is also not without problems. Total conductance is a function of ventilation, perfusion and diffusion conductances and is smaller than the smallest individual value (conductances in series add as reciprocals, i.e. as resistances). Both ventilation and perfusion conductances can be high, and therefore increase total conductance, if very large volumes of medium and blood are pumped. A high diffusion conductance requires a large gas exchanger. In fact conductance does not seem to be a good measure of efficiency in itself because the costs to the animal of high conductance are themselves inevitably high. In a later, somewhat modified view, Piiper (1989) suggested that efficiency was reflected in the degree of overlap in arterial and expired P_{O_2} so that fish gills and bird lungs were potentially more efficient than mammalian lungs. Such overlap could be an important component in keeping the overall P_{O_2} difference in the exchanger to a minimum.

The metabolic costs to the animal, both in maintaining structure in the widest sense and in the mechanical work of pumping medium and blood, ought to be taken into account in assessing efficiency (Hughes & Shelton, 1962; Shelton & Boutilier, 1982). These costs must be important in shaping the evolution of the total transfer system. Clearly an efficient system would be one in which the necessary quantities of respiratory gas could be transferred at minimal metabolic cost. This definition relates output (the transfer of O_2) and input (the metabolic cost) in a way that has some parallels with efficiency determinations in simple machines.

Metabolic costs of the passive, diffusion-driven stages in both gas exchanger and cells are very difficult to measure. The volumes and areas of the tissues involved can be determined by stereology (Weibel, 1984) and, in gas exchangers at least, some estimate of the size and metabolic rate arrived at for the gas-exchange structure itself. However, many gas exchangers have more than one function; gills, for example, are important in ion transport. It becomes more difficult to assign structural costs to the specific, gas-transport function. These difficulties are much greater at the final diffusion stage located in the general tissues and cells of the body. In addition, other much less obvious costs, such as the effects that the dimensions of these diffusion sites have on general body size, shape and motility, will certainly influence survival and therefore evolution of the structure of the diffusion stages. These effects are extremely difficult to assess and are rather remote from the central problem.

The mechanical work involved in pumping medium and blood is perhaps easier to isolate from the other components in the gas-transfer process and from most other complications affecting the assessment of efficiency. It is probably the major metabolic cost in gas transport and can also be examined profitably in simple models. The volume of respiratory fluid pumped in unit time is a variable in the basic equations 1 and 3 and, when multiplied by the pressures found in the fluid system under consideration, gives a measure of the mechanical power dissipated. This power will be related to the O_2 consumption of the pumps and so can be used to give a measure of the metabolic costs of ventilation and perfusion to the animal. The precise nature of the relationship between power output and metabolic cost is not easy to measure and, even when it has been measured, the results have been quite variable in different species and different experimental situations. These variations are not too important in examining the general relationships in simple models though, ultimately, they must be resolved if the concept of metabolic cost is to be of value.

Applying the basic equations to aquatic gas exchange, the volumes of water and blood pumped through a gill with an infinitely large diffusion conductance and different ventilation/perfusion conductance ratios can be calculated for both counter-flow and parallel-flow arrangements (Fig. 7, where the other assumptions in the model, based on a 1 kg trout, are given). In counter-flow (Fig. 7*A*), when $G\text{vent} = G\text{perf}$, arterial blood leaves the gill at the same P_{O_2} as inspired water and expired water at the same P_{O_2} as venous blood. The total fluid pumped (i.e. water and blood) is at a minimum and increases if $G\text{vent} > G\text{perf}$ or if $G\text{vent} < G\text{perf}$. Even though $G\text{diff}$ is infinitely large, inspired P_{O_2} is greater than arterial P_{O_2} when $G\text{vent} < G\text{perf}$, and the rate of ventilation remains constant as perfusion rate increases. Similarly, when $G\text{vent} > G\text{perf}$, expired P_{O_2} is greater than venous P_{O_2} and perfusion rate remains constant whilst ventilation rate changes (Fig. 7*A*). The effect of such relationships on ventilation power and perfusion power is that the former is low and constant at conductance ratios below unity whilst the latter becomes low and constant at ratios above unity (Fig. 8*A*). The sum of these two output powers must have its minimum when $G\text{vent} = G\text{perf}$.

The advantage that a counter-flow gill has over one working with parallel-flow when the diffusion conductance ($G\text{diff}$) is infinitely large can be illustrated by repeating the calculations for a parallel-flow model (Fig. 7*B*). In this model, because of the infinitely high diffusion conductance, $P\text{E}_{O_2}$ and $P\text{a}_{O_2}$ are always identical. When $G\text{vent} = G\text{perf}$, the $P\text{E}_{O_2} - P\text{a}_{O_2}$ equilibrium is midway between $P\text{I}_{O_2}$ and $P\text{v}_{O_2}$ and both ventilation and perfusion rates are twice as high as in the counter-flow model for the same rate of O_2 transfer (Fig. 7*B*). Increasing the $G\text{vent}/G\text{perf}$ ratio necessitates an increase in ventilation flow rate, as for a counter-current system, but also leads to a fall in perfusion flow rate. A decrease in the $G\text{vent}/G\text{perf}$ ratio has the opposite effect. Clearly, the power outputs of the two pumps must also change progressively in opposite directions as the conductance ratio varies (Fig. 8*B*). The minimum total power output does not therefore occur when the ratio is unity except in the event of it being equally costly to pump the water and blood required for O_2 transfer. In the example shown in Figs 7*B* and 8*B*, more power is needed to ventilate the gills than to perfuse them and the optimum conductance ratio is 0.8 for minimum pumping power. Conversely, the optimum ratio would be higher than 1 if perfusion power were to be higher than ventilation power. This conclusion has some importance in the ventilated pool systems of mammals, working in air. It seems likely that ventilation power in these systems will be less than perfusion power. In man the $G\text{vent}/G\text{perf}$ ratio is slightly greater than unity at rest but rises to 2.5 in heavy exercise (Table 1).

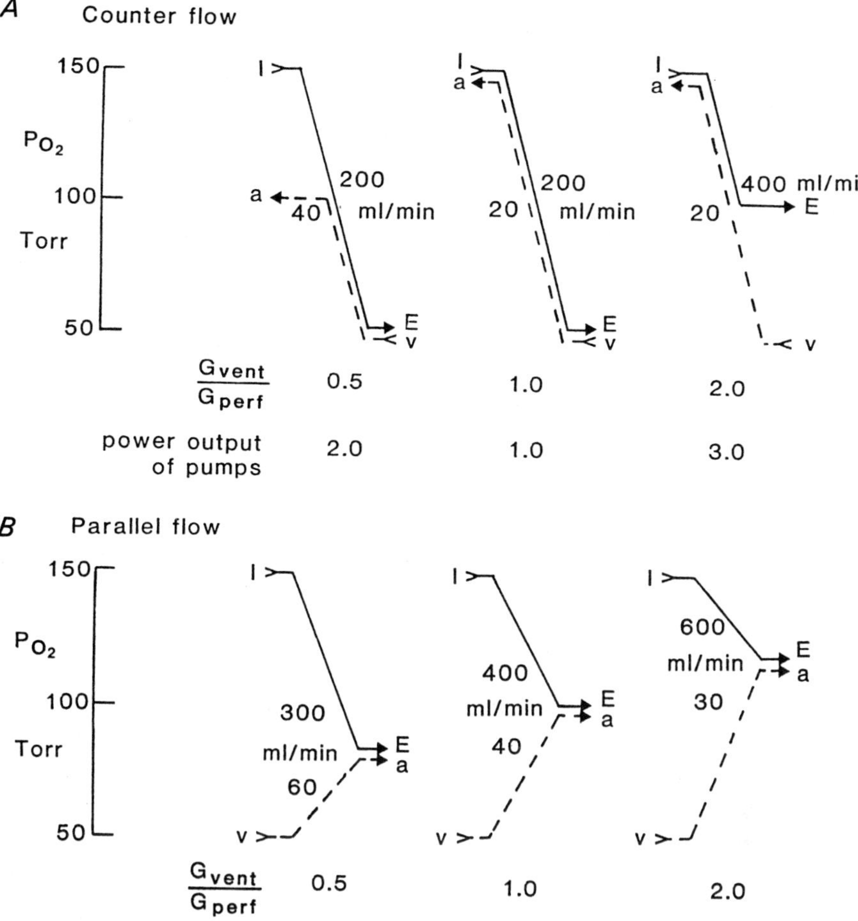

Fig. 7. Ventilation and perfusion flow rates and P_{O_2} values, calculated from the basic equations given in the text, for *A*, counter-flow and *B*, parallel-flow systems working at three different Gvent/Gperf ratios with infinite Gdiff. The calculations are based on data from 1 kg trout and assume an O_2 consumption of 1 ml min^{-1}, βwater of 0.05 ml litre^{-1} Torr^{-1}, βblood of 0.5 ml litre^{-1} Torr^{-1}, PI of 150 Torr and PE of 50 Torr. The total power output is the sum of the outputs of ventilation and perfusion pumps and is calculated by assuming that they pump into fixed resistances. It is given in relative units. The power output of 1.0 is approximately equivalent to 1.5 mW, with a ventilation output of 1.0 mW and a perfusion output of 0.5 mW, based on the data of Alexander (1967), Holeton & Randall (1967*a*), Jones & Schwarzfeld (1974) and Kiceniuk & Jones (1977).

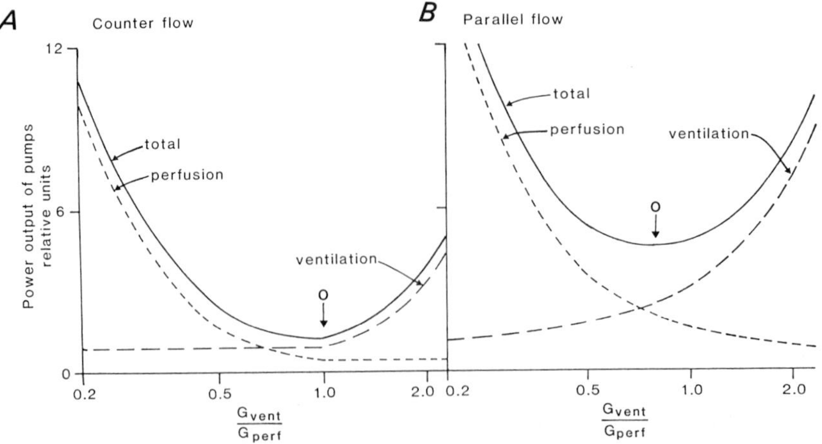

Fig. 8. The relationships between *G*vent/*G*perf ratio and power output of ventilation and perfusion pumps, calculated as described for Fig. 7, in *A*, counter-flow and *B*, parallel-flow gas exchangers. The optimum conductance ratio (*O*) for minimum total power is shown.

(a) *Experimental results and the gas-exchange model.* There are surprisingly few sets of data for fish that are sufficiently complete to allow the important prediction of conductance equality in counter-current exchange to be examined. The conductance ratio can be determined from P_{O_2} changes:

$$\frac{G\text{vent}}{G\text{perf}} = \frac{\dot{V}\text{m}\beta\text{m}}{\dot{V}\text{b}\beta\text{b}} = \frac{(P\text{a} - P\text{v})}{(P\text{I} - P\text{E})} \qquad (12)$$

The results of Holeton & Randall (1967*b*) and Kiceniuk & Jones (1977) show much larger changes of P_{O_2} in blood than in water, with conductance ratios of approximately 2.0 in resting, hypoxic and active trout (Table 1). Other data for active trout from Stevens & Randall (1967) give even higher conductance ratios that lie between 4.0 and 5.0 with extremely low utilisation (10%) of the oxygen in the ventilation stream. The results of Short *et al.* (1979) also suggest that conductance ratios are greater than unity, with a value of 2.0 in resting dogfish though this falls to 1.1 during hypoxia (Table 1). Overventilation is not universal, however, with Baumgarten-Schumann & Piiper (1968) giving a ratio of 0.4 for dogfish (Table 1) and Itazawa & Takeda (1978) even lower values of 0.1 to 0.2 for normoxic and hypoxic carp.

The discrepancy between experimental results and the predictions from the counter-current model suggests that, unless the data are unreliable,

Table 1. *Values for O_2 consumption and ventilation, perfusion and diffusion conductances in trout, dogfish and man. The diffusion conductances for the fish have been calculated from arithmetic (AM), logarithmic (LM) and harmonic (HM) mean gradients from water to blood. The diffusion conductances in man are based on determinations of carbon monoxide diffusing capacity*

	$\dot{M}_{O2}$ (μmol min^{-1} kg^{-1})	G_{vent} (μmol min^{-1} kg^{-1} Torr^{-1})	G_{perf}	G_{diff} AM	LM	HM
Trout						
Holeton & Randall (1967) 15°C						
Rest	51.0	1.28	0.64	1.06	1.12	–
Hypoxia:						
P_{IO_2} 80 Torr	58.7	3.84	1.52	1.89	1.99	–
40 Torr	47.0	6.50	2.87	2.18	2.21	–
Kiceniuk & Jones (1977) 10°C						
Rest	25.0	0.50	0.24	0.59	0.69	0.39
Exercise:						
75% maximum	84.8	1.63	0.85	1.50	1.60	–
100% maximum	193.7	3.82	1.76	3.49	3.90	–
Dogfish						
Baumgarten-Schumann & Piiper (1968) 16°C						
Rest	28.6	0.31	0.73	0.39	0.41	0.38
Short *et al.* (1979) 15°C						
Rest	36.0	0.97	0.50	0.51	0.52	–
Hypoxia						
P_{IO_2} 80 Torr	27.5	1.53	1.38	0.59	0.51	–
Man						
Ekelund & Holmgren (1964) for $\dot{M}_{O2}$						
Dejours (1981) for P_{O2}						
Weibel (1984) for G_{diff}						
Rest	195.5	5.75	3.76	19.1	–	–
Heavy exercise at 150 watts	1562.0	50.55	20.18	63.8	–	–

the model needs refinement. The experiments are difficult and cannulation of blood vessels is known to affect activity, metabolism and gill performance. Blood sampling can cause substantial changes in haematocrit. Even if the more extreme deviations from a conductance ratio of unity are due to the difficulties inherent in the experiment, the evidence suggests that G_{vent} can range at least from half to twice G_{perf} and that some inadequacies can be attributed to oversimplification in the model.

The model is clearly unrealistic in its infinitely high diffusion conductance for the gill system. In the trout, for example, Pa_{O_2} falls short of PI_{O_2} by 20–30 Torr and the differences between PE_{O_2} and Pv_{O_2} are even greater (Holeton & Randall, 1967*b*; Kiceniuk & Jones, 1977), indicating an appreciable diffusion resistance ($1/G_{diff}$). Even so, when trout are in fully aerated water, gill diffusion resistance is not sufficiently large to prevent the characteristic counter-current crossover of P_{O_2} in medium and blood, so that Pa_{O_2} is higher than PE_{O_2} (Kiceniuk & Jones, 1977). The crossover disappears in hypoxia (Holeton & Randall, 1967*b*). In dogfish diffusion limitation is somewhat greater and in some cases Pa_{O_2} is greater than PE_{O_2} but not in others (Baumgarten-Schumann & Piiper, 1968; Piiper *et al.*, 1977; Short *et al.*, 1979).

(b) *Models of gill diffusion processes*. Measuring G_{diff} is not easy and establishing reasonably trustworthy values has again been aided by some simplifying models. A physiological G_{diff} (frequently referred to as effective diffusing capacity, D_{eff}) can be determined as the quotient of O_2 consumption and the mean P_{O_2} gradient from water to blood (Eqn 7). The component gradients making up that mean cannot be measured directly and the P_{O_2} profiles along the gill are not well characterised, though some models are emerging (Scheid *et al.*, 1986). The mean gradient has therefore to be estimated from P_{O_2} values in blood and medium before and after the exchange surface. The estimates can be based on an arithmetic mean (AM) of the P_{O_2} differences at the two ends of the exchanger (Randall *et al.*, 1967), a procedure strictly applicable only when $G_{vent} = G_{perf}$ and the O_2 dissociation curve for blood is linear:

$$\mathrm{AM} = \frac{(P_I - P_a) + (P_E - P_v)}{2} \tag{13}$$

If the conduction ratio is other than unity, the P_{O_2} changes in the secondary lamellae should follow exponential profiles when the O_2 dissociation curve is linear. A logarithmic mean (LM) is then more appropriate (Hughes, 1972):

$$\mathrm{LM} = \frac{(P_I - P_a) - (P_E - P_v)}{\ln(P_I - P_a)/(P_E - P_v)} \tag{14}$$

The LM values are always smaller than the AM ones and the difference becomes more marked as the conductance ratio progressively diverges from unity, particularly when $G\text{diff}$ is high. A final estimate, taking the non-linearity of the O_2 dissociation curve into account, can be arrived at using a graphical method that is based on a modified Bohr integration (Piiper & Scheid, 1982; Piiper *et al.*, 1986). In this the O_2 concentration differences in water from $C\text{I}$ to $C\text{E}$ and in blood from $C\text{a}$ to $C\text{v}$ are divided into a number of equal parts (10 such divisions are used by the authors). The P_{O_2} gradients for each of the 10 units of concentration change from $(P\text{I} - P\text{a})$ to $(P\text{E} - P\text{v})$ are then measured on the O_2 dissociation curve and a harmonic mean (HM) taken:

$$\frac{1}{\text{HM}} = \frac{1}{n}\left(\frac{1}{(P\text{w} - P\text{b})_1} + \frac{1}{(P\text{w} - P\text{b})_2} + \ldots \frac{1}{(P\text{w} - P\text{b})_n}\right) \tag{15}$$

This method gives larger mean gradients than those that assume linear dissociation curves because of the convex shape of the actual blood curve between venous and arterial points. The more convex the curve the bigger the difference becomes. The values of AM, LM and HM calculated from the dogfish data of Piiper *et al.* (1986) are respectively 65.0, 62.9 and 72.9 Torr, and those from the trout data of Kiceniuk & Jones (1977) are 42.6, 36.3 and 63.4 Torr. It is interesting that the model incorporating most assumptions (AM) produces estimates closer to the one that is probably the most accurate (HM).

Physiologically determined values for $G\text{diff}$ ($D\text{eff}$) usually fall between those for $G\text{vent}$ and $G\text{perf}$ in both trout and dogfish at rest in well-aerated water (Table 1). All conductances increase during hypoxia or exercise, though the increase in $G\text{diff}$ is less than that of the other two conductances. During hypoxia, at least, $G\text{diff}$ ultimately becomes the lowest conductance found in both trout and dogfish (Table 1). This supports the view put forward earlier that the capacity for change is least in the diffusion component of the gas-exchange process.

Theoretically, $G\text{diff}$ can also be determined (Eqn 7) by applying morphometric techniques to measure diffusion area and thickness in the gill barrier between water and blood (Hughes, 1972; Piiper *et al.*, 1986). These measurements are tedious to make and prone to sampling, shrinkage and sectioning error but the data are very important. The conductances ($D\text{barrier}$) derived this way are higher than the physiological values ($D\text{eff}$). The reason for at least some of this discrepancy is clear. The water-filled space between secondary lamellae is about 10–100 times greater than the thickness of the tissue barrier. Since the Krogh diffusion constants for water and tissue differ only by a factor of 2 at the most, some diffusion conductance must be located in water passing through the lamellae ($D\text{w}$) and morphometric estimates take no account of it. The O_2 in this water is transported both by diffusion and by convection so that the contribution $D\text{w}$ makes to the total conductance is very difficult to determine. Scheid & Piiper (1971, 1976) have produced an ingenious model at the level of individual gill lamellae that allows some assessment to be made. With the considerable assumptions of a constant P_{O_2} at the secondary lamella wall ($P\text{lam}_{O_2}$) and an established laminar water flow, they calculate P_{O_2} profiles along the interlamellar space as O_2 is removed from water entering the system with a set $P\text{I}_{O_2}$. The $P\text{E}_{O_2}$ of water leaving the gill can then be derived and turns out to be a function of interlamellar distance, lamellar length, water velocity and O_2 diffusion coefficient. The ratio $(P\text{E}_{O_2} - P\text{lam}_{O_2})/(P\text{I}_{O_2} - P\text{lam}_{O_2})$ defines equilibration inefficiency (E) in the system. A further simplifying assumption that, instead of a parabolic flow profile, water flow between the lamellae consists of a totally stagnant zone on either side of a totally mixed central zone, allows the diffusing capacity of the water ($D\text{w}$) to be calculated. In this model the equilibration inefficiency is a simple exponential function of $G\text{vent}$ and $D\text{w}$, and rearrangement gives:

$$D\text{w} = \dot{V}\text{w} \times \beta\text{w} \times \ln(E) \qquad (16)$$

Thus, morphometric data on lamellar dimensions and experimental measurements of flow rate can be combined to arrive at $P\text{E}_{O_2}$ and equilibration inefficiency and ultimately to give diffusion conductance in the water.

The only comprehensive set of experimental and morphometric data available for full analysis is that for the dogfish. Conductance values for a resting 2.2 kg animal are $D\text{w}$, 2.10; $D\text{barrier}$, 3.87; and $D\text{eff}$, 0.83 μmol $\text{min}^{-1}\,\text{Torr}^{-1}$ (Piiper *et al.*, 1986). Water conductance is clearly lower than that of the gill barrier and the sum of the two, adding them as resistances in series, is still substantially higher at 1.36 μmol $\text{min}^{-1}\,\text{Torr}^{-1}$ than $D\text{eff}$. Morphometric conductances that are larger than physiological ones are also found in mammalian (Weibel, 1973, 1984) and other lungs and are not unexpected. Weibel (1984) suggests that morphometric data should be compared with physiological conductances from maximally active animals since, at rest, the physiological system is operating with a great deal in reserve. The difference is also thought to be due to inefficiencies in the working system, due to ventilation–perfusion inhomogeneities, shunts, haemoglobin–O_2 reaction times, etc., that are expressed in physiological determinations but do not appear in morphometric measurements.

The gill model also gives some insight into the nature of conductance changes that can occur during exercise and hypoxia in fish and, to some extent, in other animals (Table 1). If, during exercise, the conductance of any component of the O_2 transport cascade fails to increase at the same rate as does O_2 consumption, the P_{O_2} difference over that component must go up to sustain the required levels of O_2 movement (Fig. 3). In the total system, including O_2 movement into the cells, a change in P_{O_2} difference in one component can only be achieved at the expense of the remaining components. In hypoxic conditions the P_{O_2} differences are themselves restricted by the environmental changes and all conductances have to increase proportionately if O_2 consumption is to be maintained with the same balance as before between components.

Because the capacitance coefficient for O_2 in the medium (βm) is not controllable by the animal, Gvent increases solely because ventilation volume ($\dot{V}$m) goes up. In fish, during both exercise and hypoxia, the increase is large enough to maintain O_2 consumption with the same, or even a slightly decreased proportion of the total P_{O_2} gradient (Table 1). The same is true of Gperf but in this case the changes are due not only to augmented flow ($\dot{V}$b) but also to an increase in capacitance coefficient (βb). This is due to movement of the venous blood point down the steepest part of the O_2 dissociation curve, causing the slope of the line between venous and arterial points to go up. In hypoxia, and to a lesser extent in exercise, the arterial blood point also moves onto the steeper parts of the curve, further increasing βb.

Changes in Gdiff (data in Table 1 refer to physiologically determined Gdiff, or Deff) were first reported by Randall *et al.* (1967) who suggested that they were due to modifications in the pattern of blood flow in gills during hypoxia and exercise. The possibility of a non-respiratory pathway through which venous blood could be shunted to varying degrees (Steen & Kruysse, 1964; Richards & Fromm, 1969) is now largely discounted. There is more convincing evidence that lamellar recruitment can occur, increasing the functional surface of the gills (Wood, 1974; Booth, 1978) and causing real increases in Gdiff. There may also be improvements in blood flow patterns in those lamellae most exposed to high water flow, correcting inhomogeneities as blood pressures and flow rates increase (Randall, 1982). Such changes would improve Deff and bring the physiological conductances closer to morphometric estimates. Finally, an increase in water velocity through the lamellae will decrease the thickness of the stagnant water layer in the gill lamella model. Diffusion conductance attributable to the water (Dw) was 2.10 μmol min^{-1} Torr^{-1} in resting dogfish and 3.00 μmol min^{-1} Torr^{-1} in a separate experimental group of active (and somewhat larger) dogfish (Piiper *et al.*, 1986). Conductance of the epithelial barrier (Dbarrier) was 3.87 and 4.42 μmol min^{-1} Torr^{-1}, respectively, in the two groups, a change reflecting size difference only.

A disquieting consequence in the dogfish of the changes outlined above is that the effects on physiological conductance are substantial but the influence on the morphometric assessments is relatively minor so that the former becomes larger than the latter. For example, in the change from rest to exercise, physiological conductance increased more than twofold from $D_{eff} = 0.83$ to $D_{eff} = 1.95\ \mu mol\ min^{-1}\ Torr^{-1}$, whereas the sum of $D_{barrier}$ and D_{w} only changed from 1.36 to 1.79 $\mu mol\ min^{-1}\ Torr^{-1}$. Since morphometric techniques assess the maximum gill area, they should always produce larger conductances than those determined physiologically for reasons already explained. Piiper *et al.* (1986) carefully reconsidered all the assumptions and sources of inaccuracy in the model without resolving this discrepancy. In view of the difficulties in making both physiological and morphometric measurements, the agreement that the model gives between the two sets of data is remarkable and encouraging. More data from different fish species are now needed to test and refine the hypotheses.

(c) *Pumping costs and diffusion conductance.* The simple assessments of mechanical work involved in ventilating and perfusing a gill can be re-examined to see how they are affected by more realistic G_{diff} assumptions. Because part of the total P_I to $P_{\bar{v}}$ gradient is now associated with the diffusive transfer from water to blood (Eqn 2), that available for the ventilation and perfusion components of O_2 movement must diminish. The volume of medium and blood pumped must increase and the pumping costs go up. In addition, as diffusion conductance decreases, the differences between counter- and parallel-flow systems are progressively reduced (Fig. 6) and the unique properties of counter-flow seen in Figs 7 and 8 are lost. The overlap between P_E and P_a eventually disappears as does the constant ventilation flow when $G_{vent} < G_{perf}$ or constant perfusion flow when $G_{vent} > G_{perf}$. Exactly as in a parallel-flow system, both medium and blood flow rates change continuously and in opposite directions as the G_{vent}/G_{perf} ratio varies. The minimum total power required for convective pumping does not necessarily occur when G_{vent} and G_{perf} are equal. For example, if the simple trout model already considered is now given a 50 Torr gradient (arithmetic mean) from water to blood (Fig. 9), it becomes identical to the parallel-flow system, shown in Fig. 8*B*, with no diffusion resistance (or with diffusion resistance small enough to leave no detectable $P_E - P_{\bar{v}}$ gradient at the end of the exchange process). The minimum pumping power with a 50 Torr gradient therefore occurs when $G_{vent}/G_{perf} = 0.8$. If it is assumed that perfusion requires twice the power of ventilation at base levels (the converse of that in Fig. 9), then minimum total power occurs at $G_{vent}/G_{perf} = 1.25$. Depending on the magnitude of the physiological diffusion resistance and on relative power requirements

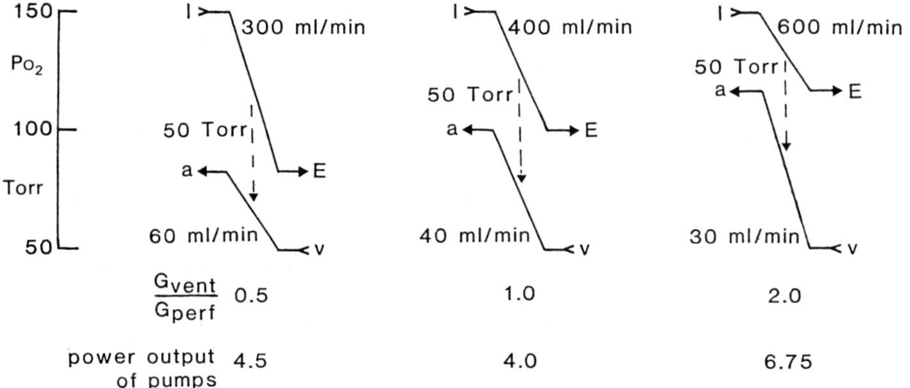

Fig. 9. Ventilation and perfusion flow rates and P_{O_2} values, calculated from the basic equations given in the text, for a counter-flow system with a 50 Torr (arithmetic mean) gradient from water to blood, working at three different *G*vent/*G*perf ratios. The other assumptions are specified in the caption to Fig. 7.

for ventilation and perfusion, the optimum *G*vent/*G*perf ratio can be larger or smaller than unity in counter-current systems.

Caution is needed before concluding that the simple model characteristics can be applied to experimental data from different species because they also show *G*vent/*G*perf ratios different from unity (Table 1). The results are extremely variable and it is difficult to reconcile ratios of 0.4 and 1.9 from a single dogfish species, or of 0.2 to 5.0 in different species of teleost, with the small deviations suggested by the model. Such huge variations in conductance ratio are very difficult to understand in terms of differences in the metabolic cost. It seems unlikely that the metabolic costs of pumping water will change dramatically unless the type of pump changes as, for example, between sharks and teleosts or between pump ventilation and ram ventilation. One set of results on dogfish suggests high ventilation rates with low O_2 extractions (Short *et al.*, 1979) whilst another set suggests the converse (Baumgarten-Schumann & Piiper, 1968). Trout seem consistently to ventilate at high levels (Holeton & Randall, 1967*b*; Kiceniuk & Jones, 1977). The metabolic costs of pumping blood are also unlikely to vary substantially since the pumps are similarly structured in all fish. The most likely variable is that of blood pressure and Johansen (1972) has suggested that perfusion work in dogfish is about half that in trout because of lower blood pressures. Again, such differences are not always found by different investigators (Short *et al.*, 1979; Kiceniuk & Jones, 1977). Comparisons are clearly bedevilled by the inconsistencies in results.

As with the gill diffusion models, the need now is for more experimental data to allow the metabolic costs of ventilation and perfusion to be determined and to see whether the relative costs influence the design and operation of the gas exchanger. A further need is for refinement of the model that has been used to consider the power relationships of the pumps. The assumptions of linear P_{O_2}–O_2 concentration relationships and of fixed $P\text{v}_{O_2}$ are gross simplifications. Clearly, O_2 concentration cannot increase continuously at high P_{O_2} values. As blood flow and $P\text{a}_{O_2}$ vary in satisfying different Gvent/Gperf ratios, so the P_{O_2} gradient from blood to cells must change. If the movement of O_2 into the cells is to remain constant then the conductance from blood to cells (Gcell) must also change. However, the model, even in its simplest form, serves to emphasise that the relationships between ventilation and perfusion conductances cannot be determined by the simple rules that appeared to be decisive, at least for counter-current devices. It also shows up the difficulties inherent in the existing data and emphasises that further experimentation is probably needed before more modelling along these lines can be profitable.

Models of aerial gas exchange

The three types of aerial gas exchanger, i.e. infinite pool (which also operates in water), cross-flow and ventilated pool systems, have already been described (Fig. 4). In all of these models, in contrast to the counter-current system, it is assumed that the medium in contact with the exchange surface remains constant in composition whilst the P_{O_2} (and P_{CO_2}) of capillary blood changes from venous to arterial values as the blood flows past the exchange surface. In the infinite pool the constancy of the medium is a consequence of the open design of the air or water compartments and infinite Gvent. In the ventilated pool, composition of alveolar gas is treated as being uniform throughout the lung and the equilibrium between it and pulmonary capillary blood is exactly as in the infinite pool model. However, alveolar gas and inspired air differ in composition because of dead space in the conducting zone of the lung and long diffusion pathways. Gvent is finite. Finally, in the cross-flow model of bird lungs, gas composition is different in different parts of the gas-exchanging system. Each cross-sectional element of the parabronchial tubes within the lung is assumed to contain gas of constant composition that comes into equilibrium with local pulmonary capillary blood. As a result of this gas exchange, the next cross-sectional element in the parabronchus is lower in O_2 and higher in CO_2 and establishes equilibrium with its local capillary blood. Thus, the blood from each parabronchial element is of different composition and arterial blood is a mixture of end-capillary blood from the total length of the parabronchi (Fig. 4*B*). It may be higher or lower in P_{O_2} than end-parabronchial gas, depending largely on the values for Gdiff, the diffusion conductance of the lung.

Diffusion processes in lungs. The transfer of O_2 by diffusion from a constant composition medium into capillary blood can be modelled in a greatly simplified fashion, as in the gill models, by assuming a linear O_2 dissociation curve and a barrier with a finite conductance or diffusing capacity ($G\text{diff} = D$), separating blood and medium (Piiper & Scheid, 1982; Piiper, 1990). The size of the gradient at the end of the pulmonary capillary ($P\text{A} - P\text{a}$) as a fraction of the initial gradient ($P\text{A} - P\text{v}$) is then an exponential function of the $D/G\text{perf}$ ratio (Piiper, 1990), as shown in Fig. 10:

$$\frac{(P\text{A} - P\text{a})}{(P\text{A} - P\text{v})} = e^{-D/G_{\text{perf}}} \quad (17)$$

By rearrangement, the diffusing capacity can be derived:

$$D = G\text{perf} \ln \frac{(P\text{A} - P\text{v})}{(P\text{A} - P\text{a})} \quad (18)$$

When D is very large compared to Gperf, the $P\text{A}_{O_2} - P\text{a}_{O_2}$ difference becomes very small in the model (Fig. 10), as it is in the actual alveolar lung of most mammals. In practice this prevents the measurement of Deff in normal steady-state conditions because the P_{O_2} values measured in arterial and venous blood cannot be used to arrive at an accurate mean capillary P_{O_2}. The diffusing capacity has to be measured in hypoxic conditions so that the $P\text{A}_{O_2} - P\text{v}_{O_2}$ difference decreases and the $P\text{A}_{O_2} - P\text{a}_{O_2}$ difference becomes large and measurable. The mean capillary P_{O_2} and the alveolar–capillary gradient can then be estimated. A difficulty with this technique is that hypoxia affects Deff in fish (Table 1) and may also do so in other animals, including mammals, by causing the dilation and recruitment of lung capillaries and possibly by reducing inhomogeneities. The method may therefore change that which it is attempting to measure. Diffusing capacity can also be determined by using carbon monoxide which has an extremely high capacitance coefficient in blood. If a relatively small quantity of CO is inspired in a single breath which is then held, P_{CO} in both arterial and venous blood remains at zero, so that:

$$D_{CO} = \frac{\dot{M}_{CO}}{P\text{A}_{CO}} \quad (19)$$

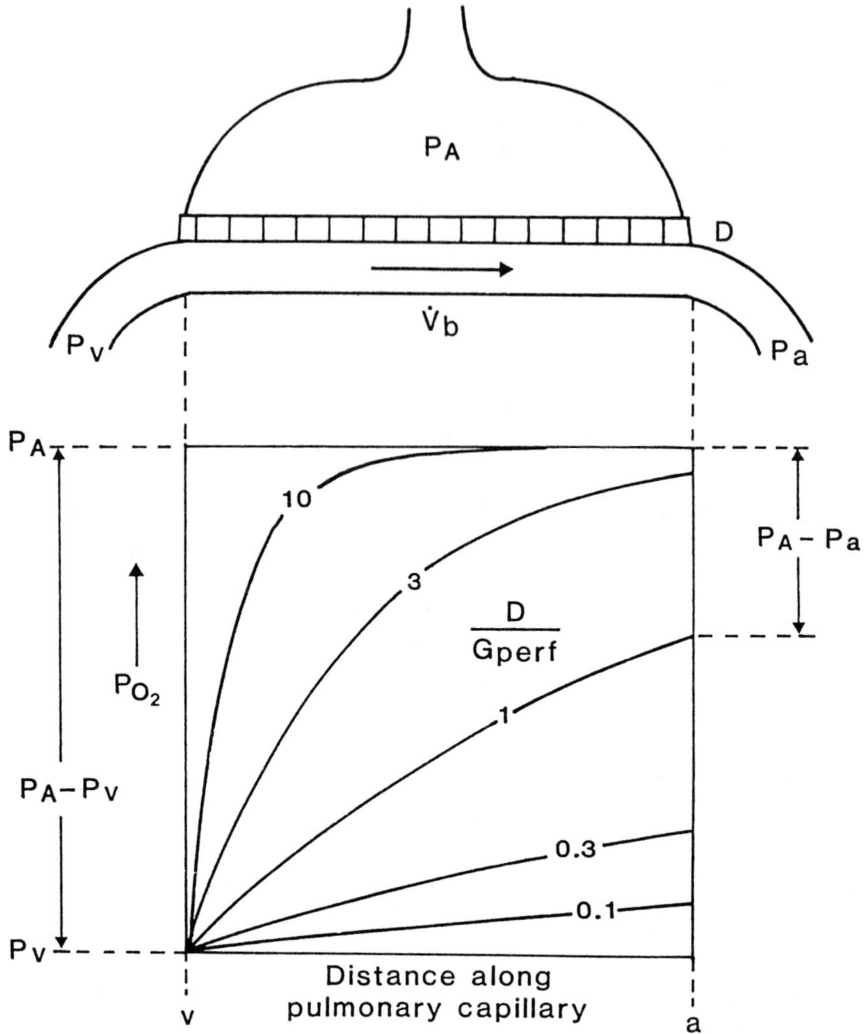

Fig. 10. Model of O_2 diffusion process in alveolar lung to show dependence of final equilibrium between alveolar gas and arterial blood, measured as a fraction of the initial gradient: $(P_A - P_a)/(P_A - P_v)$, on the D/G_{perf} (G_{diff}/G_{perf}) ratio. G_{perf} is the product of $\dot{V}_b$ and β_b, the slope of the O_2 dissociation curve, which is assumed to be constant (after Piiper, 1990).

where $\dot{M}_{CO}$ is the uptake of CO from the breath and is easily measured, as is alveolar P_{CO}. The diffusing capacity thus measured has to be corrected for differences in diffusivity and solubility between CO and O_2. The method also has the disadvantage of disturbing breathing and the steady-state conditions in the lung so, again, is not ideal. In spite of these methodological difficulties, measurements can be made and show that, in resting animals, Deff is very much larger in alveolar lung systems than either Gvent or Gperf (Table 1). This confirms the high Gdiff/Gperf ratio that the model shown in Fig. 10 suggested was necessary for complete equilibrium between medium and blood before the end of the pulmonary capillary.

During heavy exercise in man and other mammals, O_2 consumption increases about 10-fold above resting levels and all the gas-exchange conductances rise (Table 1). As in fish, the change in Gvent is attributable entirely to an increased ventilation rate that can be large enough to cause the $P\text{I}_{O_2} - P\text{E}_{O_2}$ difference to decrease. The change in Gperf is due to increases in both perfusion rate and capacitance coefficient (βb) that are more or less equal in magnitude. A considerable drop in venous P_{O_2} is responsible for the increased βb and it also increases $P\text{a}_{O_2} - P\text{v}_{O_2}$ difference. The increased O_2 convection in blood is therefore achieved by increased cardiac output, βb, and P_{O_2} difference, as it is in fish. The fall in $P\text{v}_{O_2}$ diminishes the gradient for O_2 transfer into the cells and necessitates an increase in Gcell in both fish and mammals.

However, an important difference between fish and mammal emerges from consideration of the changes in diffusion conductances. In mammals, Gdiff is so large that P_{O_2} differences between alveolar gas and arterial blood are usually insignificant even during heavy exercise; reserve capacity is inherent in a parallel-flow or ventilated pool system in that equilibrium can be reached progressively further down the pulmonary capillary as the perfusion rate increases without any change in diffusion conductance. In fact, Gdiff does increase during exercise, owing to the recruitment of pulmonary capillaries and reduction of inhomogeneities, but it does not go up to the same extent as O_2 consumption, Gvent or Gperf (Table 1). However, the lung still manages to transfer more O_2 and retain alveolar–arterial equilibrium by extending the P_{O_2} gradient to the full length of the pulmonary capillary. The mean gradient for O_2 transfer is thereby increased. In the model the Gdiff/Gperf ratio falls from 5.4 to 2.9 without changing $P\text{a}_{O_2}$ substantially.

Gdiff may increase even more during exercise in fish than it does in mammals, though it still does not go up as much as O_2 consumption. A larger P_{O_2} gradient is therefore required and the effect, in a counter-flow system, is to decrease $P\text{a}_{O_2}$. In fish both $P\text{a}_{O_2}$ and $P\text{v}_{O_2}$ decrease during exercise and the difference between them remains relatively constant. In the mammal only Pv_{O_2} falls so that the arterial–venous difference increases and contributes to O_2 transfer.

Non-steady-state models of oxygen transport

The evolutionary change from water-breathing fish to air-breathing birds and mammals, for all of which the steady-state models of O_2 transport are applicable with appropriate modifications, has taken place through a number of intermediate stages, some of which can be seen in present-day forms. The conditions, such as temperature, shade and hypoxia, that lead to the evolution of air breathing, occur frequently in freshwater environments and are sufficiently common for air breathing to have evolved a large number of times in groups of fish totally unrelated to the tetrapod ancestors and to one another. In spite of their many differences, these air breathers have some things in common, one of these being the inadequacy of steady-state models of gas exchange to describe the complexity of their respiratory and circulatory physiology (Shelton & Boutilier, 1982; Shelton *et al.*, 1986). Three major modifications are needed to the simple O_2 cascade models represented in Figs 2 and 3 if they are to be useful in studying the non-steady states found in these animals. The features that need to be incorporated in any satisfactory model are:

(a) *Bimodal gas exchange.* Air-breathing fish and amphibians have two gas-exchange surfaces, one working in air and the other in water, and each having different characteristics for the exchange of O_2 and CO_2. Elimination of CO_2 exceeds acquisition of O_2 at the gill or skin surface exchanging gas in water with the converse being found at the air-breathing organ (ABO) or lung exchanging gas in air.

(b) *Intermittent ventilation.* The ABO or lung is intermittently and usually arrhythmically ventilated. The habit clearly arose as aquatic animals were driven to supplement their O_2 uptakes from poorly oxygenated water by periodically coming to the water surface to gulp air and immediately resubmerging. A consequence of intermittent ventilation is that oscillations occur in the respiratory gas concentrations in all parts of the transport system (Boutilier & Shelton, 1986*b*). Such ventilation is found in all air-breathing fish, amphibians (Fig. 11) and even in many reptiles in spite of their unimodal, air-breathing habit (Burggren & Shelton, 1979; Shelton *et al.*, 1986).

(c) *Variable perfusion.* The circulatory connections to the ABO or lung are extraordinarily diverse in different groups of air-breathing fish. In all groups, however, the blood supply tends to be in parallel with some part

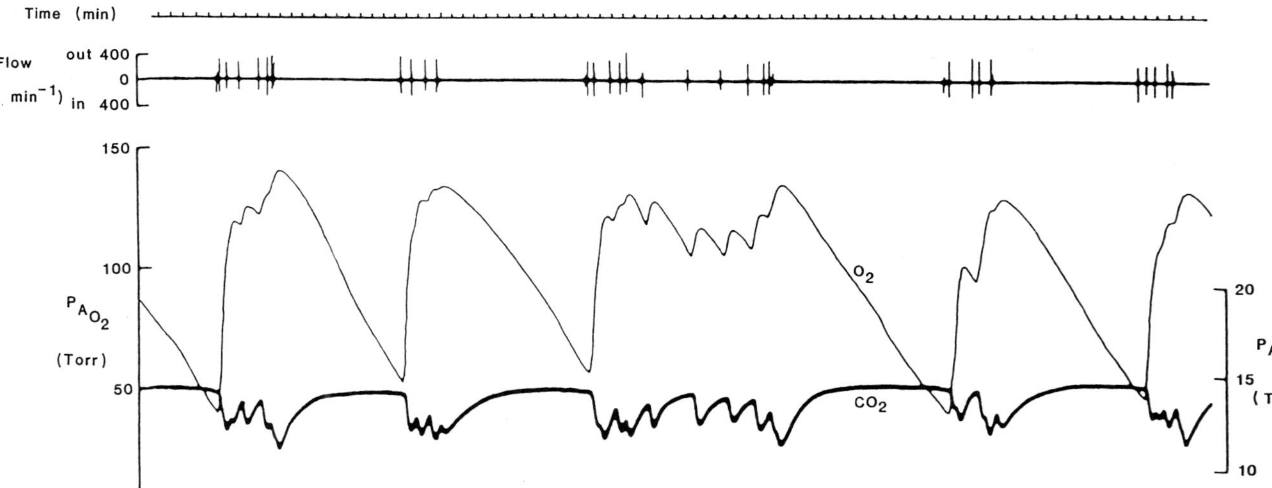

Fig. 11. Continuous recordings of lung ventilation rates (shown as vertical deflections caused by an animal breathing air at a blow-hole pneumotachograph) and partial pressures of lung gases ($P_{A_{O_2}}$, $P_{A_{CO_2}}$: measured by electrodes in an extracorporeal loop) in the amphibian *Xenopus laevis* diving and surfacing freely in air-saturated water (25 °C).

of the systemic circulation so that blood flow to the air-breathing structure can be varied independently of systemic flow. Even in amphibians and reptiles, in which the pattern of connection in the circulatory system is recognisable as one from which the totally divided, double circulation of birds and mammals has evolved, an incompletely divided heart with central shunts (Shelton, 1970, 1976, 1985) again allows perfusion of the lung to be varied quite independently of blood flow to the rest of the body. This situation is in marked contrast to that found in unimodal fish, birds and mammals in which the single gas exchanger and the body are connected in series by the circulatory system. In these animals blood arriving at the cells is necessarily that which has just been through the gas-exchange system. This is not obviously the case in animals that have the possibility of variable lung perfusion. The advantages of variable lung perfusion are that it allows pumping costs to be reduced during apnoea, it can act as a mechanism for regulating O_2 movement between stores and it allows the blood and tissue stores to be recharged as quickly as possible when the animal breathes. It has been argued (Shelton, 1985) that the need for high levels of perfusion of the lung with venous blood to recharge O_2 stores constitutes the most important selection pressure in the evolution towards a completely divided circulation.

Simple models of intermittent breathing and bimodal gas exchange

Clearly, the relationships between ventilation, perfusion and the movements of O_2 through non-steady-state systems with the characteristics outlined above are not going to be easy to study. Experiments are difficult because the measurement of constantly changing respiratory variables, using techniques that are time-dependent, limits the accuracy of any data. Models offer real possibilities of studying the dynamic properties of such systems though, at the moment, only very simplified versions have been produced (Shelton, 1985; Shelton & Croghan, 1988).

Storage capacity for O_2 and CO_2 in the component parts of the total transport system becomes a parameter of substantial importance in determining the time-dependent changes in non-steady-state systems. It can best be quantified in terms of capacitance (B: units, mM Torr^{-1} or ml Torr^{-1}), which is the increase in amount of gas in a store per increase in partial pressure and is clearly the product of volume of the storage medium, such as water, gas, blood, tissue fluid, etc. (Vstore) and the capacitance coefficient (β) of that medium:

$$B_{O_2} = \frac{\partial M_{O_2}}{\partial P_{O_2}} = V\text{store}\beta\text{medium} \tag{20}$$

The product of capacitance and partial pressure change will give the change in volume of respiratory gas moving into or out of the store. Using the electrical analogy already introduced, this movement is equivalent to the change in charge on a condenser discharging or charging through a resistance. The amount of O_2 contained in a store will fall exponentially with time, as does the charge on a condenser, if allowed to flow through a fixed resistance (R = conductance^{-1}: units, Torr min mM^{-1} or Torr min ml^{-1}), so that at time t:

$$M_{O_2} = M_{O_2}0\,e^{-t/BR} \tag{21}$$

where $M_{O_2}0$ is the amount of O_2 contained in the store at zero time. The product BR ($= BG^{-1}$) is the time constant of the system and in $t = BR$, the O_2 store will diminish to 37% of its initial value. The relationships between capacitances and conductances in the various stores of the body are important in determining the amounts of O_2 that move through the system and the time courses of their changes. Models of gas exchange, store and use having only two compartments have been described by Shelton & Croghan (1988). The compartments represent O_2 stores in lung and blood and the models have bimodal gas exchange, show intermittent ventilation and have the potential for variable perfusion but not in a convenient form. The model shown in Fig. 12 is based on data from *Xenopus* (100 g), the clawed toad. Capacitance in the lung and blood stores is plotted against P_{O_2} in those locations and the volume of O_2 contained in the stores is therefore represented as an area in Fig. 12. The shape of the blood store, whose capacitance falls as P_{O_2} increases, is due to the change in βb_{O_2} at different values of P_{O_2}. In the model the dissociation curve for the chemically bound O_2 is assumed to be hyperbolic:

$$\beta b_{O_2} = C_{O_2}max\frac{P_{50}}{(P_{50} + Pb_{O_2})^2} + \beta phys_{O_2} \tag{22}$$

where $C_{O_2}max$ is the concentration of chemically bound O_2 in fully saturated blood, P_{50} is the P_{O_2} required for half-saturation and $\beta phys_{O_2}$ is the physical solubility of O_2 in blood.

The rate of movement of O_2 from atmospheric air to lung store ($\dot{M}lung_{O_2}$) is regulated by the conductance Gvent:

$$G\text{vent} = \dot{V}_A \beta gas_{O_2} = \frac{\dot{M}lung_{O_2}}{(P_{I_{O_2}} - P_{A_{O_2}})} \tag{23}$$

The equation refers to the rate of alveolar ventilation ($\dot{V}_A$) in order to simplify the relationship between lung gas and blood. Dead space is small in the primitive air breathers and the difference between alveolar gas and that expired from the lung is negligible. O_2 input into the lung from the

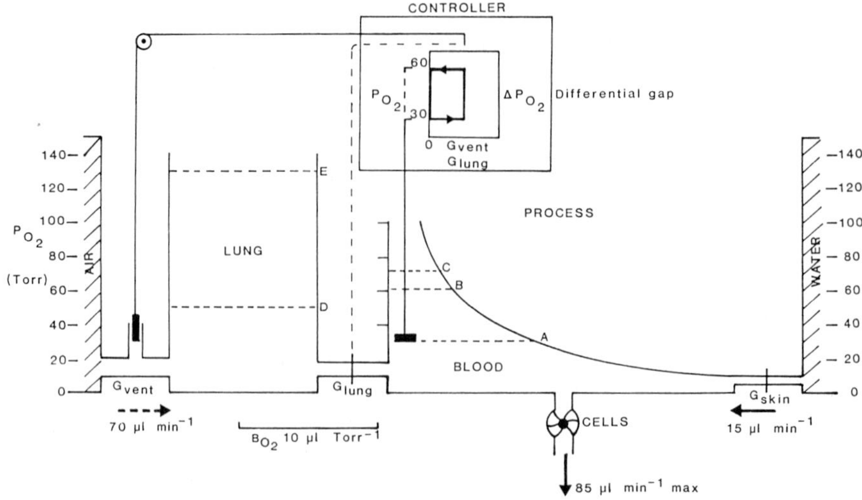

Fig. 12. Two-compartment model of gas exchange, storage and use in a bimodal system based on 100 g *Xenopus*. The equations for the model appear in the text. Volume of O_2 stored in each compartment is the product of capacitance (B) and P_{O_2} in the store and appears as an area in the figure. The model uses a hyperbolic O_2 dissociation curve (capacity: 10 vol%; P_{50}: 27 Torr). The controller initiates breathing when P_{O_2} in blood falls to 30 Torr (level A) and ends it when it rises to 60 Torr (level B). Lung store oscillates between levels D and E and blood store rises to C at the expense of lung store after breathing has ended.

environment exceeds its removal into blood when breathing occurs and Gvent is high. The converse is true during apnoea when Gvent is zero. The transfer of O_2 from lung to blood is regulated by a composite conductance Glung. This replaces Gdiff and Gperf in the more comprehensive models of steady-state gas exchange because, in the two-compartment model, blood is treated as a uniform pool with no arterial–venous O_2 difference.

$$G\text{lung} = \frac{\dot{M}\text{b}_{O_2}}{(P\text{A}_{O_2} - P\text{b}_{O_2})} \tag{24}$$

where $\dot{M}\text{b}_{O_2}$ is the rate of movement of O_2 from lung to blood and $P\text{b}_{O_2}$ is the P_{O_2} in the uniform pool of blood. The values adopted for Glung are based on estimates derived from experimental measurements of Gdiff and Gperf. Changes in Gperf, known to occur during the alternating cycles of breathing and apnoea, can be incorporated in the model at Glung as indicated by the broken line in Fig. 12. A model with more blood compartments and a separate circulation to the lung is needed to cope adequately with these relationships.

The rate of change in the amount of O_2 in the lung depends on inflow from environment and outflow to blood:

$$B\mathrm{gas}_{O_2}\frac{\partial P\mathrm{A}_{O_2}}{\partial t} = G\mathrm{vent}(P\mathrm{I}_{O_2} - P\mathrm{A}_{O_2}) - G\mathrm{lung}(P\mathrm{A}_{O_2} - P\mathrm{b}_{O_2}) \tag{25}$$

Similarly, the rate of change of O_2 in the blood depends on inflows from lungs and skin and consumption by the cells ($\dot{M}\mathrm{cell}_{O_2}$). The transfer of O_2 from environmental water to blood through the skin has been modelled in the infinite pool system by Piiper *et al.* (1976) already discussed, the rate of transfer being determined by diffusion and perfusion conductances. These are represented in Fig. 12, for the same reasons as before, by the composite conductance $G\mathrm{skin}$:

$$G\mathrm{skin} = \frac{\dot{M}\mathrm{skin}_{O_2}}{(P\mathrm{water}_{O_2} - P\mathrm{b}_{O_2})} \tag{26}$$

The rate of change of O_2 in the blood is thus:

$$B\mathrm{b}_{O_2}\frac{\partial P\mathrm{b}_{O_2}}{\partial t} = G\mathrm{lung}(P\mathrm{A}_{O_2} - P\mathrm{b}_{O_2}) + $$

$$+ G\mathrm{skin}(P\mathrm{water}_{O_2} - P\mathrm{b}_{O_2}) - \dot{M}\mathrm{cell}_{O_2} \tag{27}$$

Air-breathing fish and amphibians can rely on anaerobiosis when access to the surface is restricted but, in conditions that impose no restriction on surfacing, metabolism is predominantly aerobic (Shelton & Boutilier, 1982; Boutilier & Shelton, 1986*b*). Nevertheless, in a system showing wide oscillations, there must be some variation in the capacity of the cells to remove O_2 from the supply cascade. The conditions required for transition from aerobiosis to anaerobiosis have not been established for tissues with normal blood supplies (Boutilier *et al.*, 1986). In the model it is assumed that the rate of removal is limited at low P_{O_2} levels by a Michaelis–Menten relationship:

$$\dot{M}\mathrm{cell}_{O_2} = \dot{M}\mathrm{max}_{O_2}\frac{P\mathrm{b}_{O_2}}{(P\mathrm{b}_{O_2} + K\mathrm{m})} \tag{28}$$

where $\dot{M}\mathrm{max}_{O_2}$ is the rate of O_2 consumption when P_{O_2} is not limiting and $K\mathrm{m}$ is the P_{O_2} at which O_2 consumption is half $\dot{M}\mathrm{max}_{O_2}$. It is difficult to ascribe a value to $K\mathrm{m}$, but it is assumed to be low, at 2–3 Torr, in the model.

Both animal and model require appropriate control systems to initiate and end breathing movements, i.e. to increase $G\mathrm{vent}$ (and possibly $G\mathrm{lung}$)

and then to restore apnoea. The precise nature of the controller and its input signals are not well understood in air-breathing fish and amphibians (Shelton *et al.*, 1986). The levels of P_{O_2} in the blood are known to be important though not the only signals. In the simple control system shown in Fig. 12, the controller switches lung ventilation on at Pb_{O_2} levels of 30 Torr and off at Pb_{O_2} of 60 Torr.

With the controller operating to switch breathing on and off, Eqns 20–28 can be solved simultaneously for $\dot{M}cell_{O_2}$, O_2 stored in blood and for P_{O_2} in lung and blood, using a microcomputer. The results are plotted in Fig. 13*A* and *B*. With the parameter values that are given in the captions to Figs 12 and 13, the model predicts that a breathing burst goes on for 4–5 min and that the breathing–apnoea cycles last for 18–21 min. These times are well within the range of breathing behaviour shown by undisturbed *Xenopus* (Fig. 11), as are the oscillations in lung and blood P_{O_2}. As Fig. 13 shows, steady states do not occur in the model, even after prolonged lung ventilation. The relationship between lung and blood stores is affected by their respective time constants so that the highest levels in the latter are reached after breathing has stopped. In a model with more blood compartments, the movement of oxygen between stores will become more complicated and the pattern of breathing will depend not only on the set P_{O_2} levels but also on the possible locations of sensors in different blood compartments.

Experimental results and the two-compartment model

The limitations of the two-compartment model can be illustrated by showing a more comprehensive outline of the O_2 transport and storage system in *Xenopus* (Fig. 14). The correct circulatory relationships between the lung and four blood stores, the O_2-consuming cells, and the O_2 inputs through lungs and skin, are summarised in Fig. 14*B*. The lung and blood stores are shown in Fig. 14*A*, using the same axes as in the simple model. Data from actual O_2 dissociation curves are used to produce the blood store outlines so that maximum capacitances occur at P_{O_2} values of 21–23 Torr at which the real curves are steepest (Boutilier & Shelton, 1986*a*). The effect of falling lung volume, as O_2 is removed and not replaced by equivalent volumes of CO_2, is also taken into account in the shape of the lung store. The levels of O_2 in the stores are shown at 5 min intervals during a 30 min dive. All the information is derived from the measurements of lung gas and blood samples taken during many breathing–apnoea cycles in freely diving *Xenopus*. Figure 14 is a convenient, static summary of many experimental results and represents the simple type of descriptive model mentioned in the introduction. It shows

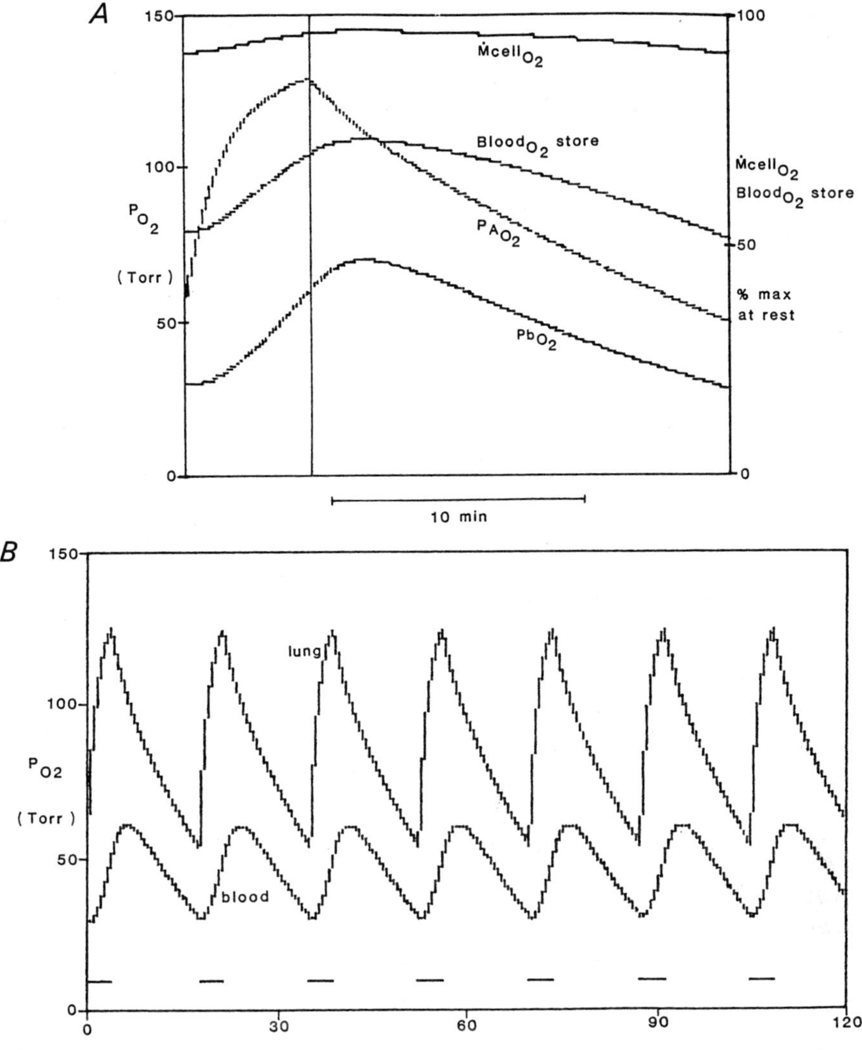

Fig. 13. Performance of the model shown in Fig. 12, based on 100 g *Xenopus*, with equations in the text. *A* shows the effect of a single breathing cycle on P_{O_2} in lung (PA_{O_2}) and blood (Pb_{O_2}), on cell O_2 consumption ($\dot{M}cell_{O_2}$) and volume of O_2 stored in blood ($Blood_{O_2}$ store), both given as percentages of maximum levels. *B* shows the effect of repetitive breathing cycles, indicated by horizontal marker lines, on lung and blood P_{O_2} levels. Values for parameters used in Eqns 20–28 are: $G\text{vent} = 0$ (during breath hold), $6.5\,\mu l\,min^{-1}\,Torr^{-1}$ (during breathing), $G\text{lung} = 1.6\,\mu l\,min^{-1}\,Torr^{-1}$, $G\text{skin} = 0.12\,\mu l\,min^{-1}\,Torr^{-1}$, $V\text{lung} = 8.0\,ml$, $V\text{blood} = 13\,ml$, $\dot{V}max_{O_2} = 85\,\mu l\,min^{-1}$, $Km = 2.5\,Torr$.

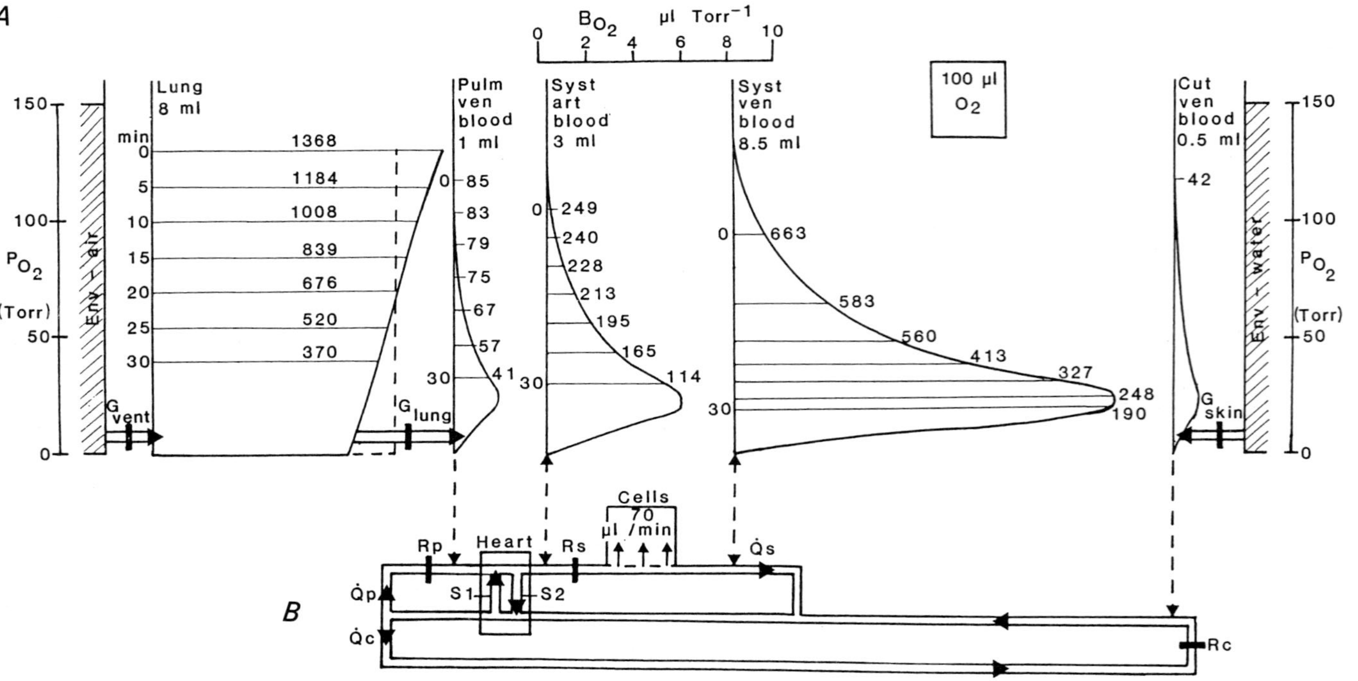

A
B
B_{O_2} µl Torr^{-1}
0 2 4 6 8 10
100 µl O_2
P_{O_2} (Torr)
150
100
50
0
Env - air
Env - water
Lung 8 ml
min
0 5 10 15 20 25 30
1368
1184
1008
839
676
520
370
G_{vent}
G_{lung}
Pulm ven blood 1 ml
85 83 79 75 67 57 41
Syst art blood 3 ml
249 240 228 213 195 165 114
Syst ven blood 8.5 ml
663 583 560 413 327 248 190
Cut ven blood 0.5 ml
42
G_{skin}
Cells 70 µl /min
Rp
Heart
Rs
S1
S2
Q̇s
Q̇p
Q̇c
Rc

the compartments and connections that need to be incorporated in the dynamic computer model for it to predict the relationships between time-dependent changes in ventilation, perfusion, O_2 movement and storage in actual multi-compartment systems as found in animals.

One of the striking differences between the models and animals is that the former produce totally regular breathing bursts whereas animals, even when undisturbed, show many irregularities (Figs 11 and 13) and may change, in a diurnal cycle for example, to quite different patterns. Some of the spontaneous fluctuations suggest that sensors in different parts of the transport system operate together with the controller in a probabilistic fashion. This type of fluctuation could easily be incorporated in the model. However, the major changes in pattern must be due to other inputs, e.g. from other sense organs such as light receptors, from an internal clock or even from higher centres of the brain, that override the basic systems incorporated in the model and take them beyond the scope of this chapter.

The process of O_2 transport from environment to cells is a part of comparative animal physiology that, since the time of Krogh, has attracted and responded to the modelling approach. The range of models surveyed in this chapter show the technique in its many forms, from simple summaries of existing knowledge to analytical models that rely on simplification and estimation where relationships are complex or data are inadequate. Dynamic models are of particular value in analysing non-steady-state phenomena in gas transfer. The relationships that such models predict are, in many cases, susceptible to experimental testing on

---

Fig. 14. Diagram summarising *A*, oxygen exchange and storage and *B*, oxygen transport and consumption determined experimentally in 100 g *Xenopus laevis* (20–25 °C). In *A* arrows indicate direction of O_2 movement in system, quantities transferred being determined by conductances and P_{O_2} differences as in Figs 12 and 13. O_2 stores in lung, pulmonary venous (Pulm ven), systemic arterial (Syst art), systemic venous (Syst ven) and cutaneous venous (Cut ven) blood shown as areas (100 μl calibration square). Store capacitance values (B_{O_2}) derived as products of volumes given above each store and capacitance coefficients (β) for gas and blood (the latter from experimentally measured dissociation curves). O_2 contents of each store (in μl) are given for 5 min intervals during a 30 min dive. In *B* large arrows indicate pulmonary ($\dot{Q}_P$), cutaneous ($\dot{Q}_C$) and systemic ($\dot{Q}_S$) blood flows, whose rates are determined in part by respective peripheral resistances (R_P, R_C and R_S). In the heart, arrows show right-to-left (S_1) and left-to-right (S_2) shunts that occur in the undivided ventricle.

real physiological systems and so offer the prospect of scientific advancement. It is also possible that models, because of their ability to respond rapidly to a variety of small changes in their various parameters, will also be useful in understanding physiological adaptations and evolutionary change.

References

Alexander, R.McN. (1967). *Functional Design in Fishes.* London: Hutchinson.

Baumgarten-Schumann, D. & Piiper, J. (1968). Gas exchange in the gills of resting, unanaesthetised dogfish (*Scyliorhinus stellaris*). *Respiration Physiology* **5**, 317–25.

Booth, J.H. (1978). The distribution of blood in the gills of fish: application of a new technique to rainbow trout (*Salmo gairdneri*). *Journal of Experimental Biology* **73**, 119–29.

Boutilier, R.G., Emilio, M.G. & Shelton, G. (1986). Aerobic and anaerobic correlates of mechanical work by gastrocnemius muscles of the aquatic amphibian, *Xenopus laevis. Journal of Experimental Biology* **122**, 223–35.

Boutilier, R.G. & Shelton, G. (1986*a*). Respiratory properties of blood from voluntarily and forcibly submerged *Xenopus laevis. Journal of Experimental Biology* **121**, 285–300.

Boutilier, R.G. & Shelton, G. (1986*b*). Gas exchange, storage and transport in voluntarily diving *Xenopus laevis. Journal of Experimental Biology* **126**, 133–55.

Brackenbury, J.H. (1973). Respiratory mechanics in the bird. *Comparative Biochemistry and Physiology* **44A**, 599–611.

Brett, S.S. & Shelton, G. (1979). Ventilatory mechanisms of the amphibian, *Xenopus laevis*: the role of the buccal force pump. *Journal of Experimental Biology* **80**, 251–69.

Bretz, W.L. & Schmidt-Nielsen, K. (1972). The movement of gas in the respiratory system of the duck. *Journal of Experimental Biology* **56**, 57–65.

Burggren, W.W. & Shelton, G. (1979). Gas exchange and transport during intermittent breathing in chelonian reptiles. *Journal of Experimental Biology* **82**, 75–92.

Dainty, J. (1960). Electrical analogues in biology. In *Models and Analogues in Biology.* Symposia of the Society for Experimental Biology, Vol. 14, pp. 140–51. Cambridge: Cambridge University Press.

De Jongh, H.J. & Gans, C. (1969). On the mechanism of respiration in the bullfrog, *Rana catesbeiana*: a reassessment. *Journal of Morphology* **127**, 259–90.

Dejours, P. (1981). *Principles of Comparative Respiratory Physiology*, 2nd edition. Amsterdam: Elsevier North Holland.

Ekelund, L.G. & Holmgren, A. (1964). Circulatory and respiratory adaptation, during long term, non steady-state exercise, in the sitting position. *Acta Physiologica Scandinavica* **62**, 240–55.

Gans, C. & Clark, B. (1976). Studies on the ventilation of *Caiman crocodilus*. *Respiration Physiology* **26**, 285–301.

Gans, C. & Hughes, G.M. (1967). The mechanism of lung ventilation in the tortoise, *Testudo graeca*. *Journal of Experimental Biology* **47**, 1–20.

Holeton, G.F. & Randall, D.J. (1967*a*). Changes in blood pressure in the rainbow trout during hypoxia. *Journal of Experimental Biology* **46**, 297–305.

Holeton, G.F. & Randall, D.J. (1967*b*). The effect of hypoxia upon the partial pressure of gases in the blood and water afferent and efferent to the gills of rainbow trout. *Journal of Experimental Biology* **46**, 317–27.

Hughes, G.M. (1972). Morphometrics of fish gills. *Respiration Physiology* **14**, 1–25.

Hughes, G.M., Knights, B. & Scammel, C.A. (1969). The distribution of P_{O_2} and hydrostatic pressure changes within the branchial chambers of the shore crab, *Carcinus maenas* L. *Journal of Experimental Biology* **51**, 203–20.

Hughes, G.M. & Shelton, G. (1958). The mechanism of gill ventilation in three freshwater teleosts. *Journal of Experimental Biology* **35**, 807–23.

Hughes, G.M. & Shelton, G. (1962). Respiratory mechanisms and their nervous control in fish. In *Advances in Comparative Physiology and Biochemistry*, Vol. 1, ed. O.E. Lowenstein, pp. 275–364. London: Academic Press.

Itazawa, Y. & Takeda, T. (1978). Gas exchange in the carp gills in normoxic and hypoxic conditions. *Respiration Physiology* **35**, 263–9.

Johansen, K. (1972). Heart and circulation in gill, skin and lung breathing. *Respiration Physiology* **14**, 193–210.

Jones, D.R. & Schwarzfeld, T. (1974). The oxygen cost to the metabolism and efficiency of breathing in trout (*Salmo gairdneri*). *Respiration Physiology* **21**, 241–54.

Kiceniuk, J.W. & Jones, D.R. (1977). The oxygen transport system in trout (*Salmo gairdneri*) during sustained exercise. *Journal of Experimental Biology* **69**, 247–60.

Krogh, A. (1922). *Anatomy and Physiology of Capillaries*. New Haven: Yale University Press.

Malte, H. & Weber, R.E. (1985). A mathematical model for gas exchange in the fish gill based on non-linear blood gas equilibrium curves. *Respiration Physiology* **62**, 359–74.

Piiper, J. (1982). Respiratory gas exchange at lungs, gills and tissues: mechanisms and adjustments. *Journal of Experimental Biology* **100**, 5–22.

Piiper, J. (1989). Gas exchange efficiency of fish gills and bird lungs. In *Physiological Function in Special Environments*, ed. C.V. Paganelli & L.E. Farhi, pp. 159–71. New York: Springer-Verlag.

Piiper, J. (1990). Modeling of gas exchange in lungs, gills and skin. In *Advances in Environmental Physiology*, Vol. 6, ed. R.G. Boutilier, pp. 15–44. Berlin: Springer-Verlag.

Piiper, J., Gatz, R.N. & Crawford, E.C. (1976). Gas transport characteristics in an exclusively skin breathing salamander, *Desmognathus fuscus* (Plethodontidae). In *Respiration of Amphibious Vertebrates*, ed. G.M. Hughes, pp. 339–56. London: Academic Press.

Piiper, J., Meyer, M., Worth, H. & Willmer, H. (1977). Respiration and circulation during swimming activity in the dogfish *Scyliorhinus stellaris*. *Respiration Physiology* **30**, 221–39.

Piiper, J. & Scheid, P. (1972). Maximum gas transfer efficacy of models for fish gills, avian lungs and mammalian lungs. *Respiration Physiology* **14**, 115–24.

Piiper, J. & Scheid, P. (1975). Gas transport efficacy of gills, lungs and skin: theory and experimental data. *Respiration Physiology* **23**, 209–21.

Piiper, J. & Scheid, P. (1977). Comparative physiology of respiration: functional analysis of gas exchange organs in vertebrates. In *International Review of Physiology, Respiratory Physiology II*, Vol. 14, ed. J.G. Widdicombe, pp. 219–53. Baltimore: University Park Press.

Piiper, J. & Scheid, P. (1982). Physical principles of respiratory gas exchange in fish gills. In *Gills*. Society for Experimental Biology Seminar Series 16, ed. D.F. Houlihan, J.C. Rankin & T.J. Shuttleworth, pp. 45–61. Cambridge: Cambridge University Press.

Piiper, J. & Scheid, P. (1989). Gas exchange. Theory, models and experimental data. In *Comparative Pulmonary Physiology: Current Concepts*, Vol. 39, ed. S.C. Wood, pp. 369–416. New York: Marcel Dekker.

Piiper, J., Scheid, P., Perry, S.F. & Hughes, G.M. (1986). Effective and morphometric oxygen diffusing capacity of the gills of the elasmobranch *Scyliorhinus stellaris*. *Journal of Experimental Biology* **123**, 27–41.

Pringle, J.W.S. (1960). Models of muscle. In *Models and Analogues in Biology*. Symposia of the Society for Experimental Biology, Vol. 14, pp. 41–68. Cambridge: Cambridge University Press.

Randall, D.J. (1982). The control of respiration and circulation in fish during exercise and hypoxia. *Journal of Experimental Biology* **100**, 275–88.

Randall, D.J., Holeton, G.F. & Stevens, E.D. (1967). The exchange of oxygen and carbon dioxide across the gills of rainbow trout. *Journal of Experimental Biology* **46**, 339–48.

Richards, B.D. & Fromm, P.O. (1969). Patterns of blood flow through filaments and lamellae of isolated–perfused rainbow trout (*Salmo gairdneri*) gills. *Comparative Biochemistry and Physiology* **29**, 1063–70.

Riggs, D.S. (1963). *The Mathematical Approach to Physiological Problems.* Baltimore: Williams & Wilkins.

Rosenberg, H.I. (1973). Functional anatomy of pulmonary ventilation in the garter snake, *Thamnophis elegans. Journal of Morphology* **140**, 171–84.

Scheid, P., Hook, C. & Piiper, J. (1986). Model for analysis of countercurrent gas transfer in fish gills. *Respiration Physiology* **64**, 365–74.

Scheid, P. & Piiper, J. (1971). Theoretical analysis of respiratory gas equilibration in water passing through fish gills. *Respiration Physiology* **13**, 305–18.

Scheid, P. & Piiper, J. (1976). Quantitative functional analysis of branchial gas transfer: theory and application to *Scyliorhinus stellaris* (Elasmobranchii). In *Respiration of Amphibious Vertebrates*, ed. G.M. Hughes, pp. 17–38. London: Academic Press.

Scheid, P., Slama, H. & Piiper, J. (1972). Mechanisms of unidirectional flow in parabronchi of avian lungs: measurements in duck lung preparations. *Respiration Physiology* **14**, 83–95.

Shelton, G. (1970). The effect of lung ventilation on blood flow to the lungs and body of the amphibian, *Xenopus laevis. Respiration Physiology* **9**, 183–96.

Shelton, G. (1976). Gas exchange, pulmonary blood supply, and the partially divided amphibian heart. In *Perspectives in Experimental Biology*, ed. P. Spencer Davies, pp. 247–59. Oxford: Pergamon.

Shelton, G. (1985). Functional and evolutionary significance of cardiovascular shunts in the Amphibia. In *Cardiovascular Shunts.* Alfred Benzon Symposium 21, ed. K. Johansen & W.W. Burggren, pp. 100–20. Copenhagen: Munksgaard.

Shelton, G. & Boutilier, R.G. (1982). Apnoea in amphibians and reptiles. *Journal of Experimental Biology* **100**, 245–73.

Shelton, G. & Croghan, P.C. (1988). Gas exchange and its control in non steady-state systems: the consequences of evolution from water to air breathing in the vertebrates. *Canadian Journal of Zoology* **66**, 100–23.

Shelton, G., Jones, D.R. & Milsom, W.K. (1986). Control of breathing in ectothermic vertebrates. In *Handbook of Physiology*, Section 3, *The Respiratory System*, Vol. II, *Control of Breathing*, ed. N.S. Cherniack & J.G. Widdicombe, pp. 857–909. Bethesda, MD: American Physiological Society.

Short, S., Taylor, E.W. & Butler, P.J. (1979). The effectiveness of oxygen transfer during normoxia and hypoxia in the dogfish (*Scyliorhinus canicula* L.) before and after cardiac vagotomy. *Journal of Comparative Physiology B* **132**, 289–95.

Steen, J.B. & Kruysse, A. (1964). The respiratory function of teleostean gills. *Comparative Biochemistry and Physiology* **12**, 127–42.

Stevens, E.D. & Randall, D.R. (1967). Changes in gas concentration in blood and water during moderate swimming activity in rainbow trout. *Journal of Experimental Biology* **46**, 329–37.

Taylor, C.R. & Weibel, E.R. (1981). Design of the mammalian respiratory system. I. Problem and strategy. *Respiration Physiology* **44**, 1–10.

Templeton, J.R. & Dawson, W.R. (1963). Respiration in the lizard, *Crotaphytus collaris*. *Physiological Zoology* **36**, 104–21.

van Dam, L. (1938). On the utilisation of oxygen and regulation of breathing in some aquatic animals. Dissertation, University of Groningen.

Vitalis, T.Z. & Shelton, G. (1990). Breathing in *Rana pipiens*: the mechanism of ventilation. *Journal of Experimental Biology* **154**, 537–56.

Weibel, E.R. (1973). Morphological basis of alveolar–capillary gas exchange. *Physiological Reviews* **53**, 419–95.

Weibel, E.R. (1984). *The Pathway for Oxygen*. Cambridge, MA: Harvard University Press.

Wells, M.J. & Smith, P.J.S. (1985). The ventilation cycle in *Octopus*. *Journal of Experimental Biology* **116**, 375–83.

Wells, M.J. & Wells, J. (1982). Ventilatory currents in the mantle of cephalopods. *Journal of Experimental Biology* **99**, 315–30.

West, N.H. & Jones, D.R. (1975). Breathing movements in the frog *Rana pipiens*. I. The mechanical events associated with lung and buccal ventilation. *Canadian Journal of Zoology* **53**, 332–44.

Wilkens, J.L. & McMahon, B.R. (1972). Aspects of branchial irrigation in the lobster, *Homarus americanus*. I. Functional analysis of scaphognathite beat, water pressures and currents. *Journal of Experimental Biology* **56**, 469–79.

Wood, C.M. (1974). A critical examination of the physical and adrenergic factors affecting blood flow through the gills of rainbow trout. *Journal of Experimental Biology* **60**, 241–65.

R.McN. ALEXANDER and I.S. YOUNG

Dynamic models of breathing

Introduction

The theory of vibration is an important branch of engineering and is the subject of many textbooks, for example Thomson (1981). It analyses such problems as the vibrations of unbalanced machinery; the behaviour of the suspension systems of vehicles; the resonances of aeroplanes and other structures; and the design of instruments such as accelerometers and seismometers. Vibrations are regularly repeated movements. Breathing involves regularly repeated movements, albeit at lower frequencies than most of the vibrations that interest engineers. We can expect the theory of vibrations to be useful in the study of breathing.

This theory, as generally used, takes a very simple view of the world. It treats everything as being built from just three types of element: springs, dashpots and masses.

Suppose that one end of a spring is fixed. Initially, no force acts on it, but the free end is then moved through a distance z, stretching the spring by that amount. The force F that is needed to do this is proportional to the displacement z,

$$F = kz \tag{1}$$

The constant of proportionality k is known as the stiffness or spring constant.

Dashpots are generally imagined as oil-filled cylinders with loose-fitting pistons: such devices are used as shock absorbers in some cars. Movements of the piston along the cylinder are resisted by the viscosity of the oil. The force required is not proportional to the displacement but to the *rate* of displacement, $\mathrm{d}z/\mathrm{d}t$.

$$F = c \cdot \mathrm{d}z/\mathrm{d}t \tag{2}$$

The constant c is known as the damping constant.

Finally, the force required to move a mass m is proportional to the acceleration $\mathrm{d}^2z/\mathrm{d}t^2$

$$F = m \cdot \mathrm{d}^2z/\mathrm{d}t^2 \tag{3}$$

Society for Experimental Biology Seminar Series 51: *Oxygen Transport in Biological Systems*, ed. S. Egginton & H.F. Ross.

Forced vibrations

One of the standard problems in the theory of vibrations considers a mass mounted on a spring and a dashpot (Fig. 1*A*). Sinusoidally fluctuating forces make the mass vibrate up and down. Let the force at time t be $F_0 \sin \omega t$. Here F_0 is the force amplitude, and ω is known as the circular frequency. (If the frequency is f the circular frequency is $2\pi f$: remember that there are 2π radians in a revolution.) The forces required to move the spring, dashpot and mass are given by Eqns 1–3, so the equation of motion of the system is

$$F_0 \sin \omega t = kz + c \cdot \mathrm{d}z/\mathrm{d}t + m \cdot \mathrm{d}^2z/\mathrm{d}t^2 \tag{4}$$

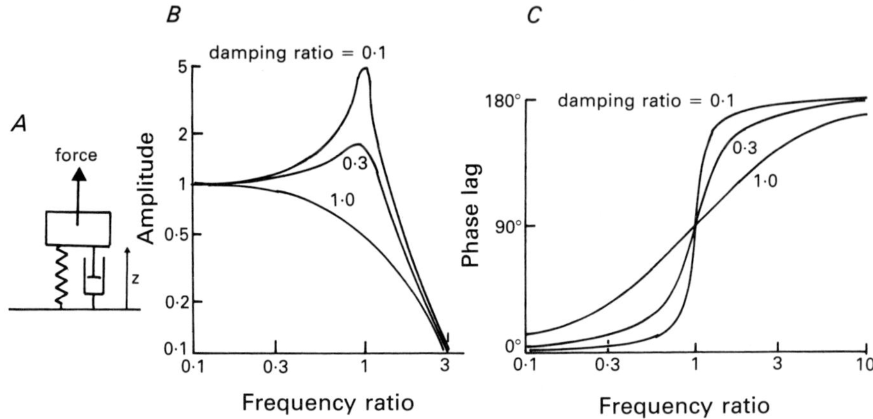

Fig. 1. *A*, Forced vibration of a mass mounted on a spring and dashpot. *B*, *C*, The steady-state response of the system shown in *A* to sinusoidally fluctuating forces. *B* shows the amplitude (expressed in dimensionless form, as explained in the text) and *C* shows the phase lag, plotted against the ratio (forcing frequency)/(undamped natural frequency).

We need not concern ourselves here with the build-up of the vibration from rest but only with the final steady state. The steady-state solution of Eqn 4 is

$$z = Z \sin(\omega t - \phi)$$

$$\text{where} \quad Z = (F_0/k)\{[1-(m\omega^2/k)]^2 + [c\omega/k]^2\}^{-\frac{1}{2}}$$
$$\text{and} \quad \omega = \arctan\{(c\omega/k)/[1-(m\omega^2/k)]\} \tag{5}$$

(See Thomson, 1981.) Notice that if there were no damping ($c = 0$) and if the circular frequency were k/m (i.e. if $m\omega^2/k = 1$) the amplitude of vibration Z would be infinite. The corresponding frequency f_n is known as the undamped natural frequency of the system.

$$f_n = (1/2\pi)(k/m)^{\frac{1}{2}} \tag{6}$$

Figure 1*B* shows how the steady-state amplitude of vibration depends on the frequency, for three different degrees of damping. The amplitude is expressed in dimensionless form, as Zk/F_0; and the frequency is shown as a multiple of the undamped natural frequency. The degree of damping is expressed in the conventional way, in terms of the dimensionless damping ratio $c/4\pi m f_n$. As this damping ratio rises towards 1, the maximum amplitude falls, as does the frequency at which it occurs. When it exceeds 1, the maximum occurs at zero frequency.

Figure 1*C* shows how the phase angle (ϕ in Eqn 5) varies with frequency. It is zero at low frequencies, 90° at the undamped natural frequency and approaches 180° at higher frequencies.

Hot dogs

Crawford (1962) found that dogs pant at very constant frequencies: 5.3 Hz in the case of his 12 kg mongrels. He suggested that the panting frequency might be the natural frequency of vibration of the respiratory system. His hypothesis implied that the system behaved like the simple model of Fig. 1, and was not too heavily damped. If that were the case, small force fluctuations could make it vibrate with large amplitudes, at frequencies near the natural frequency. More important, perhaps, at the undamped natural frequency the displacements would be 90° out of phase with the forces: in other words, the *rate* of displacement would be in phase with the force so the muscle would always be shortening while exerting force. At no stage in the respiratory cycle would the muscles be forcibly stretched, so as to act like brakes, degrading mechanical energy to heat. Because the function of panting is to dissipate heat, heat production in the respiratory muscles should be minimised.

Crawford (1962) anaesthetised his dogs and ventilated their lungs at various frequencies in a whole-body respirator. He recorded oesophageal pressure and the rate of flow of air into the lungs, and found the frequency at which they were in phase: this was the undamped natural frequency. He found that it was 5.3 Hz, the same as the panting frequency. His view of the respiratory system was, of course, highly simplified. He treated it as having only one degree of freedom of movement, whereas we now know that the relative amplitudes of rib cage movements and abdominal wall movements of dogs vary with the frequency of ventilation (Boynton *et al.*, 1989). His insight was nevertheless valuable.

Periodic forces

The pattern of force that acts on a system, making it vibrate, may not be simple harmonic but may nevertheless be periodic: that is to say, it may repeat itself at regular intervals. Any periodic pattern of force can be represented as a Fourier series,

$$F = a_0/2 + a_1 \sin(\omega t - \phi_1) + a_2 \sin(2\omega t - \phi_2) + + a_3 \sin(3\omega t - \phi_3) \ldots \quad (7)$$

Here $a_0/2$ is the mean force and the terms that follow represent sinusoidally fluctuating forces at the fundamental frequency and at 2, 3, 4, etc., times that frequency. The factors a_1, a_2, etc., are the amplitudes of the harmonic components and ϕ_1, ϕ_2, etc., are phase lags. The response of the system shown in Fig. 1*A* to a periodically fluctuating force is the sum of its responses to all the harmonic components, which can be calculated individually by means of Eqns 5 (Thomson, 1981).

Galloping horses I

Bramble & Carrier (1983) showed that, when horses and some other mammals gallop, they take exactly one breath per stride. They suggested that the running movements drive breathing, and proposed three possible mechanisms. We have investigated their feasibility further.

Alexander (1989) made a preliminary analysis. He wrote models based on the theory of vibration and tested them against the data for two short sequences of galloping that were published by Bramble & Carrier (1983). We have since been enabled to obtain better, more extensive data through the cooperation of Professor P. J. Butler and Dr A. J. Woakes of the University of Birmingham and Dr L. Anderson of the Animal Health Trust, Newmarket. They were investigating the respiratory physiology of racehorses galloping on a large treadmill, using flow meters in a mask over the nostrils to measure respiratory flow rates. We took cine films simultaneously, using a synchronisation device so that we could identify the point on the flow record corresponding to each frame of the film.

Figure 2 shows an example of our data. At this stage we are concerned with the footfall patterns, shown at the top, and the flow rate record. Notice that expiration occurs while the fore feet are on the ground.

Bramble & Carrier (1983) had suggested that air was driven out of the lungs by the force transmitted to the thorax by the fore legs, when the feet hit the ground.

This suggested the model shown in Fig. 3*A*. In it, *m* is the mass of the fore legs and chest wall, and *M* is the mass of as much of the rest of the body, as is effectively supported by the fore legs. This model is more complex than the one in Fig. 1*A*, but Alexander (1989) argued that it could be treated as a combination of the simple models show in Fig. 3*B* and *C*. Figure 3*B* represents the entire fore quarters of the horse (including the legs and chest wall), mounted on a spring that represents the combined compliance of the legs and thorax; Fig. 3*C* represents the legs and chest wall vibrating relative to the rest of the body. If the natural frequencies of these two systems were sufficiently different, there would be little interaction between them and their vibrations could be considered separately.

Alexander (1989) argued that the natural frequencies of the two systems must indeed be very different. Because mass *m* is much smaller than mass *M*, the natural frequency of the system in Fig. 3*C* must be much higher than that of the system in Fig. 3*B*. Further, he argued that the time for which the fore feet were in contact with the ground would be between half and one period of vibration of the system represented by Fig. 3*B*. This implied that the stride frequency must be much lower than the natural frequency of the system shown in Fig. 3*C*.

Figure 4 shows how this system would behave if the stride frequency were 0.3 times its natural frequency. The force exerted on it by the fore feet is assumed to have the time course shown at the bottom of the graph. The Fourier series that describes this force pattern has been used to calculate the response of the system for three different damping ratios. With low damping ratios (for example, 0.2) mass *m* would rise (squeezing air out of the lungs) while the force on the feet was increasing, fall while the force was decreasing and continue vibrating while the feet were off the ground. Only with high damping ratios (for example, 5) would the mass continue rising during most of the period of ground contact, as required to simulate the observed pattern of ventilation. The graph has been drawn for one particular ratio of the stride frequency to the natural frequency, but the conclusion does not depend critically on the choice of ratio: the proposed mechanism can explain the observed relationship between ventilation and running movements only if the system is very heavily damped.

This implies that large quantities of work would be required to ventilate the lungs. The energy cost of ventilation would be a much larger

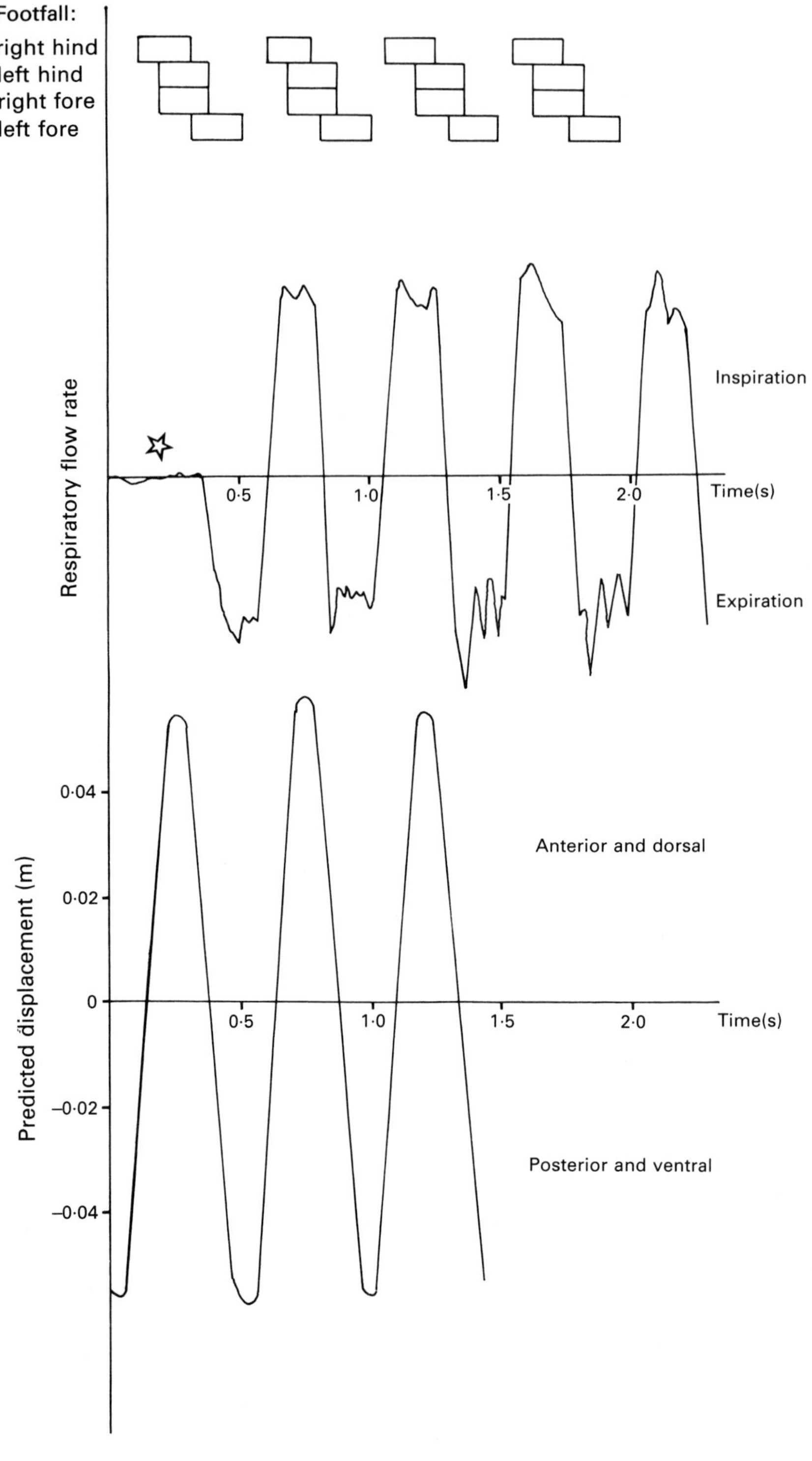
Footfall:
right hind
left hind
right fore
left fore
Respiratory flow rate
Inspiration
Expiration
0·5
1·0
1·5
2·0
Time(s)
Predicted displacement (m)
0·04
0·02
0
−0·02
−0·04
Anterior and dorsal
Posterior and ventral
0·5
1·0
1·5
2·0
Time(s)

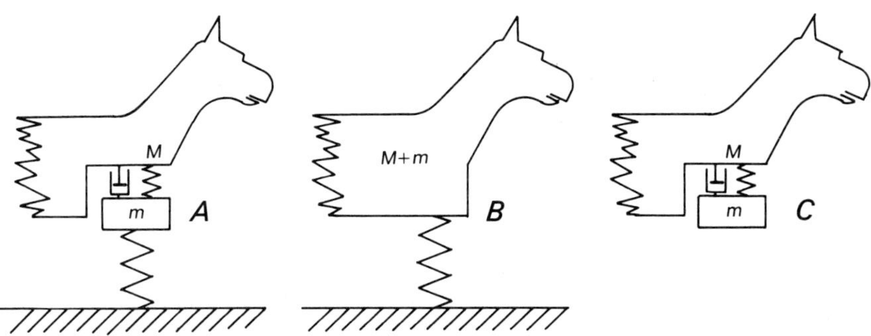

Fig. 3. *A*, A model of a possible mechanism, whereby ventilation could be driven by the forces on the fore feet. The behaviour of this system can be estimated by considering systems *B* and *C* separately. From Alexander (1989), by permission of the Zoological Society of London.

fraction of the total energy cost of running than appears from the work of Lafortuna & Saibene (1991) (also Young *et al.*, 1992). We therefore think it unlikely that this mechanism is important.

Accelerometers and seismometers

Figure 5*A* represents another standard problem in the theory of vibrations. A mass m is mounted on a spring of stiffness k and a dashpot of damping constant c in a box. The box is vibrated so that at time t its displacement, along a line parallel to the spring and dashpot, is $s_0 \sin \omega t$; and the displacements z of the mass relative to the box are observed. The acceleration of the box is $-s_0\omega^2 \sin \omega t$, so the acceleration of the mass is $\mathrm{d}^2z/\mathrm{d}t^2 - s_0\omega^2 \sin \omega t$, and the equation of motion of the system is

$$kz + c \cdot \mathrm{d}z/\mathrm{d}t + m(\mathrm{d}^2z/\mathrm{d}t^2 - s_0\omega^2 \sin \omega t) = 0 \tag{8}$$

The steady-state solution to this equation is

$$z = Z \sin(\omega t - \phi)$$

where $Z = (s_0 m\omega^2/k)\{[1 - (m\omega^2/k)]^2 + [c\omega/k]^2\}^{-\frac{1}{2}}$

and $\phi = \arctan\{(c\omega/k)/[1 - (m\omega^2/k)]\}$ (9)

Fig. 2. The footfall pattern of a galloping horse, air flow through its nostrils and predicted displacements of the visceral piston, from the data of Young *et al.* (1992). The horse was cantering on a treadmill at $9\,\mathrm{m\,sec^{-1}}$. Zero air flow is registered at the star, when the horse swallowed.

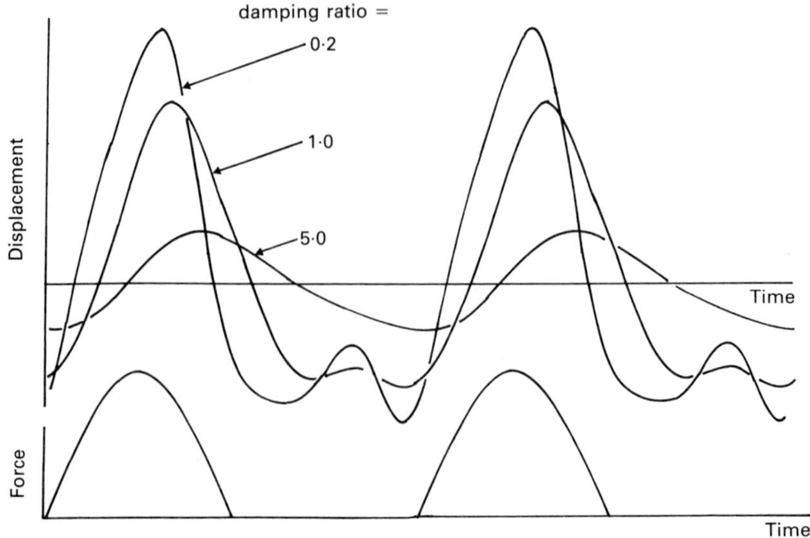

Fig. 4. If the pattern of force (shown below) were applied to the system shown in Fig. 3*C*, it would vibrate as shown in the graphs of displacement (above). The stride frequency is assumed to be 0.3 times the undamped natural frequency of the system, and displacements are shown for three different damping ratios.

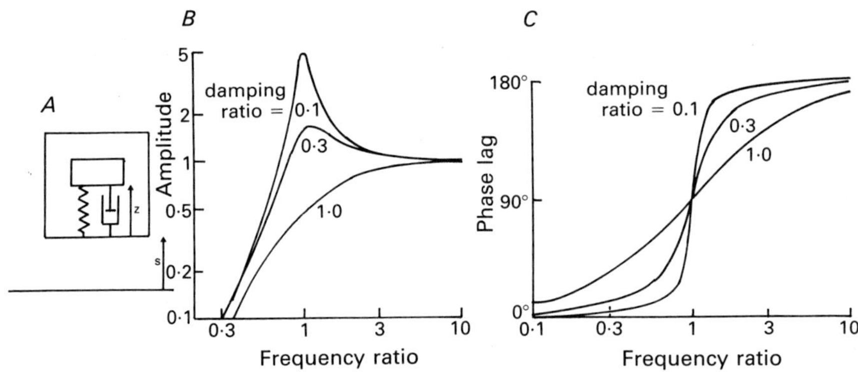

Fig. 5. *A*, A model that may represent either an accelerometer or a seismometer. *B*, *C*, The steady-state response of the model to simple harmonic vibration. *B* shows the amplitude (expressed in dimensionless form) and *C* shows the phase lag, plotted against the ratio (vibration frequency/undamped natural frequency).

(Thomson, 1981). Figure 5*B* and *C* show how the amplitude *Z* and the phase difference ϕ depend on the frequency. The undamped natural frequency is given by Eqn 6. At this frequency $m\omega^2/k = 1$, the phase difference is 90°, and if there were no damping the amplitude *Z* would be infinite.

Figure 5*A* may represent either an accelerometer or a seismometer. Accelerometers are built with stiff springs and small masses, so as to have high natural frequencies. Their displacements *z* are proportional to the acceleration of the box, provided that the box moves at a frequency much lower than the instrument's natural frequency. Seismometers are built with large masses and soft springs, so as to have low natural frequencies. If the box vibrates at a frequency much higher than the instrument's natural frequency, the mass remains almost stationary while the box vibrates around it: in other words, *z* is almost equal and opposite to *s*.

Vibrations are not generally simple harmonic, but may nevertheless be periodic. Such motions can be described by Fourier series, as explained above for periodic force fluctuations. The response of the system shown in Fig. 4 to periodic motion is the sum of its responses to all the harmonic components, so is easily calculated (Thomson, 1981).

Galloping horses II

We have already discussed and rejected one of the three mechanisms suggested by Bramble & Carrier (1983), whereby the galloping movements of horses might drive breathing. Another of their suggestions was the visceral piston mechanism, illustrated by Fig. 6. The piston represents the abdominal viscera, which are displaced posteriorly as the diaphragm flattens in inspiration, and anteriorly in expiration. The spring represents the elastic properties of the respiratory system and the dashpot represents its damping. The galloping horse does not move at constant velocity, but accelerates and decelerates in each stride, and these periodic movements must tend to make the piston (the viscera) vibrate within its cylinder (the body cavity). Thus, galloping movements may drive breathing.

The amplitude and phase of breathing movements driven by locomotion would depend on the natural frequency and damping factor of the system. If the spring were very stiff, making the natural frequency much higher than the stride frequency, the system would behave as an accelerometer: displacements of the viscera would be proportional to the acceleration of the body, and would be small. If, on the other hand, the spring were very compliant, making the natural frequency much lower than the stride frequency, the system would behave as a seismometer: the viscera would travel with almost constant velocity while the body

Fig. 6. A diagram of the visceral piston mechanism.

accelerated and decelerated around them. The respiratory movements would be much larger than in the case of the accelerometer but the system would nevertheless be unsatisfactory, because very large displacements of the viscera relative to the body wall would occur when the animal accelerated rapidly from rest. A third possibility is that the spring may have just the right stiffness to match the natural frequency of the system to the stride frequency (Bramble, 1986; Alexander, 1987). In that case the amplitude of the visceral movements would depend critically on the damping, and might be very large if the damping factor were low.

We now return to the experiment in which we filmed horses galloping on a treadmill while recording air flow through their nostrils (Fig. 2). A piece of adhesive plaster had been attached to the horses' sides, lateral to the anterior lumbar vertebrae. We measured the Cartesian coordinates of this marker in successive frames of the film and calculated the horizontal and vertical components of its acceleration. The diaphragm of the horse makes an angle of about 45° with the horizontal, so we modelled the respiratory system as a cylinder and piston tilted at 45° (Fig. 6), and calculated the component of the body's acceleration in this direction. In steady galloping this was periodic, so we were able to represent it as a Fourier series and calculate the movements of the visceral piston for chosen natural frequencies and damping factors. We assumed that the natural frequency equalled the stride frequency and chose a damping factor that gave realistic amplitudes of diaphragm movement. An example of our results is shown at the bottom of Fig. 2.

The amplitude of the calculated movements of the visceral piston tells us nothing because the damping ratio was chosen to make it realistic. The phase relationship between these movements and the movements of galloping can, however, be used to test the visceral piston hypothesis. Figure 2 shows that the visceral piston is not predicted to start moving forward until the second half of the period in which air was observed to flow out through the nostrils, and is not predicted to move backwards until late in the observed period of inspiration. The phase relationship between its movements and the movements of galloping, in this and other records, is inconsistent with the suggestion that ventilation is driven by the visceral piston.

We have eliminated two of the three mechanisms proposed by Bramble & Carrier (1983), but the third remains. The lumbosacral joint flexes at one stage of the galloping stride, tending to push the viscera forwards, and extends at another, tending to draw them back. We found that expiration occurs while the back is bending and inspiration while it is extending, so these movements may help to ventilate the lungs. We have discussed this mechanism more fully elsewhere (Young *et al.*, 1992) but we discuss it no further in this chapter as its elucidation did not depend on the theory of vibration.

Conclusion

We have been concerned throughout this chapter with a simple assembly that recurs frequently in the theory of vibrations: a mass mounted on a spring and a dashpot. We started by considering the response of this assembly to a sinusoidally fluctuating force, and used the result to interpret the panting of dogs. Next, we showed how the analysis can be extended to the more general case of periodic forces, and used it to show that in galloping horses, expiration is probably not driven by the forces on the fore feet. Finally, we considered the response of the assembly to vibration, and used this to show that in galloping horses, ventilation cannot be driven by the visceral piston mechanism. A third mechanism, flexion of the lumbosacral joint, remains as the most likely explanation of the linkage between breathing and galloping in horses. In this example the theory of vibrations has played a negative role, enabling us to eliminate false hypotheses, but in a similar analysis of wallabies (*Macropus* spp.) its role has been more positive. It seems that when wallabies hop, their breathing movements may be driven by a visceral piston mechanism (Alexander, 1989). The theory of vibrations can also be applied to more complex problems, but these examples are sufficient to demonstrate its usefulness in studies of ventilation of the respiratory tract.

Acknowledgements

I.S.Y. thanks the Science and Engineering Research Council for support. We are also most grateful to an anonymous referee, who helped us to detect and correct a serious error.

References

Alexander, R.McN. (1987). Wallabies vibrate to breathe. *Nature* **328**, 477.

Alexander, R.McN. (1989). On the synchronization of breathing with running in wallabies (*Macropus* spp.) and horses (*Equus caballus*). *Journal of Zoology* **218**, 69–85.

Boynton, B.R., Glass, G., Frantz, I.D. & Fredberg, J.J. (1989). Rib cage vs. abdominal displacement in dogs during forced oscillation to 32 Hz. *Journal of Applied Physiology* **67**, 1472–8.

Bramble, D.M. (1986). Biomechanical and neuromotor factors in mammalian locomotor–respiratory coupling. In *Nonlinear Oscillations in Biology and Chemistry*, ed. H.G. Othmer, pp. 139–49. Berlin: Springer-Verlag.

Bramble, D.M. & Carrier, D.R. (1983). Running and breathing in mammals. *Science* **219**, 251–6.

Crawford, E.C. (1962). Mechanical aspects of panting in dogs. *Journal of Applied Physiology* **17**, 249–51.

Lafortuna, C.L. & Saibene, F. (1991). Mechanics of breathing in horses at rest and during exercise. *Journal of Experimental Biology* **155**, 245–59.

Thomson, W.T. (1981). *Theory of Vibrations with Applications*, 2nd edn. London: Allen & Unwin.

Young, I.S., Alexander, R.McN., Woakes, A.J., Butler, P.J. & Anderson, L. (1992). The synchronization of ventilation and locomotion in horses (*Equus caballus*). *Journal of Experimental Biology* **166**, 19–31.

STEVEN F. PERRY

Morphometry of vertebrate gills and lungs: a critical review

Introduction

In the first chapter of this volume, G. Shelton introduced several models currently used by respiratory physiologists in order to explain their experimental results. The underlying general principle of these models, as applied to the gas-exchange organ, is illustrated in Fig. 1. Here the respiratory medium (water for the gill, air for the lung) is convectively delivered to the water/air–blood barrier of the organ. The rate (in moles per unit time) of O_2 delivery to the diffusion barrier depends on the ventilation ($\dot{V}$), on the O_2 partial pressure (P_{O_2}) of the respiratory medium and on its capacitance coefficient (βm). Similar parameters define the rate of O_2 uptake on the blood side of the diffusion barrier, whereby the convection factor is termed perfusion ($\dot{Q}$) rather than ventilation and the capacitance coefficient (βm) is equivalent to the effective slope of the O_2 dissociation curve. The properties relevant to the diffusion membrane itself are contained in Krogh's diffusion constant (K_{O_2}) for O_2 in lung tissue, connective tissue or cytoplasm-like aqueous solutions (Bartels, 1971).

In addition to this general model, Piiper & Schied (1972, 1975) have recognised at least four different specific models for gas exchange in vertebrate gills and lungs (Fig. 2). Three of these are designed to represent highly evolved and very effective gas-exchange organs: fish gills and the lungs of mammals and birds. Inspection of Fig. 2 reveals that these specific models do not differ in general principle: each demonstrates an external medium, a blood side and a membrane of finite thickness separating the two. Rather, they differ in the interrelationship of the mass flow directions of the external medium and blood.

The question arises: how good are these different specific models at exchanging gas? One measure of 'goodness' is the gas-exchange efficacy, which is defined as the total conductance of the model (Piiper & Scheid, 1972, 1975; Piiper, 1982):

$$G_{tot} = \dot{M}(P_i - P_v)^{-1} \qquad (1)$$

Society for Experimental Biology Seminar Series 51: *Oxygen Transport in Biological Systems*, ed. S. Egginton & H.F. Ross.

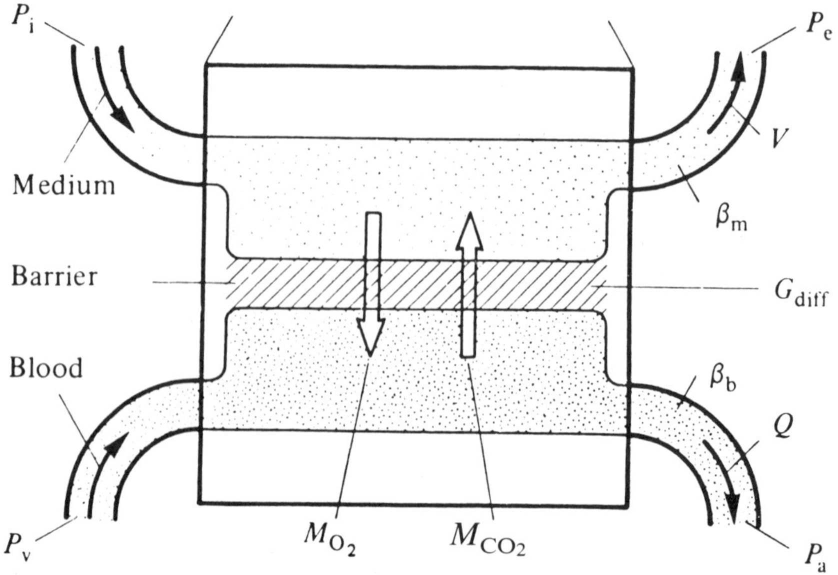

Fig. 1. General model of vertebrate gas-exchange organs. O_2 is supplied to the diffusion membrane (Barrier) which separates the respiratory Medium and the Blood. O_2 and CO_2 content in the medium are determined by their respective inspired (Pi) and expired (Pe) partial pressures and by the capacitance coefficient (βm) of the medium for the appropriate gas, while the O_2 and CO_2 content in the blood is determined by the venous (Pv) and arterial (Pa) partial pressures and by the capacitance coefficient (βb) of blood. $\dot{V}$ and $\dot{Q}$ are rates of ventilation and perfusion, $\dot{M}_{O_2}$ and $\dot{M}_{CO_2}$ are the rates of O_2 consumption and CO_2 output, respectively. From Piiper (1982), courtesy of Cambridge University Press.

where Gtot is the total conductance for a designated gas, $\dot{M}$ is the transfer rate of that gas and (Pi − Pv) is the partial pressure difference between inspired gas and venous efflux. In general, the efficacy of the specific models decreases in the order counter-current > cross-current > ventilated pool = infinite pool. The difference in efficacy between models increases as the surface area-to-barrier thickness ratio (ADF; see below) increases (Piiper, 1990).

Another measure of the 'goodness' of a gas exchanger is its efficiency. This can be expressed as the ratio of the physiologically measured diffusing capacity for oxygen (D_{O_2}; see below) to $G\text{tot}_{O_2}$ or to any other measure of the maximum possible oxygen conductance of the gas-

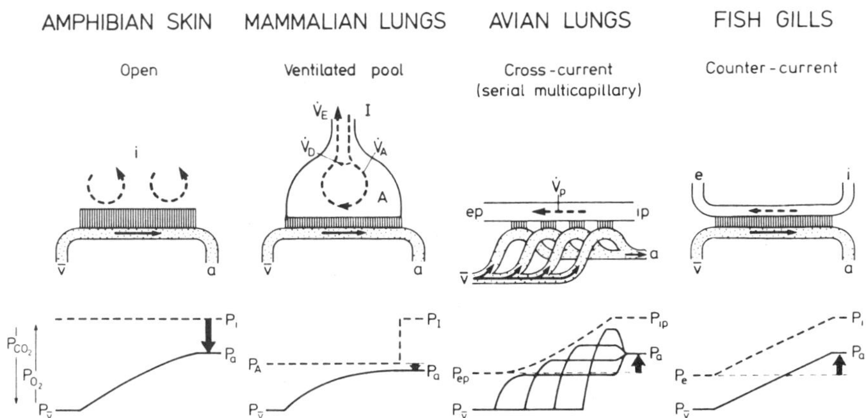

Fig. 2. Specific models for vertebrate gas-exchange organs. The Open model is exemplified by amphibian skin; the Ventilated pool model, by the mammalian lung; the Cross-current model, by the avian lung; and the Counter-current model, by the fish gill. Low-performance lungs may combine the first three models to varying degrees. A, alveolar gas; a, arterial blood; e, expired respiratory medium; ep, expired parabronchial air; I, end inspiratory air in the bronchoalveolar lung; i, inspired respiratory medium; ip, inspired parabronchial air; $\dot{V}$ is the ventilation rate of the alveoli ($\dot{V}$A), of dead space ($\dot{V}$D) or of both ($\dot{V}$E). Thin arrows indicate the flow direction of blood (solid lines) or medium (broken lines) and thick arrows, the medium–blood partial pressure differences. From Piiper (1990), courtesy of Springer-Verlag.

exchange organ. It is also possible to express the efficiency of gas exchange at the whole-animal level as the rate of oxygen consumption ($\dot{M}_{O_2}$) per unit diffusing capacity. Here again, a measure of the maximum possible diffusing capacity is necessary; there are, however, a number of technical difficulties associated with such measurements (see Shelton, this volume).

Since stereological morphometry (see Mayhew, this volume) can provide precise measurements of the lengths, surface areas and volumes necessary for determination of diffusing capacity, it was hoped that morphometry could be the gold standard against which physiological results could be measured. The concept of diffusing capacity would thus allow free translation between morphology and physiology. It is now clear, however, that the morphometrically accessible model is not completely congruent with the physiological one and that neither truly represents the actual gas-exchange situation in the living animal.

Diffusing capacity and the morphometrically accessible model

For both the morphometrist and the physiologist, oxygen diffusing capacity (D_{O_2}) – the physical capacity for oxygen transfer per unit time and driving pressure (Dejours, 1981) – is approached in terms of Fick's law of diffusion, the respiratory physiologist's equivalent of Ohm's law:

$$I = 1/R \cdot U \tag{2}$$

$$\dot{M}_{O_2} = D_{O_2} \cdot \Delta P_{O_2} \tag{3}$$

where I is current, R is resistance, U is potential difference (voltage), $\dot{M}_{O_2}$ is the oxygen consumption rate and ΔP_{O_2} is the mean partial pressure gradient for O_2, theoretically the alveolar–capillary P_{O_2} difference across the diffusion membrane. As we see, D_{O_2} is by analogy the reciprocal of the resistance to oxygen transfer.

In order to determine the total resistance of an electrical circuit in series, the resistances are added. The total conductance (diffusing capacity) of the gas-exchange organ, however, is given by the reciprocal of the sum of the reciprocals of the individual components:

$$1/D(\mathrm{G;L})_{O_2} = \sum_{i=1}^{n} 1/D\mathrm{i}_{O_2} \tag{4}$$

where $1/D(\mathrm{G;L})_{O_2}$ is the total diffusing capacity of the gill or lung and $D\mathrm{i}_{O_2}$ is the oxygen diffusing capacity of each consecutive layer of the diffusion barrier which oxygen must traverse from the external medium to the haemoglobin of the erythrocyte.

These individual components, while readily discernible anatomically, are physiologically separable only into two components: the 'membrane' diffusing capacity and the rate of oxygen combination with capillary blood, $\theta\mathrm{c}$ (Cotes, 1968). Morphometrically, however, it is possible to break down the 'membrane' diffusing capacity into its epithelial, interstitial, endothelial and plasma layers as follows:

$$D\mathrm{i}_{O_2} = K\mathrm{i}_{O_2}(S\mathrm{i}/\tau\mathrm{hi}) \tag{5}$$

where $D\mathrm{i}_{O_2}$, $K\mathrm{i}_{O_2}$ and $S\mathrm{i}$ are the O_2 diffusing capacity, Krogh's diffusion constant and the surface area of the appropriate layer, respectively. $\tau\mathrm{hi}$ is the harmonic mean distance across the barrier in question (Weibel & Knight, 1964). The harmonic mean (the reciprocal of the mean of the reciprocals) weights short distances in a skewed distribution. Since short distances functionally predominate, it is a more attractive measure of central tendency than is the arithmetic mean in order to indicate the mean diffusion distance: it expresses mean tissue 'thinness' rather than thickness (Weibel & Knight, 1964).

Thus, knowing only a single physical constant ($K\text{i}_{O_2}$) for each tissue type encountered, the morphometrist can calculate the diffusing capacity not only of the whole gill or lung, but also of each successive layer which lies in the intrapulmonary oxygen diffusion path (for placental diffusion: see Mayhew, this volume). In addition, the proportional contribution of any desired anatomical subunit *j* (e.g. gill arch or lung lobe) of the whole gas-exchange organ can be calculated separately:

$$D(\text{G;L})_{O_2} = \sum_{j=1}^{n} D_j \tag{6}$$

This method has been successively employed in order to compare quantitatively various regions within the reptilian lungs (Perry, 1983, 1990*b*).

In spite of this promising beginning, close agreement has been reached between morphometric and physiological diffusing capacities only in the case of the unperfused visceral pleura of the dog (Magnussen *et al.*, 1974), the salamander skin, and the shark gill (Piiper *et al.*, 1986). The reasons for this lack of agreement could be several (Fig. 3). First, the membrane diffusing capacity as a physiological entity includes diffusion through a layer of stagnant air, a concept termed stratified inhomogeneity (Piiper & Scheid, 1971). Indeed, the matching of morphometrical and physiological diffusing capacities in the shark gill was achieved only by taking into account a boundary layer on the respiratory surface (Piiper *et al.*, 1986).

Secondly, neither the P_{O_2} at the alveolar surface nor that in the pulmonary capillary bed is directly measurable physiologically, and must be estimated from values for end-expiratory air, mixed venous blood and arterial blood. Furthermore, shunts and occlusions of all types are unavoidable physiologically, but inaccessible morphometrically. On the other hand, although stereological morphometry can easily provide estimates of alveolar and capillary surface area and of air–blood diffusion distances, the distance travelled by a red blood cell in the capillary bed of an average alveolus is much more difficult, and less accurate, to derive. These arguments apply in similar measure to gills. Thus, although respiratory physiology and morphology are merely two different ways of looking at the same phenomenon, the approaches are so different that their combination at best remains an approximation rather than a calculation of the 'true' diffusing capacity of a lung.

Thirdly, using Eqns 4 and 5 alone, it is only conditionally possible to

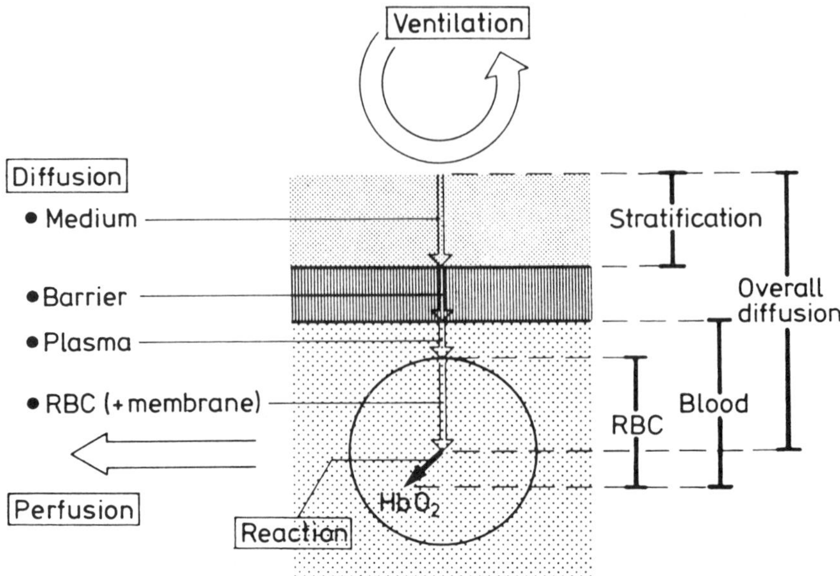

Fig. 3. Stages on O_2 passage from the respiratory medium to the red blood cell (RBC). HbO_2, oxygenated haemoglobin. From Piiper (1990), courtesy of Springer-Verlag.

calculate $D(G;L)_{O_2}$ for all animals. Even assuming that the 'true' surface area and barrier thickness for gas diffusion can be precisely determined, the application of physical constants in order to allow the calculation of diffusing capacity is problematic. To begin with, Ki_{O_2} has been determined for only a limited number of tissues and these include neither gills nor lower vertebrate and avian lungs. In addition, it has not been established that convective gas movement is negligible at all levels of the morphometrically accessible model. This may be particularly true within the plasma layer and the red blood cells (Piiper, 1990). The relative contribution of convective mixing may become important in comparison of such animals as the dog and the goat, which differ greatly in erythrocyte size, haematocrit and harmonic mean thickness of the plasma layer (see p. 64).

Finally, the fixation procedures commonly employed for lung morphometry do not guarantee physiologically meaningful haematocrits (Bur *et al.*, 1985) and blood pressures. Thus, significant variations in intravascular dimensions may occur from investigator to investigator for the same species, as has been demonstrated in the parakeet (Dubach, 1981; Maina, 1989).

In conclusion, even should a perfect agreement between physiological and morphometric diffusing capacity be achieved, the meaning of this agreement is not certain. How, then, can morphometry help us to understand better the function of gills and lungs as gas-exchange organs?

Morphometry as an aid to understanding the function of gas-exchange organs

Although the comparison of morphometrically and physiologically derived data has serious limitations (see above), morphometry can be used to advantage for comparison with other morphometric data, and the results of these analyses can be presented as an hypothesis to be tested physiologically.

The numerical characterisation of respiratory organs, however, is complicated by certain procedural differences outlined in an earlier review (Perry, 1989), as well as by the tendency of workers in different fields to report their data in ways befitting the functional interpretation in the respective animal group. Most students of mammalian and avian lungs, for example, report morphometric diffusing capacities which include measured 'diffusion' distances through the plasma layer of pulmonary capillaries as well as θ, the oxygen-binding capacity of the erythrocytes (for further references see: Weibel, 1970/1; Weibel *et al.*, 1981; Lechner, 1984; Maina *et al.*, 1989). Those studying reptiles and fish, however, tend to report the anatomical diffusion factor (ADF), from which the diffusing capacity of the air/water–blood tissue barrier can be estimated as follows:

$$\mathrm{ADF} = S\mathrm{AR}/\tau\mathrm{ht} \qquad (7)$$

$$D\mathrm{t}_{O_2} = \mathrm{ADF} \cdot K\mathrm{t}_{O_2} \qquad (8)$$

where $S\mathrm{AR}$ is the respiratory surface area of the gill or lung, $\tau\mathrm{ht}$ is the harmonic mean thickness of the air/water–blood tissue barrier, $D\mathrm{t}_{O_2}$ is the oxygen diffusing capacity of the air/water–blood tissue barrier and $K\mathrm{t}_{O_2}$ is Krogh's diffusion coefficient for oxygen in the appropriate diffusion barrier at the appropriate temperature.

Still other investigators report only surface areas (Tenney & Remmers, 1963; Tenney & Tenney, 1970), barrier thicknesses (Goniakowska-Witalinska, 1980; Meban, 1980; Welsch & Müller, 1980) or capillary dimensions (Czopek, 1955, 1965; Jakubowski, 1989). The *a posteriori* combination of these surface areas and barrier thicknesses to yield an ADF is risky, since the sampling location is usually not given. In the Nile crocodile failure to appropriately area-weight the barrier thickness measurements leads to a 30% underestimate of ADF for the whole lung (Perry, 1990*a*).

A further pitfall in the interspecific comparison of morphometric data is the failure of some investigators to conduct and/or to publish exhaustive analyses of error. In the lesser spotted dogfish (*Scyliorhinus canicula*), for example, van Dieken (1989) has recently shown that traditional estimates of gill surface area obtained from projected whole mounts of secondary lamellae contain a systematic error resulting in values that are from 9 to 33% too small.

In poikilothermic vertebrates, particularly in those with a three-chambered heart, only limited significance can be attached to calculated Dt_{O_2} values. Changes in body temperature of 20 °C are not unusual in desert reptiles and the possibility of intracardiac blood mixing means that the θ value for erythrocytes in the lung could vary within a single blood sample (Hicks & Wood, 1989). In addition, for a basking animal there is no assurance that the temperature is uniform within the lung. The temperature not only affects the blood chemistry, it also causes a 1% change in Kt_{O_2} per °C (Bartels, 1971). We shall thus rely on purely morphological parameters for the main interspecies comparisons and, taking account of all caveats, proceed to compare Dt_{O_2} in selected fish, reptiles, mammals and birds.

Morphometric comparison of gills and lungs

Anatomical diffusion factor (ADF)

Fish gills are not innately anatomically inferior to lungs as gas-exchange organs. Even taking into account a possible 30% underestimation of secondary lamellar surface area would change little in Figs 4 and 5: the tuna gill remains a high-performance gas-exchange organ that is anatomically comparable with the mammalian lung. The body mass-normalised ADF of fish gills ranges over two orders of magnitude, the surface area and barrier thickness values completely overlapping those of reptilian lungs. This wide range suggests that the structure of fish gills may more closely reflect the metabolic strategy of the animal. A rewarding field for future research would be to quantify the 'pathway for oxygen' from gill to mitochondrion in the tuna, trout and dogfish, all of which have been extensively studied physiologically (for references, see Randall *et al.*, 1967; Piiper *et al.*, 1977; Brill, 1979, 1987; Gooding *et al.*, 1981; Graham & Laurs, 1982; Wood & Perry, 1985; Bushnell & Brill, 1990).

From the comparative anatomical point of view, it is interesting to compare the position of the tuna in Fig. 6 relative to that of other fish with the position of birds compared with that of reptiles and mammals. The tuna differs from the trout, for example, by having both a six-times

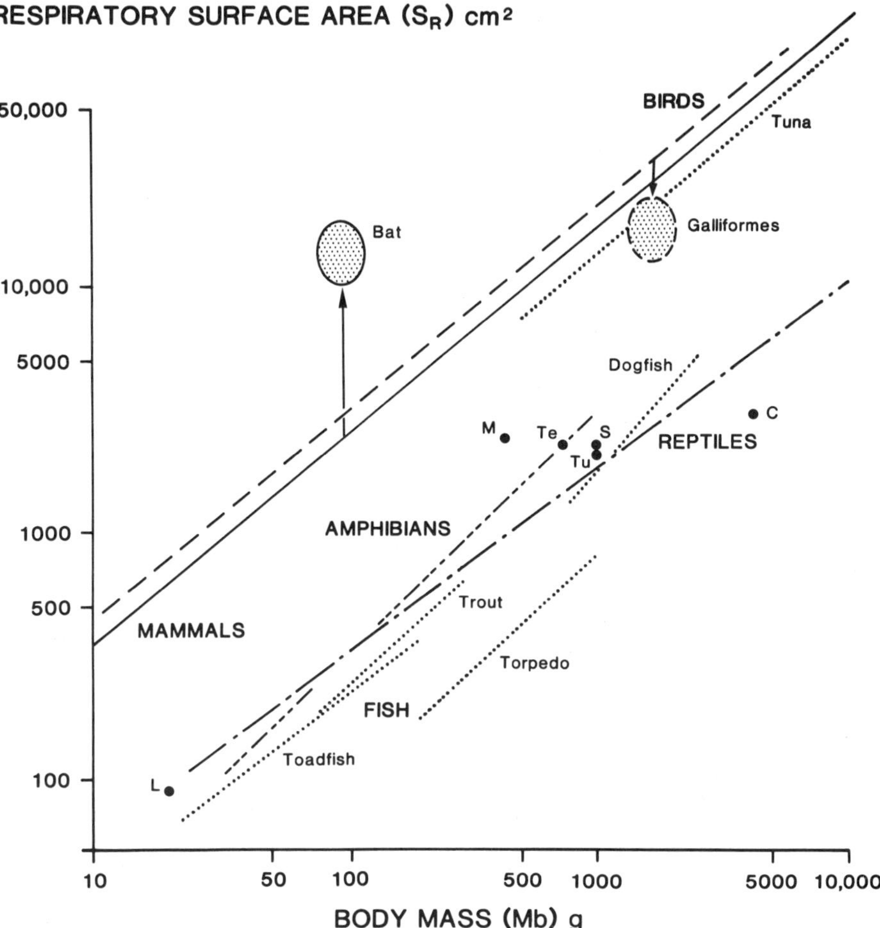

Fig. 4. Double logarithmic plot of respiratory surface area against body mass in selected species of fish (· · · · ·), amphibian (— - - - —), reptile (— · — and ●), bird (– – – –) and mammal (———). The regression lines are calculated for animals between 10 and 10 000 g and do not include the values for bats or galliform birds. Fish: 'Dogfish', the elasmobranch lesser spotted dogfish *Scyliorhinus canicula*; 'Toadfish', the teleost, *Opsanus tau*; 'Trout', the teleost rainbow trout *Oncorhynchus mykiss*; 'Tuna', the teleost yellowfin, *Thunnus albacares*. Reptiles: 'C', the Nile crocodile *Crocodylus niloticus*; 'L', the European lizard *Lacerta* sp.; 'M', the savanna monitor lizard *Varanus exanthematicus*; 'S', the pine gopher snake *Pituophis melanoleucus*; 'Te', the teju lizard *Tupinambis nigropunctatus*; 'Tu', the slider turtle *Trachemys scripta*. After Perry (1990*b*), courtesy of Springer-Verlag.

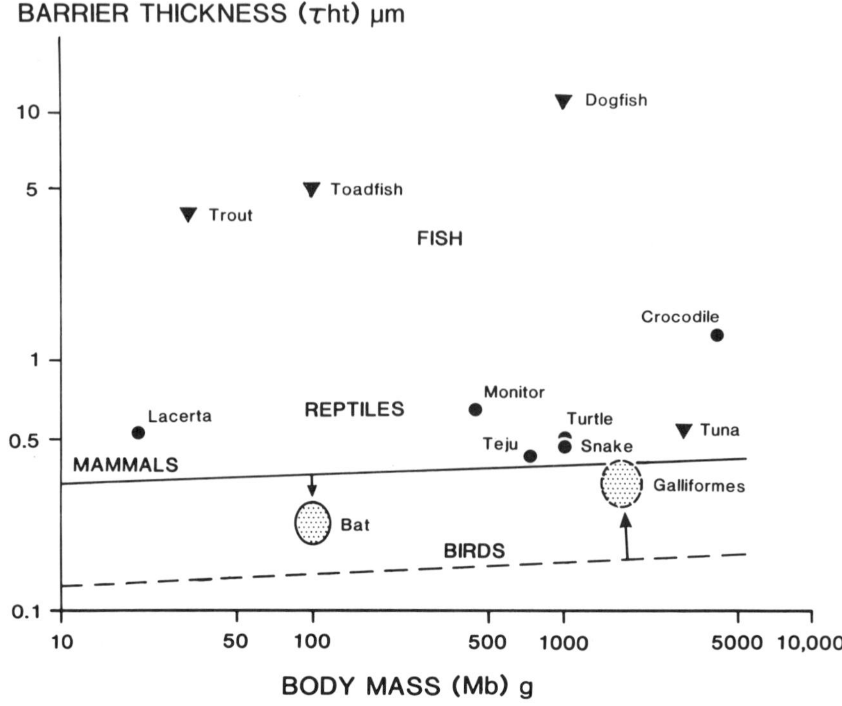

Fig. 5. Double-logarithmic plot of respiratory medium–blood barrier thickness in selected species of fish, reptile, bird and mammal. Values for fish are 'preferred channel' or harmonic means. All others are harmonic means. Species as indicated in Fig. 6. From Perry (1990*b*), courtesy of Springer-Verlag.

thinner water–blood barrier and a nearly eight-times greater secondary lamellar surface area for a hypothetical 100 g fish (Hughes, 1984*a*). Similarly, birds tend to differ from their sauropsid cousins, the reptiles, by displaying both a much greater pulmonary surface area and a much thinner air–blood barrier. Most birds also have a thinner air–blood barrier than do mammals but the respiratory surface area per unit body mass (*M*b) is about the same in both groups. Thus, both the tuna among fish and the birds among terrestrial vertebrates achieve a relatively large surface area for gas exchange, but the real breakthrough appears to be the extreme thinness of the diffusion barrier. This may be related to the rigidity of the gas-exchange organs in each group, and to the concomitant lack of necessity for cytoplasmic reserve in gas-exchange epithelium that is not subjected to extensive stretching.

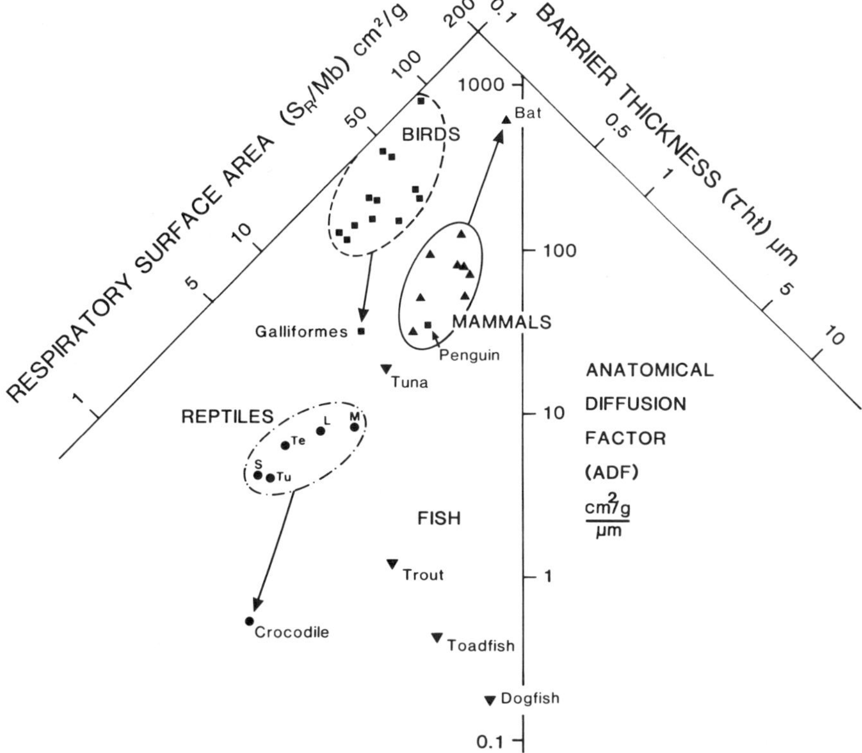

Fig. 6. Double-logarithmic plot of respiratory surface area and respiratory medium–blood barrier thickness in selected species of fish, reptile, bird and mammal between 10 and 10 000 g body mass. Each point can be read off perpendicularly against each of the three axes. The vertical middle axis gives the anatomical diffusion factor (ADF), which is surface area per unit barrier thickness. Abbreviations and symbols as in Fig. 4. From Perry (1990*b*), courtesy of Springer-Verlag.

Ratio of diffusing capacity to consumption ($D_{\mathrm{tO_2}}/\dot{M}_{\mathrm{O_2}}$)

By multiplying the ADF values obtained in Fig. 5 by the best available estimate for the appropriate Krogh's diffusion constant (Bartels, 1971), one obtains a value for $D_{\mathrm{tO_2}}$ of the gas-exchange organ. This value always exceeds the physiological $D_{\mathrm{O_2}}$: in 'resting' reptiles, birds and mammals by a factor of 13–25 (Piiper *et al.*, 1969; Perry, 1983; Maina *et al.*, 1989), but in the dogfish only by a factor of 3 (Piiper *et al.*, 1986).

Inspection of Fig. 7 shows that for terrestrial vertebrates, the Dt_{O_2} per unit exercising $\dot{M}_{O_2}$ tends to rise only slightly as one proceeds from birds to mammals and then to reptiles. In other words, at the order-of-magnitude level permitted by this graphical comparison, the capability of the lung tissue to deliver oxygen to the blood matches the near-maximal oxygen demand by the body to approximately the same degree in all three classes of terrestrial vertebrates. Comparing 'resting' values, however, one sees that the diffusing capacity of a reptilian lung appears to be exploited to a greater degree than that of a mammal or bird, in sequence. This tendency is particularly clear when one compares the values for the monitor lizard, the rat and the pigeon: all relatively active species in the 200–500 g range of Mb.

Extending the comparison to include fish gills, the tuna and the trout conform as if they were reptiles. The gills of the toadfish and the dogfish, on the other hand, display a very low $Dt_{O_2}/\dot{M}_{O_2}$ ratio. The difference is not attributable to phylogenetic considerations alone since the toadfish, like the trout and the tuna, is a teleost. Both the toadfish and the dogfish, however, are relatively sluggish, bottom-dwelling species with large secondary lamellae (Hughes, 1984*a*). It is possible that for these species the efficiency imparted by the counter-current model is of particularly great importance. The fact that Dt_{O_2} in the dogfish is only three times greater than the physiologically measured D_{O_2} lends some credence to this hypothesis. To my knowledge Dt_{O_2}/D_{O_2} in the toadfish and tuna has not been determined.

Applications of interspecific morphometric comparisons

The power of morphometry lies in the fact that, unlike descriptive morphology, it allows the objective and quantitatively reproducible dissection of a gas-exchange organ. With the use of appropriate sampling techniques and operating within a set of unambiguous definitions, stereological methods reduce the lung to a set of nested compartments, such that all defined components regardless of their absolute dimensions are accounted for. Thus, it is possible to calculate not only alveolar volume per unit lung volume but also erythrocyte volume or mitochondrial volume per unit lung volume, should these values be required. Similarly, surface areas, line lengths and particle numbers can be expressed for any desired reference volume. Of particular relevance as an indicator of the degree of specialisation for gas exchange is the surface area-to-volume ratio (surface density) within the lung parenchyma.

The changes in surface density values as one precedes from the pigeon ($3053\,cm^{-1}$) to the rat ($750\,cm^{-1}$), the emerald lizard ($130\,cm^{-1}$) and the

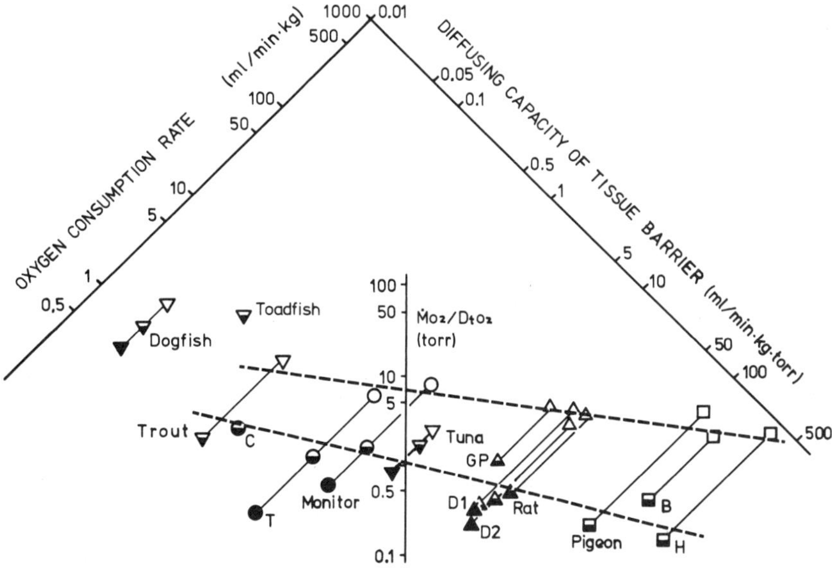

Fig. 7. Double-logarithmic plot of O_2 consumption rate ($\dot{M}_{O_2}$) and diffusing capacity of the respiratory medium–blood tissue barrier (Dt_{O_2}) in selected species of fish (inverted triangles), reptiles (circles), birds (squares) and mammals (triangles). $\dot{M}_{O_2}/Dt_{O_2}$, an indicator of the degree of exploitation of the gas-exchange organ, is given on the central axis. All data points are read off perpendicularly against any of the three axes. $\dot{M}_{O_2}$ at maximal exercising rate is given by open symbols, half-filled symbols represent resting but alert $\dot{M}_{O_2}$ rates and filled symbols give basal rates. The upper broken line is for exercising animals, and the steeper, lower one is for resting animals. Both lines exclude 'Dogfish' and 'Toadfish'. Fish and reptilian species as in Fig. 4. Birds: B, budgerigar (parakeet) *Melopsittacus undulatus*; H, Violet-eared hummingbird *Colibri coruscans*; Pigeon, *Columba livia*. Mammals: D1 and D2, respectively small and large dogs *Canis familiaris*; GP, guinea pig *Cavia porcellus*; Rat, laboratory rat *Rattus norvegicus*. Data from diverse sources given in Perry (1983); additionally, Jones & Randall (1978); Hughes (1984*a*); Brill (1987), Brill & Bushnell (1991). The conventional unit of diffusing capacity, ml O_2/min torr, is converted to the SI unit, ml O_2/s mbar, by multiplying by the conversion constant 1.2501×10^{-2}. The conventional unit for $\dot{M}_{O_2}/Dt_{O_2}$, torr, is converted to the SI unit, mbar, by multiplying by the conversion constant 1.333. Modified and supplemented after Perry (1983), courtesy of Springer-Verlag.

red-eared turtle ($18\,\text{cm}^{-1}$) are dramatic. Although it is intuitively obvious that packing a large surface area into a small space should lead to an effective gas-exchange organ, the physiological disadvantages presented by highly specialised gas-exchange organs, such as increased airway resistance, decreased elastic compliance, susceptibility to blockage and difficulties in maintaining the proper ventilation/perfusion ($\dot{V}/\dot{Q}$) ratio in the lung parenchyma, are less obvious.

In reptiles, a low surface density actually appears to correlate with the ability to increase the oxygen extraction to levels in excess of 25% of $C\text{I}_{\text{O}_2}$, presumably because the gas-exchange surfaces are more readily ventilated than in species that display a high parenchymal surface density (Perry, 1992). Furthermore, morphometric estimates of surface density reflect only the extended state of the parenchyma in fully inflated lungs. While this may be relevant during the non-ventilatory period in reptiles, amphibians and lungfish, the surface density may change drastically during the ventilatory cycle (McGregor *et al.*, 1992).

Morphometric comparisons of gas-exchange organs in different species can be brought to bear upon two major areas: (i) basic research on the functional morphology of the oxygen transport pathway from the external medium to the mitochondrion and (ii) the evolution of oxidative metabolic strategies. In the first case, differences among species are used in order to deduce the underlying principles without emphasising the systematic position of the animals. In fact, for the sake of accuracy, the comparisons should be carried out within a group of closely related species (Weibel *et al.*, 1987). Tantamount to this approach is the hypothesis of 'Symmorphosis' (see below). In the second case, gas-exchange organs are compared quantitatively against the backdrop of systematic zoology. A wide variety of species from divergent taxa is desirable. All functions of the gill or lung are of interest as they relate to the general metabolic strategy of the animals, but the gas-exchange function is most readily quantified.

Basic research: the 'Symmorphosis' hypothesis

In a major thrust beginning in 1981 (Weibel & Taylor, 1981), an international group of physiologists and morphometrists have combined in an attempt to test in mammals the following hypothesis (Weibel & Taylor, 1981): 'no more structure is formed and maintained than is required to satisfy functional needs; this is achieved by regulating morphogenesis during growth and during maintenance of structures'. For the sake of convenience, they chose to test this hypothesis in the respiratory system in its broadest sense: from the lung surface (major oxygen supplier) to the skeletal muscle mitochondrion (major oxygen consumer) and carefully analysed the system for superfluity.

In 1987, Taylor *et al.* found a close match between supply and demand only in the smallest mammals. In larger ones, there is always a superfluity on the supply side. In 'athletic' animals such as the pony and dog, there is less reserve on the supply side than in 'less athletic' animals of similar M_b: the calf and goat, respectively. The differences are attributed to adaptive factors such as capillary haematocrit, allometric factors such as heart rate and factors which are both adaptive and allometric such as the ΔP_{O_2} in the lung. The authors now conclude: 'The hypothesis of symmorphosis is thus not completely supported, certainly not in a simple way. Because of its stringency, it has, however, shown considerable heuristic value in advancing insight into how structure and function are interdependent in an integrated functional system' (Weibel *et al.*, 1992).

The evolution of gas-exchange organs

Reconstruction of the evolution of the respiratory system is confounded by the virtual lack of fossil lungs. Imprints of gills, however, are present for some palaeozoic fishes (for references, see Jarvik, 1980), but the record is still very fragmentary. In order to fill the holes in the fossil record, the evolutionary biologists traditionally resort to comparative embryology, comparative anatomy and – more recently – molecular biology. Functional morphology has not been heavily relied upon because of the risk of introducing analogous rather than haemologous structure into zoological systematics. Knowing the relationship between a phylogenetically primitive and a more derived species, however, it is possible to use principles of functional morphology in order to construct hypothetical intermediate structures.

A case in point is the reconstruction of the origin of the avian lung airsac system with its unidirectional air flow in the palaeopulmo. First, morphometric data were employed in order to test the hypothesis that the crocodilian lung processes the anatomical prerequisites for the cross-current model (Perry, 1989). It was then possible to determine that the common ancestor of birds and crocodiles also could have possessed the cross-current model. The structural and functional principles of the avian lung are relatively well known. Thus, knowing these endpoints, lung types that are intermediate between that of the avian/crocodilian ancestors and avian types were conceptually constructed, that are consistent with what is known about the thoracic structure and life style of protoavian animals (Ostrom, 1976; Perry, 1992). Using mathematical models of the type employed by Kuethe (1988) for the bird lung, it should now be possible to determine at what level of archosaurian evolution the unidirectional flow pattern of the protoavian lung became established.

Similarly, knowledge of the functional morphology of gills provided by the combination of morphometric and physiological data could provide crucial information about the importance of the counter-current model in early fish (see p. 67). Knowing the stage at which counter-current exchange may have evolved, one can intensify the search for the structural prerequisites in available palaeozoic impressions.

A final pitfall: double-logarithmic plots

The analysis of double-logarithmic plots, such as Figs 4–7, will remain an important method for studying the growth, differentiation, adaptation and evolution of respiratory organs. Various authors (e.g. Heusner, 1982, 1983; Hughes, 1984*b*) have pointed out the pitfalls in regression lines which incorporate different species over a wide range of body mass and energy strategies. (i) The overall slope of the regression line can be influenced by the sampling procedure, particularly by highly specialised species at the upper end of the body mass scale, such as scombrid fish (Fig. 6; Hughes, 1980) and cetacean mammals (Tenney & Tenney, 1970). (ii) The overall slope may bear little relationship to the slopes for the individual species included. Hughes (1980), for example, has demonstrated that the allometric slope of gill surface area as a function of body mass in highly active fish approaches 1 (as does lung surface area in mammals and birds: Gehr *et al.*, 1981; Maina, 1989), as opposed to nearly 0.6 in some benthic species.

Conclusion

Theoretically, the diffusing capacity of any gas-exchange organ for O_2 can be calculated morphometrically and also determined physiologically using the same general model. The free interconversion of morphometric and physiological data for the functional comparison of skin, gills and lungs, however, has remained elusive. The reason is that the general model is not accessible in practice either to physiological or to morphometric methods.

Pitfalls in the application of the morphometrically accessible model include the difficulty in applying stereological morphometry to heterogeneously distributed plate-like structures, the uncertain physiological significance of interlamellar diffusion distance in gills and of intercapillary diffusion distance in all gas-exchange organs, and the effects of morphometrically inaccessible adaptive mechanisms, such as central blood shunting, intrapulmonary temperature gradients in basking animals and breath holding. In spite of these difficulties, some quantitative comparisons of the gas-exchange organs in fish, reptiles, birds and mammals are attempted and matched with O_2 consumption rates in the respective species.

The high-performance gills of tunas are comparable with mammalian lungs in all investigated parameters: respiratory surface area per unit body mass, harmonic mean thickness of the medium–blood diffusion barrier, anatomical diffusion factor and ratio of the diffusing capacity of the medium–blood tissue barrier (Dt_{O_2}) to the O_2 consumption rate ($\dot{M}_{O_2}$). Other fish species tend to display a smaller mass-specific respiratory surface area and particularly a thick water–blood diffusion barrier. The relationship of tunas to other fish is similar to that of birds to other sauropsids. With notable exceptions, mammals and birds tend to have a similar mass-specific pulmonary surface area, but mammals have a thicker air–blood barrier. Compared with similarly sized reptiles, mammals have smaller lungs with a greater surface area, but the air–blood barrier is similar in both groups. The degree of exploitation of the respiratory organ, as indicated by the $Dt_{O_2}/\dot{M}_{O_2}$ ratio, is similar for strenuously exercising animals in all groups. In the 'resting' state, however, it tends to increase in the order bird < mammal < reptile = active fish < sluggish fish.

In this review, I have attempted to sketch the conceptual models basic to both the morphometric and the physiological approaches to vertebrate gas exchange and to point out their similarities and differences. The euphoria of the 1970s surrounding the use of diffusing capacity as a meeting point for the morphology and physiology of gas exchange has paled even among its most ardent advocates. Nevertheless, by exercising proper caution, one can profitably combine physiological and morphological data to achieve goals in basic and applied research that are inaccessible to either discipline alone.

Acknowledgements

Financial assistance from the Society for Experimental Biology is gratefully acknowledged. The Deutsche Forschungsgemeinschaft supported much of my research cited here, and Dr John E. Remmers supported me through grants from the Medical Research Council of Canada and the Alberta Heritage Foundation for Medical Research during completion of the manuscript.

References

Bartels, H. (1971). Diffusion coefficients and Krogh's diffusion constants. In *Respiration and Circulation*, ed. P.L. Altman & D.S. Dittmer, pp. 21–33. Bethesda: FASEB.

Brill, R.W. (1979). The effect of body size on the standard metabolic rate of skipjack tuna, *Katasuwonus pelamis. Fisheries Bulletin, US* **77**, 494–8.

Brill, R.W. (1987). On the standard metabolic rates of tropical tunas, including the effect of body size and acute temperature change. *Fisheries Bulletin, US* **85**, 25–36.

Brill, R.W. & Bushnell, P.G. (1991). Metabolic and cardiac scope of high energy demand teleosts: the tunas. *Canadian Journal of Zoology* (in press).

Bur, W., Bachofen, H., Gehr, P. & Weibel, E.R. (1985). Lung fixation by airway instillation: effects on capillary hematocrit. *Experimental Lung Research* **9**, 57–66.

Bushnell, P.G. & Brill, R.W. (1990). Cardiorespiratory responses of swimming yellowfin tuna (*Thunnus albacares*) and skipjack tuna (*Katasuwonus pelamis*) exposed to acute hypoxia. *Physiological Zoology* (in press).

Cotes, J.E. (1968). *Lung Function. Assessment and Application in Medicine*, 2nd edn. Oxford: Blackwell Scientific Publications.

Czopek, J. (1955). Vascularization of the respiratory surfaces in *Leiopelma hochstetteri* Fitzinger and *Xenopus laevis* (Daudin). *Acta Anatomica* **25**, 346–60.

Czopek, J. (1965). Quantitative studies on the morphology of respiratory surfaces in Amphibia. *Acta Anatomica* **62**, 196–323.

Dejours, P. (1981). *Principles of Comparative Respiratory Physiology.* Amsterdam: Elsevier.

Dubach, M. (1981). Quantitative analysis of the respiratory system of the house sparrow, budgerigar and violet-eared hummingbird. *Respiration Physiology* **46**, 43–60.

Gehr, P., Mwangi, D.K., Ammann, A., Maloiy, G.M.O., Taylor, D.R. & Weibel, E.R. (1981). Design of the mammalian respiratory system. V. Scaling morphometric pulmonary diffusing capacity to body mass: wild and domestic animals. *Respiration Physiology* **44**, 61–86.

Gooding, R.M., Neill, W.H. & Dizon, A.E. (1981). Respiration rates and low-oxygen tolerance limits in skipjack tuna, *Katasuwonus pelamis. Fisheries Bulletin, US* **79**, 31–48.

Goniakowska-Witalinska, L. (1980). Ultrastructural and morphometric changes in the lung of the newt, *Triturus cristatus carnifex* Laur. during ontogeny. *Journal of Anatomy* **130**, 571–83.

Graham, J.B. & Laurs, R.M. (1982). Metabolic rate of the albacore tuna *Thunnus albacares. Marine Biology* **72**, 1–6.

Heusner, A.A. (1982). Energy metabolism and body size. I. Is the 0.75 mass exponent of Kleiber's equation a statistical artifact? *Respiration Physiology* **48**, 1–12.

Heusner, A.A. (1983). Body size, energy metabolism and the lungs. *Journal of Physiology (Respiration, Environmental & Exercise Physiology)* **54**, 867–73.

Hicks, J.W. & Wood, S.C. (1989). Oxygen homeostasis in lower vertebrates. The impact of external and internal hypoxia. In *Comparative Pulmonary Physiology. Current Concepts*, ed. S.C. Wood, pp. 311–41. New York: Marcel Dekker.

Hughes, G.M. (1980). Morphometry of fish gas exchange organs in relation to their respiratory function. In *Environmental Physiology of Fishes*, ed. M.A. Ali, pp. 33–56. New York: Plenum Press.

Hughes, G.M. (1984*a*). General anatomy of the gills. In *Fish Physiology*, ed. W.S. Hoar & D.J. Randall, pp. 1–72. New York: Academic Press.

Hughes, G.M. (1984*b*). Scaling of respiratory areas in relation to oxygen consumption in vertebrates. *Experientia* **40**, 519–652.

Jakubowski, M. (1989). Skin vascularization in fishes compared to that in amphibians. In *Trends in Vertebrate Morphology*, ed. H. Spechtna & H. Hilgers, pp. 542–5. Stuttgart: G. Fischer-Verlag.

Jarvik, E. (1980). *Basic Structure and Evolution of Vertebrates*, Vol. 1. London.: Academic Press.

Jones, D.R. & Randall, D.J. (1978). The respiratory and circulatory systems during exercise. In *Fish Physiology*, Vol. 7, ed. W.S. Hoar & D.J. Randall, pp. 425–501. New York: Academic Press.

Kuethe, D.O. (1988). Fluid mechanical valving of air flow in bird lungs. *Journal of Experimental Biology* **136**, 1–12.

Lechner, A.J. (1984). Pulmonary design in a microchiropteran bat (*Pipistrellus subflavus*) during hibernation. *Respiration Physiology* **59**, 301–12.

McGregor, L.K., Daniels, C.B. & Nicholas, T.E. (1992). The dragon's breath: the dynamics of breathing and faveolar ventilation in agamid lizards. *Physiological Zoology* (in press).

Magnussen, H., Perry, S.G., Willmer, H. & Piiper, J. (1974). Transpleural diffusion of inert gases in excised lung lobes of the dog. *Respiration Physiology* **20**, 1–15.

Maina, J.N. (1989). The morphometry of the avian lung. In *Form and Function in Birds*, Vol. 4, ed. A.S. King & J. McLelland, pp. 307–68. London: Academic Press.

Maina, J.N., King, A.S. & Settle, G. (1989). An allometric study of pulmonary morphometric parameters in birds, with mammalian comparisons. *Philosophical Transactions of the Royal Society of London, B* **326**, 1–57.

Meban, C. (1980). Thickness of the air–blood barrier in vertebrate lungs. *Journal of Anatomy* **131**, 299–307.

Ostrom, J.H. (1976). *Archaeopteryx* and the origin of birds. *Biological Journal of the Linnean Society* **8**, 91–182.

Perry, S.F. (1983). Reptilian lungs. Functional anatomy and evolution. *Advances in Anatomy, Embryology and Cell Biology* **79**, 1–81.

Perry, S.F. (1989). Mainstreams in the evolution of vertebrate respiratory structures. In *Form and Function in Birds*, Vol. 4, ed. A.S. King & J. McLelland, pp. 1–67. London: Academic Press.

Perry, S.F. (1990*a*). Gas exchange strategy in the Nile crocodile: a morphometric study. *Journal of Comparative Physiology B* **159**, 761–9.

Perry, S.J. (1990*b*). Recent advances and trends in the comparative morphometry of vertebrate gas exchange organs. In *Advances in Comparative and Environmental Physiology*, Vol. 6. *Vertebrate Gas Exchange from Environment to Cell*, ed. R.G. Boutilier, pp. 45–71. Berlin: Springer-Verlag.

Perry, S.F. (1992). Gas exchange strategies in reptiles and the origin of the avian lung. In *Physiological Adaptations in Vertebrates*, ed. S.C. Wood, R.E. Weber, A.R. Hargens & R.W. Millard. New York: Marcel Dekker (in press).

Piiper, J. (1982). Respiratory gas exchange at lungs, gills and tissues: mechanisms and adjustments. *Journal of Experimental Biology* **100**, 5–22.

Piiper, J. (1990). Modeling of gas exchange in lungs, gills and skin. *Advances in Environmental Physiology* **6**, 15–44.

Piiper, J., Meyer, M., Worth, H. & Willmer, H. (1977). Respiration and circulation during swimming activity in the dogfish *Scyliorhinus stellaria. Respiration Physiology* **30**, 221–39.

Piiper, J., Pfeiffer, K. & Scheid, P. (1969). Carbon monoxide diffusing capacity of the respiratory system in the domestic fowl. *Respiration Physiology* **6**, 309–17.

Piiper, J. & Scheid, P. (1971). Respiration: alveolar gas exchange. *Annual Review of Physiology* **33**, 131–54.

Piiper, J. & Scheid, P. (1972). Maximum gas transport efficacy of models for fish gills, avian lungs and mammalian lungs. *Respiration Physiology* **14**, 29–41.

Piiper, J. & Scheid, P. (1975). Gas transport efficacy of gills, lungs and skin: theory and experimental data. *Respiration Physiology* **23**, 209–21.

Piiper, J., Scheid, P., Perry, S.F. & Hughes, G.M. (1986). Physiological and morphometric diffusing capacity of the gills of the elasmobranch *Scyliorhinus stellaris. Journal of Experimental Biology* **123**, 27–41.

Randall, D.J., Holeton, G.F. & Stevens, E.D. (1967). Exchange of oxygen and carbon dioxide across the gills of the rainbow trout. *Journal of Experimental Biology* **46**, 339–48.

Taylor, C.R., Karas, R.H., Weibel, E.R. & Hoppeler, H. (ed.) (1987).

Adaptive variation in the mammalian respiratory system in relation to energetic demand. *Respiration Physiology* **69**, 1–127.

Tenney, S.M. & Remmers, J.E. (1963). Comparative quantitative morphology of mammalian lungs; diffusing areas. *Nature* **197**, 54–6.

Tenney, S.M. & Tenney, J.B. (1970). Quantitative morphology of cold-blooded lungs: Amphibia and Reptilia. *Respiration Physiology* **9**, 197–215.

van Dieken, E. (1989). *Morphometrie der Haifischkieme am Beispiel von* Scyliorhinus canicula, *dem Kleingefleckten Katzenhai*. Diplomarbeit, Universität Oldenburg, Germany.

Weibel, E.R. (1970/1). Morphometric estimation of pulmonary diffusing capacity. I. Model and method. *Respiration Physiology* **11**, 54–75.

Weibel, E.R., Gehr, P., Cruz-Orive, L.M., Müller, A.E., Mwangi, D.K. & Haussener, V. (1981). Design of the mammalian respiratory system. IV. Morphometric estimation of pulmonary diffusing capacity: critical evaluation of a new sampling method. *Respiration Physiology* **44**, 39–59.

Weibel, E.R. & Knight, B.W. (1964). A morphometric study of the thickness of the pulmonary air–blood barrier. *Journal of Cell Biology* **21**, 367–84.

Weibel, E.R., Marques, L.B., Constantinpol, M., Doffey, F., Gehr, P. & Taylor, C.R. (1987). Adaptive variation in the mammalian respiratory system in relation to energetic demand: VI. The pulmonary gas exchanger. *Respiration Physiology* **69**, 81–100.

Weibel, E.R. & Taylor, C.R. (1981). Design of the mammalian respiratory system. *Respiration Physiology* **44**, 1–164.

Weibel, E.R., Taylor, C.R. & Hoppeler, H. (1992). The concept of symmorphosis: a testable hypothesis of structure–function relationship. *Science* (in press).

Welsch, U. & Müller, W. (1980). Feinstrukturelle Beobachtungen am Alveolarepithel in Reptilien unterschiedlicher Lebensweise. *Zeitschrift Mikroskopisch anatomisch Forschungsgebeit* (Leipzig) **94**, 479–503.

Wood, C.M. & Perry, S.F. (1985). Respiratory, circulatory, and metabolic adjustments to exercise in fish. In *Circulation, Respiration and Metabolism – Current Comparative Approaches*, ed. G. Gilles, pp. 2–22. Berlin: Springer-Verlag.

T.M. MAYHEW

The structural basis of oxygen diffusion in the human placenta

Introduction

Gas-exchange organs may have several different tissue layers which separate the gas delivery and gas uptake compartments. For example, the air–blood barrier of the lung has layers of alveolar epithelium, connective tissue and capillary endothelium separating air in the alveoli from blood in pulmonary capillaries. A thin barrier of similar composition separates environmental water from capillary blood in the fish gill within which a large respiratory surface is established by the presence of densely packed lamellae on the filaments of branchial or gill arches (see Perry, this volume).

In the case of the human placenta, the exchanger considered in this chapter, the principal tissue compartment interposed between the maternal and fetal vascular beds is the villous membrane. On its outer aspect, bathed by maternal blood, this membrane forms a complete syncytial layer (the syncytiotrophoblast) of variable thickness. Below this layer are two other compartments of the villous membrane which are also of inconstant thickness. A cellular layer (the cytotrophoblast) gradually diminishes during gestation and eventually becomes incomplete as its post-mitotic elements are recruited into the overlying syncytium. The second compartment (the villous stroma) can be taken to include connective tissue and fetal capillary endothelium.

At certain sites within the villous membrane, fetal capillaries may obtrude into the overlying trophoblast and all compartments of the villous membrane are attenuated. These sites are known as vasculosyncytial membranes and, by virtue of their thinness, they play an important role in transplacental exchanges which occur by passive diffusion (e.g. transfer of respiratory oxygen).

The respiratory function of the placenta (and other gas exchangers) can be interpreted in terms of its composition and physical dimensions. This must be done with caution because placental structure represents a compromise between diverse functional demands. Thus, synthesis of hor-

Society for Experimental Biology Seminar Series 51: *Oxygen Transport in Biological Systems*, ed. S. Egginton & H.F. Ross.

mones and other secretion products by the trophoblast depends in part on its volume. Since this layer is confined to a villous surface, the volume per surface area provides a measure of trophoblast thickness. For a constant surface, a greater volume implies a thicker layer. However, the passive diffusion of gases and nutrients requires a thin trophoblast because this will minimise the effective diffusion distances between maternal and fetal vascular spaces.

To determine placental diffusing capacity for oxygen (D_P), tissue sections are crucially important for two reasons. First, physiological methods for estimating D_P are technically difficult. It has been stated that they can be obtained 'only by stealth and indirection' (Barron & Meschia, 1954). Secondly, the exact description of placental microstructure in terms of volumes, surface areas and thicknesses can be achieved *only* via 'measurements' made on sectional images of tissue ingredients.

The corpus of knowledge which permits such extrapolations is a branch of morphometry known as stereology (Weibel, 1979). The validity of stereological estimates is governed by proper organ sampling. By confronting randomly sampled tissue sections with test overlays bearing regular arrays of points and lines, unbiased estimates of structural quantities can be obtained merely by counting chance events, i.e. without making any measurements at all! This makes the technique also extremely efficient (Weibel, 1979; Mayhew, 1983, 1991*a*; Gundersen & Jensen, 1987; Cruz-Orive & Weibel, 1990).

A method for estimating total D_P by stereological analysis of tissue sections has been described (for references, see Mayhew *et al.*, 1984, 1986, 1990; Mayhew, 1986). It represents an adaptation of the model used to quantify pulmonary diffusing capacity and is described elsewhere by Ewald Weibel and his colleagues (e.g. Weibel, 1970; Gehr *et al.*, 1978). Naturally, the pulmonary model must be modified to suit the different structural organisation of the placenta. To implement the method, each placenta must be sampled in a manner which meets stereological requirements for obtaining valid estimates of volumes, surface areas and thicknesses. In practice, this means that sections must be cut so as to be random in position and spatial orientation (Weibel, 1979; Mayhew & Burton, 1988; Cruz-Orive & Weibel, 1990; Mayhew, 1991*a*).

Examples of the application of stereology to estimate D_P for human placentae are presented in this chapter. Attention is focused on adaptations which occur during 'normal' gestation and during chronic maternal hypoxia associated with pregnancy at high altitude (HA).

Historical perspective

Morphological changes during gestation

Placental size (whether expressed as volume or mass) increases during gestation and correlates well with fetal mass which, in turn, influences the demands for oxygen and nutrients. In fact, the rate of placental growth reaches its peak at about 29 weeks whilst that of the fetus is maximal at 33 weeks (Bonds *et al.*, 1984). Placental growth is accompanied by villous growth and maturation. Villi can be classified like the ramifications of a tree. A trunk (stem villus) gives rise to branches (intermediate villi) and these, in turn, ramify into twigs (terminal villi). The processes of villous growth and maturation are concentrated on the terminal villi. Growth involves the elaboration of new terminal villi which arise by budding from the surfaces of stem and intermediate villi. Villous maturation involves changes within the newly generated villi which possess a thinner trophoblast and more peripheral fetal capillaries (Boyd, 1984; Jackson *et al.*, 1987*a*; Mayhew *et al.*, 1991). The resulting combination of expanded exchange surface areas and reduced intervascular distances improves considerably the opportunity for efficient transfer of oxygen and nutrients by passive diffusion.

There is confusion as to whether or not deceleration of growth and development towards term is merely an artefact of cross-sectional studies or a real phenomenon which represents placental senescence. Based on studies of placental size and DNA content, Winick *et al.* (1967) suggested that morphological development of the human placenta ceases in the third trimester. More recent studies have contradicted this view (see Sands & Dobbing, 1985). Trophoblast volume seems to increase to term and into postmaturity (Bouw *et al.*, 1978). There is some agreement that a decline in the rate of growth of villous surface area occurs after 34 weeks but disagreement about whether or not absolute surface decreases (Teasdale, 1980; Boyd, 1984; Teasdale & Jean-Jacques, 1985). The uncertainty over these morphological events obscures their possible functional (and clinical) significance. Using quantitative morphological data to assess oxygen diffusive conductances is one way of tackling this uncertainty.

Morphological changes at high altitude

One of the principal stressors of life at high altitude is hypobaric hypoxia which leads to lower partial pressures of oxygen in inspired and alveolar air. In turn, the haemoglobin in lung capillaries is less saturated with oxygen and oxygen tensions are reduced in the blood within systemic arteries, tissue capillaries and veins. To compensate for these deficiencies, the oxygen pathway must adapt in order to improve the gas conductances for alveolar ventilation, vascular convection and tissue diffusion. These adaptations are particularly important when the deficiencies are further exacerbated by pregnancy at high altitude.

Human birthweights are reduced when pregnancies occur at high altitudes (see Haas, 1980; Haas *et al.*, 1980) and there is a trend towards a decrease in fetal growth rate with altitude which suggests that growth is influenced by environmental oxygen tension (Bonds *et al.*, 1984). Since the placenta mediates oxygen transfer from the mother to the fetus, it seems reasonable to expect that changes in fetal weight might be associated with differences in placental microstructure and oxygen transport capability. Indeed, morphometry has shown that villi in HA placentae have a smaller overall surface area and length than those in low-altitude (LA) organs (Jackson *et al.*, 1987*b*). In addition, the volume of the fetal vascular bed is diminished and mean capillary diameter is less (Jackson *et al.*, 1987*c*). As during gestation, the trophoblast is thinner and fetal vessels are more eccentric in position (Jackson *et al.*, 1988*a*,*b*). Finally, the volume of the maternal inter-villous space increases greatly in HA placentae. Most, but not all, of these changes may be seen as adaptations aimed at improving or maintaining the delivery of oxygen to the highland fetus.

Functional interpretation of morphological changes

In an early attempt to assess the importance of fetal capillary location during gestation and its impact on placental efficiency, Aherne (1975) adopted techniques used by heat flow engineers. He concluded that placental transfer efficiency increases at least six-fold from 16 weeks to term. However, this approach is too simplistic to describe passive diffusion because it does not cater for a sufficiently comprehensive set of physical dimensions. Indeed, most studies have estimated vascular bed volumes and exchange surface areas but have neglected completely villous membrane thickness. Yet harmonic mean thickness has a critical impact on passive diffusion, relatively modest decreases leading to large improvements in diffusive conductance which cannot be achieved by comparable changes in exchange surface areas or vascular bed volumes (Mayhew *et al.*, 1986). Moreover, it is known that villous membrane thickness decreases during normal gestation and during pregnancy at HA.

The idea that morphological differences might lead to an increase in D_{p} for oxygen in hypoxic conditions came from studies on sheep and guineapigs. Sheep apparently maintain fetal oxygen tensions at HA (Barron *et al.*, 1964; Metcalfe *et al.*, 1967) and D_P for carbon monoxide can increase in guineapig placentae during hypoxia (Gilbert *et al.*, 1979; Bacon *et al.*, 1984). Unfortunately, experimental design in the sheep studies was complicated by interstrain differences and guineapigs may be pre-adapted to hypobaric hypoxia.

In order to appreciate better the functional significance of morphological changes, we must estimate a full complement of relevant structural quantities and relate these to a defined physiological process. Passive diffusion is an ideal candidate. In the present context, attention is confined to D_P for oxygen which expresses the volume flux of oxygen per unit of partial pressure gradient. However, the methods would apply equally well to other substances which are transported across the human placenta (or other organs) by simple diffusion. Of course, the number and nature of the morphological variables may also be modified to suit those placentae which depart from the human haemochorial plan or to suit the particular anatomy of other exchange organs.

Materials and methods

Provenance of the placentae

Gestational study

Ninety-two organs were collected from pre-term and normal term pregnancies spanning 10–41 weeks. All fetuses and neonates appeared to be normal (Boyd, 1984). Organs at 10–24 weeks were obtained from pregnancies terminated using a prostaglandin gel. Those at 25–36 weeks were from pregnancies classified as uncomplicated until premature onset of labour. There were no indications of hypertension, diabetes mellitus, pre-eclampsia or other disorders. All other pregnancies (37–41 weeks) were uncomplicated. Gestational ages were decided from the date of the last menses, with support from ultrasound measurements and clinical examination. These samples did not reveal any sex differences in fetal/neonatal weights. Placentae were collected at delivery and immediately refrigerated. Within 24 h, they were trimmed and weighed and their volumes were determined by formalin displacement (Boyd *et al.*, 1980). Later, fixed weights and volumes were determined.

Altitudinal study

Sixty-eight organs were collected from women who were born, brought up and completed their pregnancies at LA (mean altitude 400 m) or HA (3600 m) in Bolivia (Haas, 1980; Haas *et al.*, 1980). The women were from native Amerindian and non-native (European and mixed) populations which displayed altitudinal, ethnic and sex differences in birthweight but not in placental weight (see also Jackson *et al.*, 1987*b*). Average birthweights were greater at LA, greater in natives and, at least at LA, greater for males (Haas, 1980; Haas *et al.*, 1980). The samples used here reproduce all but the sex differences. Placentae were obtained from normal term pregnancies after Caesarean section (4) or after spontaneous vaginal delivery (64). The LA organs ($n = 24$) comprised 10 natives and 14 non-natives whilst the HA group ($n = 44$) consisted of 16 natives and 28 non-natives. Immediately after collection, placentae were transferred to a refrigerator and, within 8 h, examined for gross pathologies, trimmed and weighed.

Tissue sampling and stereology

Details of tissue handling, preparation and sampling have been provided in previous publications (Jackson *et al.*, 1987*b*,*c*, 1988*a*,*b*). Briefly, the aim was to give all regions of a placenta the same chance of being chosen by randomising both section location and section orientation (Cruz-Orive & Weibel, 1990; Mayhew, 1991*a*). Whilst randomising orientation is necessary for obtaining unbiased estimates of exchange surface areas and harmonic mean diffusion distances, it is sufficient to randomise location alone when estimating blood space volumes. Exactly the same requirements would apply when dealing with other exchange organs.

To achieve these objectives, tissue pieces from each placenta were sampled at a series of levels in a systematic random fashion (see Fig. 1 and Gundersen & Jensen, 1987; Mayhew & Burton, 1988). This served to

Fig. 1. Illustration of a systematic random sampling scheme. In *A*, an organ (e.g. the placenta, hatched) is scanned with an overlay comprising a systematic set of squares positioned at random. Some of these squares (solid outline) hit the organ and provide sites from which tissue pieces are sampled and processed for microscopical examination. In *B*, a prepared slice of tissue (hatched) lies on a microslide and under a coverslip. Fields of view on this slice are sampled systematically using the *x*- and *y*-axes of the microscope stage micrometers. Some fields (circles) hit the tissue and are recorded. In *C*, a randomly selected field has superimposed a transparent grid of squares which is used to generate random events for stereological estimations. *D* illustrates how each square can be seen as possessing an associated test point (P) and a length of test line (L). [Figure 2 illustrates how the test points and lines may be used to estimate the volumes, surface areas and thicknesses of tissue ingredients (e.g. of the stippled areas in *C*).]

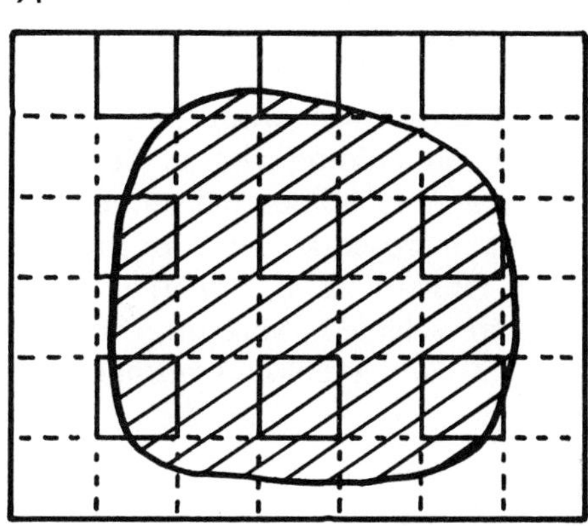

B

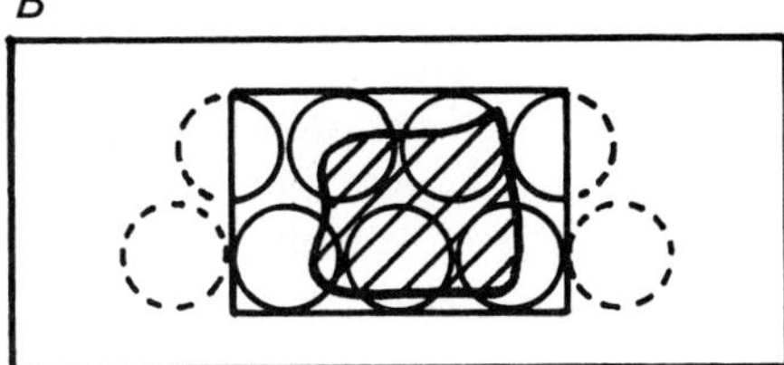

C

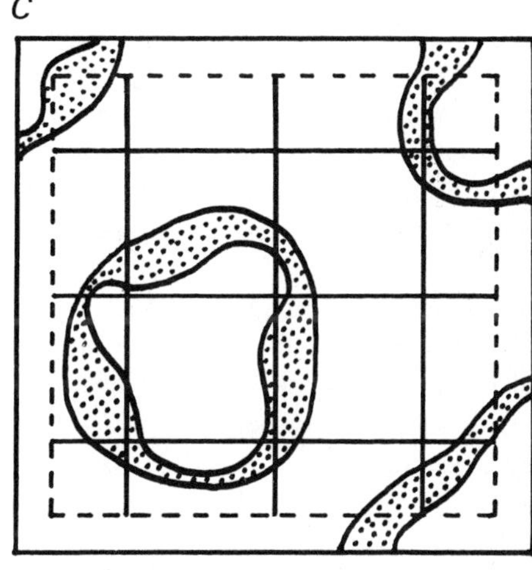

D

P

L

randomise location. Tissues were fixed in formalin and then embedded haphazardly in paraffin wax in order to randomise section orientations (Stringer *et al.*, 1982). An alternative, and generally superior, approach would be to orientate tissue pieces for vertical sectioning (for details, consult Baddeley *et al.*, 1986). Following embedding, tissues were sectioned at 3–5 μm thickness, mounted on glass microslides and stained with connective tissue stains. Under the light microscope, fields of view on microslides were chosen systematically (Burton *et al.*, 1989) so as to randomise field location. The fields so selected were prepared as micrographs (linear magnification, $M = 250$) and as colour transparencies ($M = 2000$) for morphometric analysis.

The morphometric diffusion model

A modification of the model described earlier (Mayhew *et al.*, 1984, 1986, 1990) was adopted. Oxygen released from haemoglobin within maternal erythrocytes must negotiate a set of different tissue compartments before recombining with haemoglobin within fetal erythrocytes. For present purposes, the intervascular diffusion pathway was analysed as six tissue compartments: maternal erythrocytes (me), maternal plasma (mp), trophoblast (tr), stroma (st), fetal plasma (fp) and fetal erythrocytes (fe). The compartments are connected in series (refer to Fig. 2*A*) and so were treated as oxygen diffusion resistances, *R*.

Resistances were estimated for each compartment (R_{me}, R_{mp}, R_{tr}, R_{st}, R_{fp}, R_{fe}) and then summed for each organ to obtain the total resistance, R_p. The reciprocal of R_p is overall diffusive conductance, D_p, expressed in $\mathrm{ml\,min^{-1}\,kPa^{-1}}$. Values of specific D_p ($\mathrm{ml\,min^{-1}\,kPa^{-1}\,kg^{-1}}$) were obtained for each placenta by taking into account the appropriate fetal or neonatal weight.

Each resistance (and, hence, each conductance) was calculated from physical and physicochemical data. To estimate R_{me} and R_{fe}, volumes of vascular beds (maternal inter-villous and fetal capillary spaces) were multiplied by oxygen–haemoglobin reaction rates for whole blood, θ (see references in Mayhew *et al.*, 1984). Values of θ are available for normal (non-pregnant, low-altitude) adult blood and for fetal blood. Therefore, they had to be adjusted to suit the different haematocrits found in different circumstances, e.g. the elevated haematocrits found in maternal and fetal bloods at HA (Haas, 1980; Ballew & Haas, 1986).

The four intervening resistances (R_{mp} through to R_{fp}) were estimated from exchange surface areas, harmonic mean thicknesses and Krogh's tissue diffusion coefficients for oxygen, *K* (Mayhew *et al.*, 1984). The equation relating these variables is based on Fick's equation for gas diffusion across a tissue membrane. The original equation depended on a model which envisaged the membrane as being of uniform thickness and possessing inner and outer surfaces of equal area. This simplistic model needed to be modified to cater for real placental tissue membranes. The main modifications involved estimating (i) surface areas on *both* aspects of the membrane (to cater for the possibility that these two surfaces were unequal) and (ii) harmonic mean rather than arithmetic mean thicknesses (in recognition of the fact that the real membranes are not uniformly thick). Harmonic mean thickness gives greater weighting to thinner regions of membrane, i.e. to precisely those regions where oxygen flux is expected to be greater.

In terms of diffusive conductances, D, two generic equations apply. The first defines conductances within the maternal and fetal erythrocyte compartments:

$$D = \theta V$$

where V denotes blood space volume. The second equation covers all intervening tissue compartments and takes the form:

$$D = (S_1 + S_2)K/2\text{Th}$$

where Th denotes harmonic mean thickness (equivalent to the effective diffusion distance) and S_1 and S_2 refer to the surface areas on the two (maternal- and fetal-facing) aspects of the compartment concerned.

For the appropriate physicochemical coefficients, median values of published ranges were adopted. The baseline value of θ was taken to be $13\ \text{ml ml}^{-1}\ \text{min}^{-1}\ \text{kPa}^{-1}$ for both maternal and fetal bloods. The value of K for maternal and fetal plasmas was taken to be $24 \times 10^{-8}\ \text{cm}^2\ \text{min}^{-1}\ \text{kPa}^{-1}$ and that for villous tissue was taken to be $17 \times 10^{-8}\ \text{cm}^2\ \text{min}^{-1}\ \text{kPa}^{-1}$.

Note that the stereological estimate of $D\text{p}$ provides a measure of the potential for oxygen diffusion afforded by the physical dimensions of the placenta. Certain technical limitations mean that the estimate may not have the same value as $D\text{p}$ determined by physiological methods. These include difficulties in measuring oxygen tensions at the actual site of exchange (i.e. in the inter-villous and fetal capillary spaces rather than in umbilical and uterine vessels), oxygen consumption by the placenta itself, vascular shunting and lack of regional homogeneity in the perfusion : perfusion and diffusion : perfusion ratios. Similar discrepancies between morphometric and physiological $D\text{p}$ for oxygen have been found in the human lung. In this organ the discrepancies are partly explicable in terms of its ability to respond rapidly to the increased demands placed on it by strenuous exercise. No such discrepancies are seen in the chorioallantois of the avian embryo (for references, see Mayhew *et al.*, 1984).

Stereological estimations

Vascular bed volumes were determined from volume densities estimated by point counting using overlays with quadratic patterns of test points (Weibel, 1979; Jackson *et al.*, 1987*b*). The basic procedure is illustrated in Fig. 2. Randomising section location satisfied the requirements for obtaining unbiased estimates (Fig. 2*B*). Absolute volumes were computed by multiplying volume densities by fresh placental volumes.

Surface areas were calculated from surface densities (see Fig. 2*C*). These were estimated by intersection counting using systematic patterns of parallel test lines on overlays (Weibel, 1979; Jackson *et al.*, 1987*c*). In this case, it is necessary to randomise section location *and* orientation in order to obtain valid estimates. To aim for this, tissue blocks were allowed to adopt random orientations during paraffin embedding. Absolute surfaces were calculated from fresh organ volumes after allowing for tissue processing distortions. These were assessed by using erythrocyte diameter as an internal reference.

Harmonic mean thicknesses were determined by random intercept length measurement (Jackson *et al.*, 1985). Again, overlays of parallel test lines were employed (Fig. 2*D*) and random section locations and orientations were prerequisites for achieving unbiasedness. Absolute distances were obtained after correcting for processing artefact.

Fig. 2. *A*, a random section through villi embedded in the maternal inter-villous space (ivs). Red blood cells in this space are separated from those in fetal capillaries (fc) by blood plasma, the trophoblast (tro) and villous stroma. The physical dimensions of these ingredients are estimated as shown in *B–D*. Volume densities of blood spaces within placenta are estimated by counting test points hitting blood spaces (see *B* for those hitting ivs, arrowheads) and dividing by the total hitting the placenta as a whole. The points must be randomly located and counts are made on all sections and all fields sampled from each placenta. The fraction of total test points which hit blood spaces is then an unbiased estimator of volume density. Surface densities are estimated by counting intersections between exchange surfaces (e.g. outer trophoblast in *C*, arrowheads) and test lines of known length. If the lines are isotropic in 3-D, then the intersection density per line length is an unbiased estimator of surface density. Random intercept lengths are measured in order to determine harmonic mean distances. Again, the lines must be isotropic in 3-D and distances are measured in chance directions across the tissue compartment (e.g. *some* intercepts across the trophoblast are exemplified in *D*).

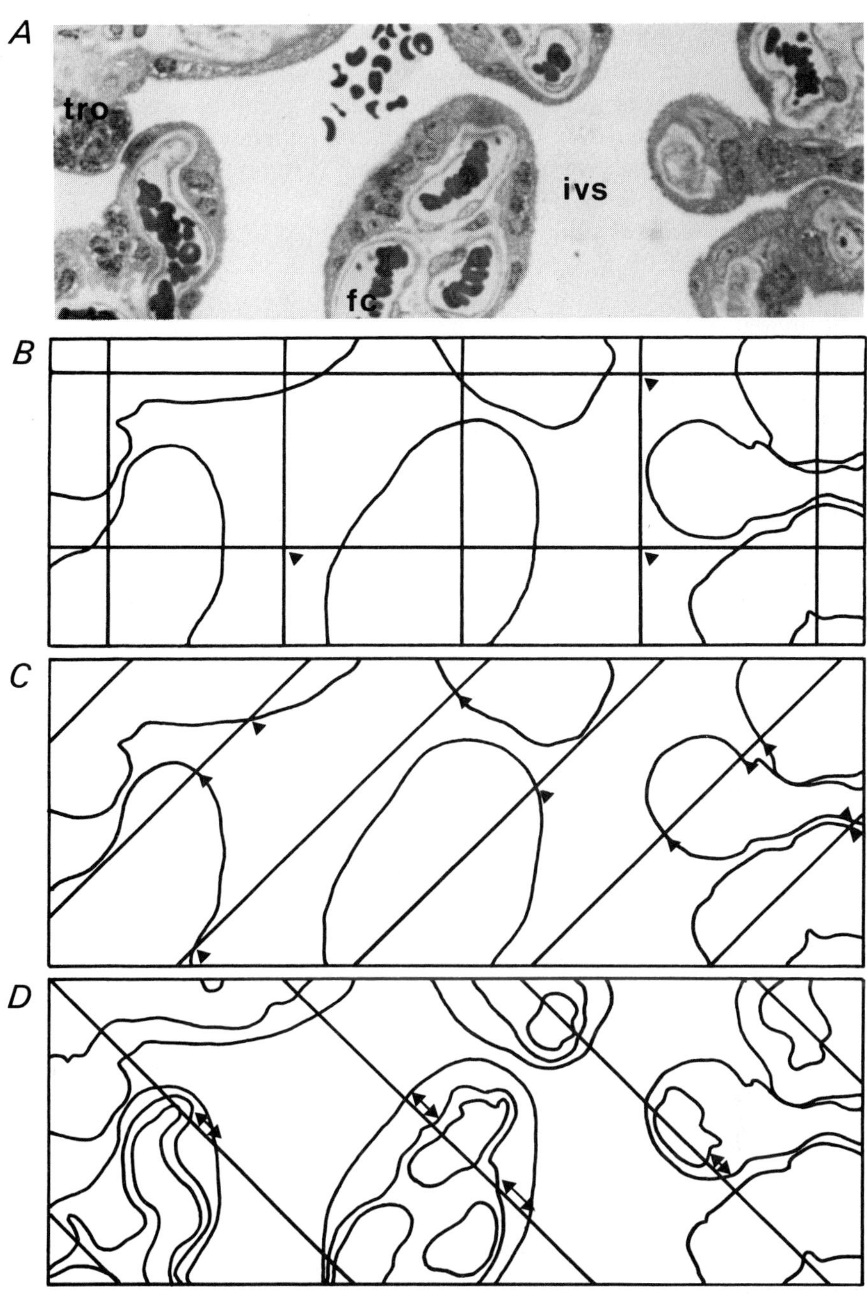
A
tro
ivs
fc
B
C
D

Statistics

All statistical calculations were based on methods described in Sokal & Rohlf (1981). Means were calculated for each group of organs and coefficient of variation (cv = standard deviation expressed as a percentage of the corresponding group mean) was used as the measure of observed between-organ variation within a given group.

For the gestational study, organs were divided into four groups (G1: 10–22 weeks; G2: 23–31 weeks; G3: 32–36 weeks and G4: 37–41 weeks) in order to approximate group sample sizes and to test whether or not there was evidence for loss of capacity near term. Apparent differences between groups were tested using two-way analyses of variance (2-way ANOVA) with age and sex of fetus/neonate as the main effects. This test generates an interaction (age × sex) term which provides a measure of whether or not age affects both sexes equally. Present findings are confined to the effects of gestational age.

Statistical comparisons between LA and HA groups were undertaken using 3-way ANOVA with altitude, ethnic grouping and sex of newborn as the main effects. This test generates three first-order interactions (altitude × ethnic group, altitude × sex, ethnic group × sex) and one second-order interaction (altitude × ethnic group × sex). Present results focus solely on the altitudinal effects.

Apparent differences were considered to be statistically significant at a probability level of $P < 0.05$ for the appropriate degrees of freedom.

Results

Gestational study

Apart from an effect on stromal thickness, no significant sex effects were detected. Mean weight of the fetus/neonate rose roughly 23-fold between G1 and G4. At 37–41 weeks, the average birthweight was 3.4 kg (cv 16%). Over the same period, placental weight increased almost sevenfold to an average of 460 g (cv 21%) during the period denoted G4.

The volumes of vascular beds expanded five-fold (maternal inter-villous space) to 15-fold (fetal capillaries), whilst exchange surface areas of villi and fetal capillaries expanded 16- to 18-fold throughout gestation. In contrast, harmonic mean diffusion distances across the trophoblast and across the villous stroma decreased by 60–70%. Near term, the average organ possessed 200 ml (cv 21%) of maternal and 50 ml (cv 38%) of fetal vascular space. Total villous surface area amounted to 10 m^2 (cv 23%) and fetal capillary surface area to 11 m^2 (41%). The harmonic mean thickness

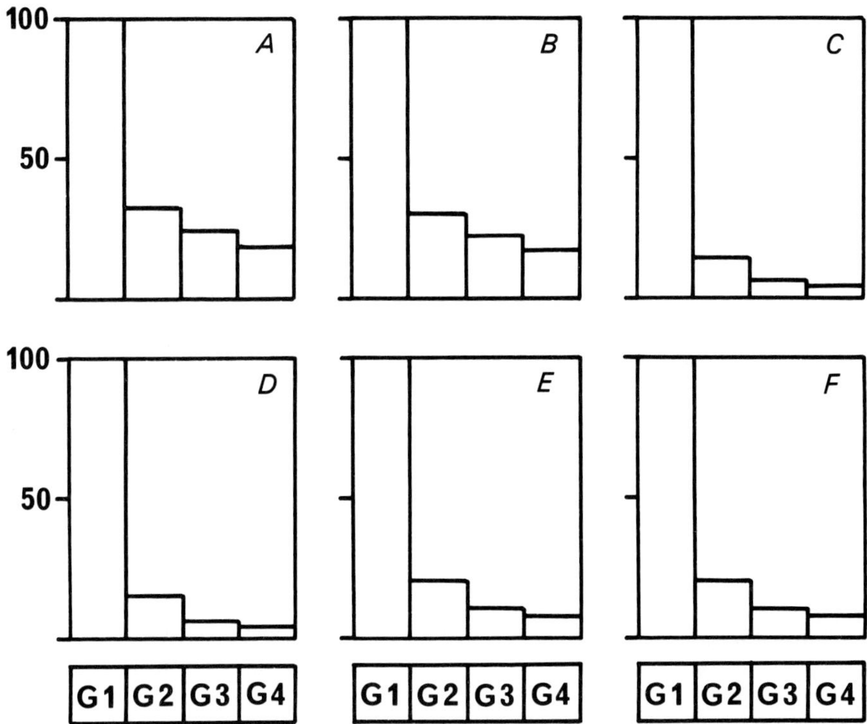

Fig. 3. Decreases in relative resistances of placental compartments from 10–22 weeks (G1) to 37–41 weeks (G4) of gestation. Each value is expressed as a percentage of the resistance found at G1 which is set arbitrarily at 100%. Six tissue compartments are identified: *A*, maternal erythrocytes; *B*, maternal plasma; *C*, trophoblast; *D*, villous stroma; *E*, fetal plasma; *F*, fetal erythrocytes. All resistances decline during gestation but the major changes involve the trophoblast and stroma which together make up at least 70% of total resistance to diffusion.

of the trophoblast was 2.0 μm (cv 19%) and that of the stroma was 1.2 μm (cv 38%).

As a consequence of these and other morphological changes, all six serial resistances of the diffusion pathway decreased progressively during gestation (Fig. 3). At each gestational age, the main contributors to total resistance were the trophoblast (55% in G1, 49% in G2 and G3, 45% in G4) and the stroma (30–31% in G1 and G2, 25–26% in G3 and G4). Together, these two compartments accounted for 70% of total resistance to diffusion. Throughout gestation, resistances on both maternal and fetal

Table 1. *Group means (cv) for placental diffusive conductances at G1 (10–22 weeks), G2 (23–31 weeks), G3 (32–36 weeks) and G4 (37–41 weeks) of gestation*

Conductance	G1 (n = 23)	G2 (n = 26)	G3 (n = 23)	G4 (n = 20)
D_{me}	140 (40%)	500 (37%)	710 (27%)	850 (22%)
D_{mp}	100 (40%)	370 (36%)	530 (25%)	640 (20%)
D_{tr}	3 (59%)	30 (42%)	60 (26%)	100 (33%)
D_{st}	5 (81%)	40 (65%)	120 (29%)	180 (46%)
D_{fp}	60 (74%)	350 (44%)	730 (28%)	950 (39%)
D_{fe}	20 (76%)	120 (44%)	240 (30%)	310 (38%)
D_P, total	2 (65%)	13 (47%)	29 (23%)	43 (33%)
D_P, specific	13 (39%)	12 (31%)	13 (20%)	13 (33%)

sides made minor contributions to total resistance but these proportions tended to increase from G1 to G4.

Total D_P increased from 2 ml min^{-1} kPa^{-1} (cv 65%) at 10–22 weeks to 43 ml min^{-1} kPa^{-1} (cv 33%) at 37–41 weeks (Table 1). Changes in D_P were commensurate with the gain in fetal weight and did not demonstrate any convincing decline at 37–41 weeks (specific D_P was roughly 13 ml min^{-1} kPa^{-1} kg^{-1} throughout gestation).

Altitudinal study

No significant sex effects or interaction terms were detected. However, there were significant main effects of altitude and ethnic grouping. Birthweights were significantly greater (by about 300 g) at LA vs HA. The differences were not attributable to differences in gestational age or placental weight. Mean birthweight at LA amounted to about 3.3 kg (cv 13%).

Though weighing the same as the LA placenta, the HA organ was substantially different in composition and microscopical dimensions. At LA, the average organ contained 180 ml (cv 33%) of maternal intervillous space, 43 ml (cv 42%) of fetal capillary bed, 7 m^2 (cv 31%) of villous surface area and almost 6 m^2 (cv 41%) of capillary surface area. Harmonic mean thicknesses were 4.2 μm (cv 20%) for trophoblast and 2.2 μm (cv 40%) for stroma.

The relative changes in the resistances to oxygen diffusion of individual compartments are shown in Fig. 4 for all LA and all HA organs. Maternal

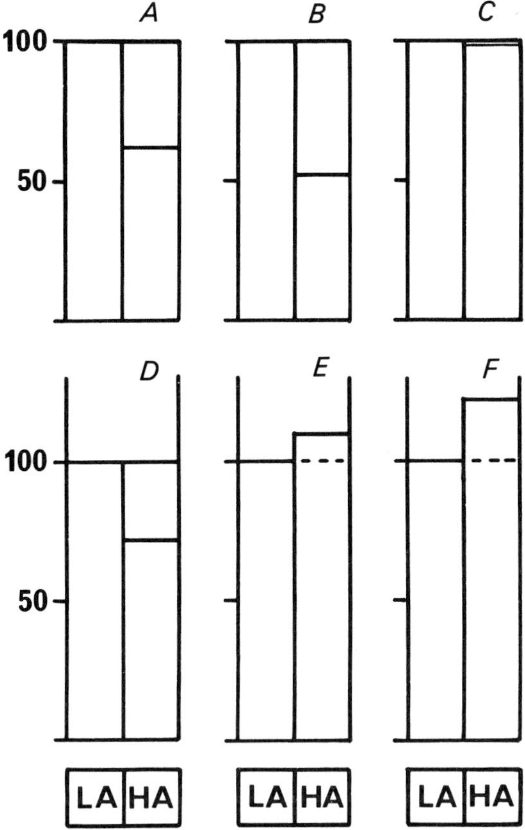

Fig. 4. Decreases in relative resistances of placental compartments from low altitude (LA) to high altitude (HA). Each value is expressed as a percentage of the resistance at LA which is set at 100%. The tissue compartments *A–D* are identified in Fig. 3. Except for the trophoblast (*C*, no change) and the fetal compartments (*E* and *F*, both increasing), resistances are diminished at HA. Important changes occur in the stroma (*D*) and, together with trophoblast, this compartment accounts for about 90% of total resistance.

and stromal resistances declined in highland placentae and there was no difference in the resistance of the trophoblast. Regardless of altitude, the principal contributors to total resistance were R_{tr} (56–62% of total) and R_{st} (25–32% of total). As with the gestational study, the diffusive resistances of maternal and fetal blood spaces made only minor contributions to total resistance. Fetal resistances were the only ones which were greater at HA.

Table 2. *Group means (cv) for placental diffusive conductances at G4 (37–41 weeks) of gestation and at term at low and high altitude*

Conductance	Gestation (n = 20)	Low altitude (n = 24)	High altitude (n = 44)
D^{me}	850 (22%)	1050 (33%)	1700 (22%)
D^{mp}	640 (20%)	620 (53%)	1190 (58%)
D^{tr}	100 (33%)	30 (35%)	30 (27%)
D^{st}	180 (46%)	50 (42%)	70 (38%)
D^{fp}	950 (39%)	730 (58%)	670 (41%)
D^{fe}	310 (38%)	290 (42%)	240 (38%)
D^{p}, total	43 (33%)	16 (36%)	18 (27%)
D^{p}, specific	13 (33%)	5 (32%)	6 (24%)

Taken together, the results suggest that total D_p is greater at HA, i.e. 18 ml min^{-1} kPa^{-1} (cv 27%) vs 16 ml min^{-1} kPa^{-1} (cv 36%). However, the apparent difference was not statistically significant, and this probably reflects the substantial impact of the resistance of the trophoblastic layer on determining overall D_p (Table 2). Significant altitudinal differences were found for the conductances of maternal and fetal erythrocytes, maternal plasma and the stroma. In HA placentae, D_{me} was about 62% greater than in LA organs, D_{mp} was about 92% greater and D_{st} 42% greater. In contrast, D_{fe} was 18% smaller in HA organs.

The highland fetus was advantaged in terms of the specific D_p of its placenta. The apparent improvement in overall D_p, combined with the reduced birthweight at HA, served to produce a significantly greater specific D_p (Table 2). This increased from 5 ml min^{-1} kPa^{-1} kg^{-1} (cv 32%) to 6 ml min^{-1} kPa^{-1} kg^{-1} (cv 24%).

Discussion

Adaptations of the functional capacity of the human placenta occur through the cumulative effects of many factors, including the microstructural and physiological variables built into the present model. The contributions made by some of these factors to the changes in D_p observed during gestation and at high altitude will now be discussed.

Gestational study

Whilst caution must be exercised when trying to draw conclusions from cross-sectional data, the present study has revealed no evidence to support the idea that placental function declines during the last trimester. Total D_p continues to rise throughout gestation and the rise appears to be commensurate with the increase in fetal weight. Without available information about the relationship between fetal weight and oxygen uptake, the precise relationship between D_p and oxygen flux remains unclear. However, it is obvious at least that morphometric D_p increases with fetal oxygen consumption.

The progressive increases in maternal and fetal vascular conductances (D_{me}, D_{mp}, D_{fp}, D_{fe}) make small contributions to overall D_p but are consistent with reported increases in perfusion rates during gestation. The changes in D_{tr} and D_{st} can be explained by the gradual growth and maturation of terminal villi which provides not only expanded exchange surface areas but also reduced diffusion distances. In previous studies, we have found that villous growth after 22 weeks is attributable almost exclusively to terminal villi (Jackson *et al.*, 1987*a*). The increase in villous volume, together with those in villous surface area, are attributable to increases in the combined length of villous branches. The increase in length is associated with a decrease in the mean diameter of villi and is consistent with continued elaboration of new terminal villi. Circumstantial support for this also comes from studies using a dimensionless coefficient which tests for isomorphic growth of villi by relating volume to the surface area raised to the power $\frac{3}{2}$ (Aherne & Dunhill, 1966; Boyd, 1984).

Apart from reduced distances across the trophoblast and stroma, maturation of villi also involves alterations in their volumetric composition (Jackson *et al.*, 1987*a*). The relative volume occupied by fetal capillaries increases but there is no increase in capillary mean diameter. Therefore, sinusoidal dilation of capillaries, although a real phenomenon (Kaufmann *et al.*, 1985), cannot be a general phenomenon. It must be compensated for by smaller vessels elsewhere. More recent findings (Mayhew *et al.*, 1991) suggest that villous maturation via trophoblast attenuation involves two processes, namely, protoplasmic redistribution within the syncytium (the dynamic component) and elaboration and margination of capillaries within villi (the mechanistic component). It appears that both processes contribute to the development of vasculosyncytial regions on the villous surface and to local improvements in gas diffusive conductance.

Altitudinal study

The present study suggests that the highland placenta near term exhibits a greater D_p per kilogram of fetal mass. This increase in specific D_p may be explained not only by differences in birthweight but also by changes in the partial conductances of the oxygen pathway. Notable are the improvements in D_{me}, D_{mp} and D_{st}. These observations are in accord with those made on the physiological conductances for carbon monoxide in guineapig placentae. In this organ, the specific D_p for carbon monoxide increases in normobaric hypoxia simulating high altitude (Gilbert *et al.*, 1979).

The altitudinal difference in D_{me} can be attributed to the more voluminous maternal blood space (Jackson *et al.*, 1987*b*) and haematocrit (Haas, 1980). Similar adaptations occur in the placentae of guinea pigs maintained under hypoxic conditions during pregnancy (Bacon *et al.*, 1984). Vascular bed volume and haematocrit also contribute to the greater D_{mp} at HA because they lead to an increase in the membrane surface area of maternal erythrocytes. The greater mass of erythrocytes must also reduce the harmonic mean distances across the maternal plasma and further increase the potential for oxygen flux from erythrocyte to trophoblast. Thinning of the maternal plasma compartment can be viewed as part of a more generalised thinning response which also includes the trophoblast and stroma (Jackson *et al.*, 1988*a*,*b*). It appears that thinning of the trophoblast is sufficient only to compensate for the reduced villous surface area and to maintain D_{tr}, whereas stromal thinning improves D_{st}.

The maternal vascular and membrane thickness changes must compensate for the poor villous growth in HA placentae (Jackson *et al.*, 1987*b*,*c*). Both the volume and surface of trophoblast are diminished at HA. Harmonic mean thickness is smaller at HA because the trophoblast is more irregular and this option is a very economical way of improving diffusive conductance (Jackson *et al.*, 1985). As during gestation, thinning seems to occur by dynamic and mechanistic processes although the balance favours the mechanistic process as the major contributor (Jackson *et al.*, 1988*b*). Attenuation of trophoblast has also been witnessed in guineapig placentae after chronic maternal hypoxia (Bacon *et al.*, 1984) and in cultured human placentae (Burton *et al.*, 1989). However, in the latter context it is probably not a truly adaptive response.

The principal mechanism underlying the increase in D_{st}, and in overall D_p, is the margination of capillaries within villi (Jackson *et al.*, 1988*b*). The other dimensions which contribute to D_{st} (i.e. the inner surface area of trophoblast and capillary surface area) do not alter significantly at HA (Jackson *et al.*, 1987*c*, 1988*b*). Thinning of the stroma probably reflects differential growth of capillaries within villi rather than their physical migration. Certainly, there is an increase in the relative length of smaller diameter villi at HA (Jackson *et al.*, 1987*c*). As during gestation, dilation does not account for the peripheralisation because the mean diameter of capillaries at HA is smaller. These differences in the fetal capillaries are similar to those observed in hypoxia in the pregnant guineapig (Bacon *et al.*, 1984) and, conceivably, the decrease in mean vessel diameter could increase resistance to blood flow and allow more time for oxygen exchange to occur between maternal and fetal vascular beds.

D_{fp} is conserved at HA, despite a decline in fetal erythrocyte surface area. Since capillary surface areas are maintained, reduced plasma distances are important and perhaps occur via the reduced capillary diameter and the elevated fetal haematocrit (Haas, 1980; Ballew & Haas, 1986; Jackson *et al.*, 1987*c*). D_{fe} is the only diffusive conductance to decrease in HA organs, so the rise in fetal haematocrit does not compensate for the smaller fetal vascular space (Jackson *et al.*, 1987*b*).

These adaptations should be viewed in the wider context of improving the supply of oxygen to the highland fetus. However, despite their occurrence HA birthweights are reduced (see Haas, 1980; Haas *et al.*, 1980). If the lower birthweight is attributable to an impoverished oxygen flux to the fetus then the present findings suggest that this probably arises as a direct consequence of lowered transplacental oxygen gradients. The structural adaptations seen in the human placenta help to conserve D_p in the face of reduced villous growth and of inadequate responses on the fetal side of the highland organ (Mayhew *et al.*, 1990; Mayhew, 1991*b*).

General remarks

These studies have revealed some common features of placental adaptation during gestation and pregnancy at HA. In both situations, between-organ variation tended to be greater on the fetal aspect of the placenta. This finding supports an earlier observation on capillary surface areas (Laga *et al.*, 1973). Similarities were found also in the mechanism of villous membrane thinning which involved contributions from both the trophoblast and stroma. These results suggest that reducing effective diffusion distances, including those across the maternal and fetal plasma, has real adaptive significance.

The contributors to thinning of the villous membrane were quantitatively different in the two studies. Between G1 (10–22 weeks) and G4 (37–41 weeks), dynamic and mechanistic components appeared to contribute equally. At HA, the emphasis was more on the mechanistic process (i.e. on the obtrusion of peripheralised capillaries into overlying trophoblast). The change of emphasis may reflect the different constraints on growth of villi in the two situations. It is known that villous growth is impoverished in HA organs (Jackson *et al.*, 1987*c*).

In highland organs, the reduced harmonic mean thickness of the trophoblast is sufficient only to conserve D_{tr} at LA values. Since D_{tr} is the major determinant of *D*p, this raises the possibility that there may be a natural limit to thinning of trophoblast, perhaps influenced by the balance between trophoblast volume and villous surface area. Unpublished findings, on which the present diffusive conductances are based, confirm that trophoblast volume in G4 is comparable to that found in LA organs in the altitudinal study. However, this volume is spread over a much more extensive villous surface and this helps to explain the lower mean trophoblast thickness near term in the gestational study (2 μm vs 4 μm).

The large discrepancy in *D*p between the two studies cannot be attributed to technical differences such as variations in birthweights, placental weights, modes of delivery or tissue processing. It appears to result mainly from differences in surface areas and diffusion distances rather than changes in vascular volumes. In the gestational study, the term organ had longer and thinner villi with longer and thinner capillaries. In this sense, they appeared to be better adapted than their LA counterparts.

These findings suggest that genetic and/or environmental factors may determine the arborisation and capillarisation of villi and so account for the observed discrepancies in surfaces and thicknesses. Future studies on placental changes in maternal diabetes mellitus and maternal smoking during pregnancy may throw further light on this. Finally, it is interesting to note that trophoblast volume at term is similar in both of the studies reported here. Trophoblast volume is related to fetal well-being and hormonal status (Vermeulen *et al.*, 1982) and continues to increase in postmaturity (Bouw *et al.*, 1978).

Acknowledgements

These studies were conducted in collaboration with Moira R. Jackson (Aberdeen), Patricia A. Boyd (Oxford) and Jere D. Haas (Cornell). My researches on the placenta have been supported by The Leverhulme Trust, The Anatomical Society of Great Britain & Ireland, The Cunningham Trust, Action Research, The Carnegie Trust for the Universities of Scotland and The Tobacco Products Research Trust.

References

Aherne, W. (1975). Morphometry. In *The Placenta and its Maternal Supply Line*, ed. P. Gruenwald, pp. 80–97. Lancaster: MTP.

Aherne, W. & Dunhill, M.S. (1966). Quantitative aspects of placental structure. *Journal of Pathology and Bacteriology* **91**, 123–39.

Bacon, B.J., Gilbert, R.D., Kaufmann, P., Smith, A.D., Trevino, F.T.

& Longo, L.D. (1984). Placental anatomy and diffusing capacity in guinea pigs following long-term maternal hypoxia. *Placenta* **5**, 475–88.

Baddeley, A.J., Gundersen, H.J.G. & Cruz-Orive, L.M. (1986). Estimation of surface area from vertical sections. *Journal of Microscopy* **142**, 259–76.

Ballew, C. & Haas, J.D. (1986). Hematologic evidence of fetal hypoxia among newborn infants at high altitude in Bolivia. *American Journal of Obstetrics and Gynecology* **155**, 166–9.

Barron, D.H. & Meschia, G. (1954). A comparative study of the exchange of the respiratory gases across the placenta. In *Cold Spring Harbor Symposia on Quantitative Biology*, Vol. XIX, *The Mammalian Fetus: Physiological Aspects of Development*, pp. 93–101. New York: Long Island Biological Association.

Barron, D.H., Metcalfe, J., Meschia, G., Huckabee, W., Hellegers, A. & Prystowsky, H. (1964). Adaptations of pregnant ewes and their fetuses to high altitude. In *The Physiological Effects of High Altitude*, ed. W.H. Weihe, pp. 115–29. Oxford: Pergamon Press.

Bonds, D.R., Mwape, B., Kumar, S. & Gabbe, S.G. (1984). Human fetal weight and placental weight growth curves. A mathematical analysis from a population at sea level. *Biology of the Neonate* **45**, 261–74.

Bouw, G.M., Stolte, L.A.M., Baak, J.P.A. & Oort, J. (1978). Quantitative morphology of the placenta. II. The growth of the placenta and the problem of postmaturity. *European Journal of Obstetrics, Gynaecology and Reproductive Biology* **8**, 31–42.

Boyd, P.A. (1984). Quantitative structure of the normal human placenta from 10 weeks of gestation to term. *Early Human Development* **9**, 297–307.

Boyd, P.A., Brown, R.A. & Stewart, W.J. (1980). Quantitative structural differences within the normal term human placenta: a pilot study. *Placenta* **1**, 337–44.

Burton, G.J., Mayhew, T.M. & Robertson, L.A. (1989). Stereological re-examination of the effects of varying oxygen tensions on human placental villi maintained in organ culture for up to 12 h. *Placenta* **10**, 263–73.

Cruz-Orive, L.M. & Weibel, E.R. (1990). Recent stereological methods for cell biology: a brief survey. *American Journal of Physiology* **258**, L148–56.

Gehr, P., Bachofen, M. & Weibel, E.R. (1978). The normal human lung: ultrastructure and morphometric estimation of diffusion capacity. *Respiration Physiology* **32**, 121–40.

Gilbert, R.D., Cummings, L.A., Juchau, M.R. & Longo, L.D. (1979). Placental diffusing capacity and fetal development in exercising or hypoxic guinea pigs. *Journal of Applied Physiology* **46**, 828–34.

Gundersen, H.J.G. & Jensen, E.B. (1987). The efficiency of systematic sampling in stereology and its prediction. *Journal of Microscopy* **147**, 229–63.

Haas, J.D. (1980). Maternal adaptation and fetal growth at high altitude in Bolivia. In *Social and Biological Predictors of Nutritional Status, Physical Growth and Neurological Development*, ed. L.S. Greene & F.S. Johnston, pp. 257–90. New York: Academic Press.

Haas, J.D., Frongillo, E.A., Stepick, C.D., Beard, J.L. & Hurtado, L. (1980). Altitude, ethnic and sex differences in birth weight and length in Bolivia. *Human Biology* **52**, 459–77.

Jackson, M.R., Joy, C.F., Mayhew, T.M. & Haas, J.D. (1985). Stereological studies on the true thickness of the villous membrane in human term placentae: a study of placentae from high-altitude pregnancies. *Placenta* **6**, 249–58.

Jackson, M.R., Mayhew, T.M. & Boyd, P.A. (1987*a*). A cross sectional study on the growth and maturation of human placental villi from 10 weeks of gestation to term. *Journal of Anatomy* **155**, 235–6.

Jackson, M.R., Mayhew, T.M. & Haas, J.D. (1987*b*). The volumetric composition of human term placentae: altitudinal, ethnic and sex differences in Bolivia. *Journal of Anatomy* **152**, 173–87.

Jackson, M.R., Mayhew, T.M. & Haas, J.D. (1987*c*). Morphometric studies on villi in human term placentae and the effects of altitude, ethnic grouping and sex of newborn. *Placenta* **8**, 487–95.

Jackson, M.R., Mayhew, T.M. & Haas, J.D. (1988*a*). On the factors which contribute to thinning of the villous membrane in human placentae at high altitude. I. Thinning and regional variation in thickness of trophoblast. *Placenta* **9**, 1–8.

Jackson, M.R., Mayhew, T.M. & Haas, J.D. (1988*b*). On the factors which contribute to thinning of the villous membrane in human placentae at high altitude. II. An increase in the degree of peripheralization of fetal capillaries. *Placenta* **9**, 9–18.

Kaufmann, P., Bruns, U., Leiser, R., Luckhardt, M. & Winterhager, E. (1985). The fetal vascularisation of term human placental villi. II. Intermediate and terminal villi. *Anatomy and Embryology* **173**, 203–14.

Laga, E.M., Driscoll, S.G. & Munro, H.H. (1973). Quantitative studies of human placenta. I. Morphometry. *Biology of the Neonate* **23**, 231–59.

Mayhew, T.M. (1983). Stereology: progress in quantitative microscopical anatomy. In *Progress in Anatomy*, Vol. 3, ed. V. Navaratnam & R.J. Harrison, pp. 81–112. Cambridge: Cambridge University Press.

Mayhew, T.M. (1986). Morphometric diffusing capacity for oxygen of the human placenta at high altitude. In *Aspects of Hypoxia*, ed. D. Heath, pp. 181–94. Liverpool: Liverpool University Press.

Mayhew, T.M. (1991*a*). The new stereological methods for interpreting functional morphology from slices of cells and organs. *Experimental Physiology* **76**, 639–65.

Mayhew, T.M. (1991*b*). Scaling placental oxygen diffusion to birth-weight: studies on placentae from low- and high-altitude pregnancies. *Journal of Anatomy* **175**, 187–94.

Mayhew, T.M. & Burton, G.J. (1988). Methodological problems in placental morphometry: apologia for the use of stereology based on sound sampling practice. *Placenta* **9**, 565–81.

Mayhew, T.M., Jackson, M.R. & Boyd, P.A. (1991). Mechanisms of villous membrane attenuation in terminal villi of human placentae during gestation. *Journal of Anatomy* **176**, 249–51.

Mayhew, T.M., Jackson, M.R. & Haas, J.D. (1986). Microscopical morphology of the human placenta and its impact on oxygen diffusion: a morphometric model. *Placenta* **7**, 121–31.

Mayhew, T.M., Jackson, M.R. & Haas, J.D. (1990). Oxygen diffusive conductances of human placentae from term pregnancies at low and high altitudes. *Placenta* **11**, 493–503.

Mayhew, T.M., Joy, C.F. & Haas, J.D. (1984). Structure–function correlation in the human placenta: the morphometric diffusing capacity for oxygen at full term. *Journal of Anatomy* **139**, 691–708.

Metcalfe, J., Bartels, H. & Moll, W. (1967). Gas exchange in the pregnant uterus. *Physiological Reviews* **47**, 782–838.

Sands, J. & Dobbing, J. (1985). Continuing growth and development of the third-trimester human placenta. *Placenta* **6**, 13–22.

Sokal, R.R. & Rohlf, F.J. (1981). *Biometry. The Principles and Practice of Statistics in Biological Research*. San Francisco: W.H. Freeman.

Stringer, B.M.J., Wynford-Thomas, D. & Williams, E.D. (1982). Physical randomisation of tissue architecture: an alternative to systematic sampling. *Journal of Microscopy* **126**, 179–82.

Teasdale, F. (1980). Gestational changes in the functional structure of the human placenta in relation to fetal growth: a morphometric study. *American Journal of Obstetrics and Gynecology* **137**, 560–8.

Teasdale, F. & Jean-Jacques, G. (1985). Morphometric evaluation of the microvillous surface enlargement factor in the human placenta from mid-gestation to term. *Placenta* **6**, 375–81.

Vermeulen, R.C.W., Kurver, P.H.J., Arts, N.F.T., Van Kressel, H., Wilson, G.R. & Klopper, A. (1982). The relationship between the surface area of the trophoblast and some placental products. *Placenta* **3**, 359–66.

Weibel, E.R. (1970). Morphometric estimation of pulmonary diffusion capacity. I. Model and method. *Respiration Physiology* **11**, 54–75.

Weibel, E.R. (1979). *Stereological Methods*, Vol. 1, *Practical Methods for Biological Morphometry*. London: Academic Press.

Winick, M., Coscia, A. & Noble, A. (1967). Cellular growth in human placenta. I. Normal cellular growth. *Pediatrics* **39**, 248–51.

ANDREA BELLELLI and GUIDO DI PRISCO

Thermodynamic and stereochemical modelling of vertebrate haemoglobin

Introduction

Haemoglobin is the protein which carries oxygen from the environment to tissues; in vertebrates, it is contained in specialised cells, the erythrocytes. Over the last century, the study of the chemical properties of haemoglobin has provided a wealth of information. Thus, it has become feasible to use this protein as a 'model' to define functional features, such as allostery and cooperativity among respiratory pigments and enzymes. The aim of this chapter is to try to introduce the most important modelling approaches to the understanding of its complex structure–function relationships.

Human and horse haemoglobins are the best characterised, and their crystallographic structure has been known for almost 30 years. However, many challenging observations came also from the study of other vertebrate haemoglobins, most notably fishes and amphibians, which may clearly display properties that are almost undetectable in human haemoglobin (e.g. the functional heterogeneity of the α and β chains). For this reason we will focus our attention upon models which were originally related to human haemoglobin, and subsequently their behaviour when verified on other haemoglobins.

Molecular properties

In higher vertebrates, haemoglobin ($\alpha_2\beta_2$; molecular weight, 64 000) is a tetramer of two pairs of identical subunits, α (141 or 142 amino acid residues) and β (146 residues). Each subunit contains one prosthetic group, the iron protoporphyrin IX (haem) whose Fe^{2+} ion reversibly binds oxygen, and is maintained in the physiologically viable reduced form by the specific interactions with the protein moiety. The oxidation of Fe^{2+} to Fe^{3+} in methaemoglobin impairs oxygen binding. The four subunits are not covalently bound to each other and upon dilution

Society for Experimental Biology Seminar Series 51: *Oxygen Transport in Biological Systems*, ed. S. Egginton & H.F. Ross.

oxyhaemoglobin splits into two symmetrical $\alpha\beta$ dimers, with a dissociation constant $K_D \simeq 1$ μM (Edelstein *et al.*, 1970).

Another protein involved in reversible oxygen binding (storage) is myoglobin (Antonini & Brunori, 1971), a single-chain haemoprotein of molecular weight 17 000, extracted from striated muscle. Although it shares the overall molecular structure with the haemoglobin subunits, its behaviour differs from that of haemoglobin, and it will be briefly discussed later.

The reaction with oxygen

The reversible reaction of oxygen binding to haemoglobin does not correspond to a simple equilibrium. In fact, the curve describing the oxygen content of blood as a function of its partial pressure has a sigmoidal, instead of hyperbolic, shape (Antonini & Brunori, 1971). This phenomenon, defined as cooperativity, characterises all vertebrate haemoglobins, the only exception so far being that from the tuatara, *Sphenodon punctatus* (Wells *et al.*, 1983).

The overall affinity of haemoglobin for oxygen is expressed as the gas partial pressure required to achieve half-saturation ($P_{1/2}$ or P_{50}); in a non-cooperative system (e.g. myoglobin) this corresponds to the equilibrium constant for the dissociation reaction. Cooperative ligand binding has important physiological consequences, since it allows oxygen to be efficiently released at relatively high oxygen partial pressures: even a 75% desaturation of haemoglobin does not require a tissue oxygen tension lower than 40 mm Hg (Fig. 1).

In the course of evolution, complex and sophisticated molecular mechanisms (which include the effect of pH, carbon dioxide, organic phosphates and temperature) have been developed to control oxygen transport by haemoglobin.

The allosteric effectors; Bohr and Root effects

In allosteric control the reactivity of a site is modulated, through conformational equilibria, by binding of ligands at other sites of the same molecule. The allosteric interactions can be homotropic, if occurring among sites which bind the same ligand (i.e. oxygen), or heterotropic, if the sites bind different ligands (protons, chloride, organic phosphates, carbon dioxide) which affect the oxygen affinity.

The decreased oxygen affinity of haemoglobin at lower values in the physiological pH range is known as the alkaline Bohr effect (Bohr, Hasselbach & Krogh, 1904, quoted by Edsall, 1980; reviewed by Riggs,

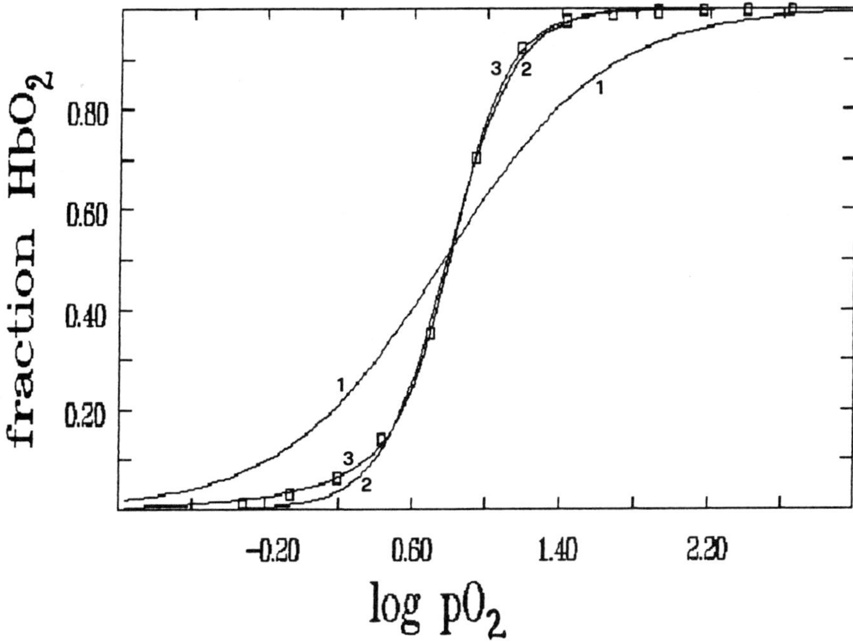

Fig. 1. Oxygen–equilibrium curve, fitted according to Eqn 2.1 with $n = 1$ (1, Hufner's theory; non-cooperative curve) and with n freely variable (2, Hill's theory), and to Eqn 17 (3, two-state model); see p. 116. The experimental data were kindly provided by Drs R. Ippoliti and E. Lendaro (Rome), whose help is gratefully acknowledged.

1988). The acid Bohr effect, i.e. an increase of oxygen affinity upon further pH decrease, occurs at pH lower than 6.0 (Fig. 2). The physiological relevance of the alkaline Bohr effect is clear when one considers that the metabolism of highly active tissues produces acid substances which enhance oxygen unloading from haemoglobin. As will be discussed later, the amino acid residues whose contribution to the alkaline Bohr effect is most relevant are His-146 on the β chains and Val-1 on the α chains; in the presence of other ligands (Cl^-, CO_2), other residues may contribute. The molecular mechanism which couples oxygen binding and proton release has been thoroughly investigated from both a thermodynamic (Wyman, 1964) and a structural standpoint (Perutz, 1970*a*,*b*; Perutz *et al.*, 1980).

Haemoglobin from many teleost fishes does not display the acid Bohr effect and, when the pH is lowered, the oxygen affinity decreases to such an extent that haemoglobins cannot be saturated even at very high press-

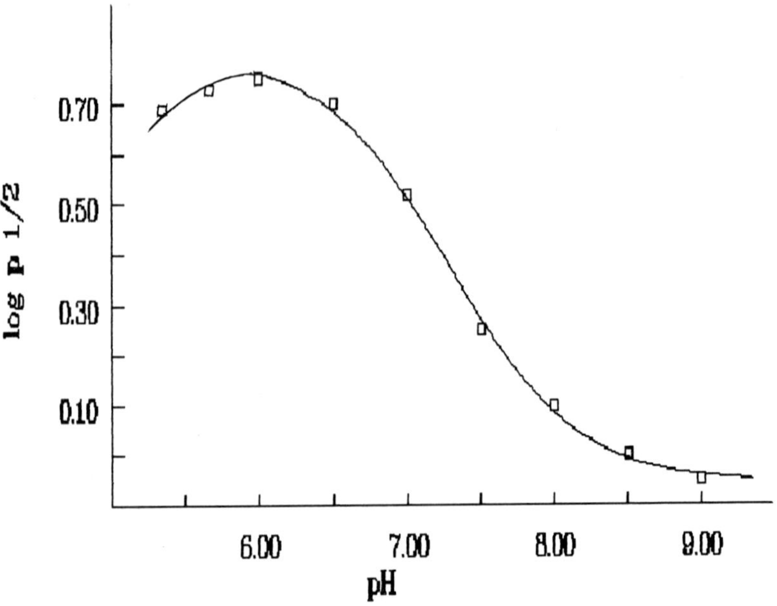

Fig. 2. The Bohr effect of human haemoglobin.

ures of pure oxygen (Fig. 3). In addition, cooperativity is totally lost and the oxygen capacity of blood may be reduced to only 50% of its value at alkaline pH. This characteristic is known as the Root effect (Root, 1931; Root & Irving, 1941; Scholander & Van Dam, 1954; reviewed by Brittain, 1987): it is an exaggerated Bohr effect and dictates to what extent the oxygen tension can be raised in acid-producing tissues. The physiological significance of Root-effect haemoglobins has been linked to the presence of at least one of two anatomical structures which require high oxygen pressure: the rete mirabile which supplies the gas gland that inflates fish swimbladders with oxygen, and the choroid rete, a vascular structure which supplies oxygen to the poorly vascularised retina (Wittemberg & Wittemberg, 1974). Although the role of the Root effect in filling the swimbladder (when this organ is present) is well established, the correlation between the presence of Root-effect haemoglobins and the choroid rete (much more common than the swimbladder), appears to be more sound. It is worth mentioning that in Antarctic fishes (all lacking the swimbladder), only the few species possessing haemoglobins without a Root effect (di Prisco *et al.*, 1991), as well as those of a family whose colourless blood has no haemoglobin at all, lack the choroid rete (Eastman, 1988).

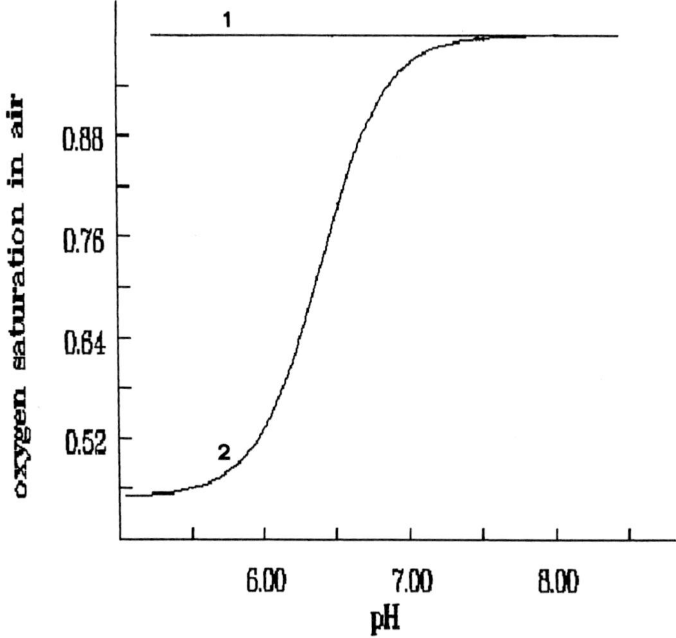

Fig. 3. Oxygen saturation as a function of pH, in air-equilibrated samples of (1) human haemoglobin and (2) Root-effect fish haemoglobin.

Characteristically, the blood of most fishes contains multiple haemoglobins, which in some instances (such as carp, a bottom feeder: Gillen & Riggs, 1972; Tan *et al.*, 1972) are functionally indistinguishable from one another and display the Root effect; in fast swimmers, on the other hand, haemoglobins show functional differences in oxygen binding, thus enabling oxygen transport even under conditions of acidosis, when the gill saturation of a Root-effect haemoglobin would be low (e.g. trout haemoglobins: reviewed by Brunori, 1975). A remarkable exception is represented by the vast majority of endemic Antarctic species: they have either a single or a major haemoglobin, accounting for about 95% of the total blood content (di Prisco *et al.*, 1991).

Haemoglobin recognises three other physiologically important allosteric effectors: chloride ions (whose effect is largely linked to the Bohr effect), carbon dioxide and organic phosphates. Carbon dioxide forms a Schiff base with the terminal amino group of each chain (Christiansen, Douglas & Haldane, 1914, quoted by Edsall, 1980; Perutz, 1990), and haemoglobin contributes to its transport; binding of CO_2 reduces the oxygen affinity (and the Bohr effect) and has the same physiological meaning as the uptake of protons. Organic phosphates (2,3-diphosphoglycerate, DPG, in mammals; ATP or GTP in reptiles, amphibians and fishes) bind to a specific site in the β chains and lower the oxygen affinity (Benesch & Benesch, 1967; Perutz, 1970*b*; Arnone, 1972).

The structure of haemoglobin as defined by X-ray crystallography

Very detailed information about the 3-dimensional structure of haemoglobin has been obtained by X-ray crystallography. The globins are made of 7 (in the α chain) or 8 (in the β chain) segments of α-helix (named from A to H; in α chains, helix D is lacking) and 7 (in α) or 6 (in β) non-helical portions, making up the corners between some of these segments (named AB, CD, etc.), and the chain N- and C-terminal regions (NA and HC).

Figure 4 shows the structure of the β subunit from human haemoglobin (Perutz, 1970*a*). It will be appreciated that helices E and F come into close contact with the haem group; slightly further, helices C, G and the CD corner border the haem pocket. The haem is wedged by Phe CD1 into a deep pocket of the polypeptide chain, with the propionate chains facing the exterior. In all haemoglobins, the iron is linked to His F8 (the proximal His) and in most it is bridged by means of the bound oxygen molecule to His E7 (the distal His: Phillips & Schoenborn, 1981). This structural arrangement is shared with the α chains, and with the chains from all other vertebrate haemoglobins and myoglobins (Perutz, 1987, 1990).

The tetrameric haemoglobin molecule may be viewed as the assembly of two perfectly symmetrical αβ dimers, $\alpha_1\beta_1$ and $\alpha_2\beta_2$; this notation serves the purpose of simplifying the description of interchain contacts, and corresponds to real chemical entities since, as noted above, $\alpha_1\beta_1$-type dimers may be easily obtained by dilution of oxyhaemoglobin (Edelstein *et al.*, 1970); by contrast, the two α or the two β subunits are in only limited contact with each other and the $\alpha_1\beta_2$ interface is easily broken (thus dimers other than the $\alpha_1\beta_1$-type cannot be isolated). The $\alpha_1\beta_1$ interface is contributed by residues of helices B, G and H and of the GH corner of both chains, which come into contact through hydrophobic interactions. The two $\alpha_1\beta_1$-type dimers are connected by the $\alpha_1\beta_2$-type interface, described in detail below.

The amino acid residues of haemoglobin are conveniently indicated by their position in the helical or non-helical segment of the polypeptide chain. In Fig. 4, residues F8 (proximal His) and E7 (distal His) are

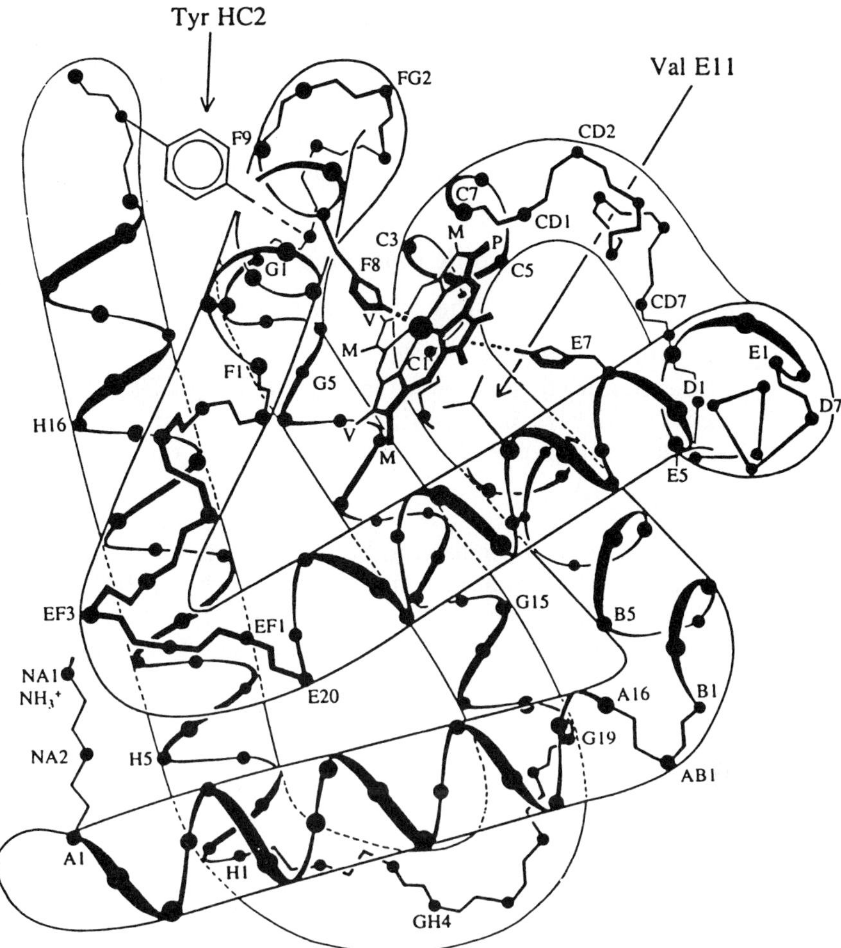

Fig. 4. Tertiary structure of the β chain of human haemoglobin. Residues F8 and E7 (the proximal and distal His) are indicated. Reproduced with permission of the author from M.F. Perutz (1990), *Mechanisms of Cooperativity and Allosteric Regulation in Proteins*, Cambridge University Press.

indicated along with Val E11, which, as well as Phe CD1, also comes into contact with the haem. The tertiary structure of the subunits is so highly conserved that, for example, the α chains (141 residues), the β chains (146 residues) and myoglobins (over 150 residues) all have the proximal and distal His at positions F8 and E7. The consequences of this finding can be carried much further and it has been possible to predict the structures of some haemoglobins of known sequence, before their crystallographic resolution, by computer-modelling techniques, using the structure of human haemoglobin as reference (Perutz & Brunori, 1982; Perutz, 1987). X-ray crystallography has also shed light on the effect of hydrogen ions and other allosteric effectors, as will be discussed below (see p. 119).

Classical thermodynamic modelling

Early descriptions of the cooperative oxygen binding

The respiratory properties of haemoglobin were first formalised at the beginning of this century by Hufner & Gansser (quoted by Edsall, 1980), who proposed that oxygen is bound to the iron and that the reaction obeys the mass law. Myoglobin does actually conform to this hypothesis:

$$\mathrm{Mb} + \mathrm{O_2} \rightleftharpoons \mathrm{MbO_2} \qquad \text{(scheme 1)}$$

Haemoglobin was soon recognised to behave in a complex way, since its titration with ligands is sigmoidal instead of hyperbolic (Bohr *et al.*, quoted by Edsall, 1980). Hill (1910) proposed that the haemoglobin molecule is polymeric and that its n equivalent sites bind oxygen all together, in the reaction:

$$\mathrm{Hb} + n\mathrm{O_2} \rightleftharpoons \mathrm{Hb(O_2)}_n \qquad \text{(scheme 2)}$$

The mass law, applied to the reaction of scheme 2, dictates:

$$K = [\mathrm{Hb(O_2)}_n]/[\mathrm{Hb}]P_{\mathrm{O_2}}{}^n \qquad (1)$$

Therefore the fractional ligand saturation Y is:

$$Y = KP_{\mathrm{O_2}}{}^n/(1 + KP_{\mathrm{O_2}}{}^n) \qquad (2.1)$$

This equation, for $n = 1$, corresponds to the equilibrium of a non-cooperative oxygen carrier, such as myoglobin. Taking the ratio of liganded and unliganded haems instead of the fractional saturation, and using a logarithmic notation, one obtains:

$$\log[Y/(1 - Y)] = \log K + n \log P_{\mathrm{O_2}} \qquad (2.2)$$

Equations 2.1 and 2.2 (the Hill equations) describe a non-hyperbolic oxygen equilibrium curve and give a value of n of approximately 2.8 for human haemoglobin; this was interpreted as the average degree of polymerisation of haemoglobin monomers in solution. It is worth mentioning that the model fails to describe the experimental data outside the fractional saturation range of $0.1 < Y < 0.9$, even though reliable measures in this region were not available at that time (line 2 in Fig. 1). A conceptual improvement over the Hill equation was suggested by Haldane, who proposed that a fast association–dissociation equilibrium between monomers and polymers characterises haemoglobin solutions, and that only monomers bind oxygen; this hypothesis generates a sigmoid ligand-binding curve, and applies strictly to the case of lamprey haemoglobin (Antonini *et al.*, 1964). The oxygen saturation of haemoglobin predicted by schemes 1 and 2 is depicted in Fig. 1; in the case of the Hill equation, the fit of the experimental data is poor in the low saturation range, as expected.

Adair (1925), using osmotic pressure techniques, demonstrated beyond doubt that the molecular weight of haemoglobin is 64 000, independently of the oxygen partial pressure, and corresponds to the aggregation of four subunits, each carrying a haem group. Adair also suggested that oxygen binding to the tetrameric haemoglobin molecule need not be an all-or-none phenomenon and that partially liganded molecular species may be present in solution, according to the multiple equilibrium:

$$\mathrm{Hb} + 4\mathrm{O}_2 \overset{K_1}{\rightleftharpoons} \mathrm{HbO}_2 + 3\mathrm{O}_2 \overset{K_2}{\rightleftharpoons} \mathrm{Hb(O_2)_2} + \\ + 2\mathrm{O}_2 \overset{K_3}{\rightleftharpoons} \mathrm{Hb(O_2)_3} + \mathrm{O}_2 \overset{K_4}{\rightleftharpoons} \mathrm{Hb(O_2)_4} \qquad \text{(scheme 3)}$$

The mass law, applied to the series of reactions of scheme 3, gives:

$$K_n = [\mathrm{Hb(O_2)}_n]/[\mathrm{Hb(O_2)}_{n-1}]P_{\mathrm{O}_2} \qquad (3)$$

Equation 3 is conveniently rewritten defining the apparent constant β_n:

$$\beta_n = [\mathrm{Hb(O_2)}_n]/[\mathrm{Hb}]P_{\mathrm{O}_2}{}^n \qquad (4)$$

The distribution of haemoglobin amongst partially liganded intermediates (the so-called binding polynomial, BP), taking the concentration of the unliganded species as unity, is given by:

$$\mathrm{BP} = 1 + \beta_1 P_{\mathrm{O}_2} + \beta_2 P_{\mathrm{O}_2}{}^2 + \beta_3 P_{\mathrm{O}_2}{}^3 + \beta_4 P_{\mathrm{O}_2}{}^4 \qquad (5)$$

Multiplying each term of the binding polynomial by the exponent of the oxygen partial pressure, the relative amount of liganded haems (HO_2) is obtained:

$$\mathrm{HO}_2 = \beta_1 P_{\mathrm{O}_2} + 2\beta_2 P_{\mathrm{O}_2}{}^2 + 3\beta_3 P_{\mathrm{O}_2}{}^3 + 4\beta_4 P_{\mathrm{O}_2}{}^4 \qquad (6)$$

The fractional saturation of haemoglobin at any given ligand activity is obtained by combination of Eqns 5 and 6 (considering that the total number of oxygen-binding sites is four times the binding polynomial):

$$Y = \frac{\beta_1 P_{\mathrm{O}_2} + 2\beta_2 P_{\mathrm{O}_2}{}^2 + 3\beta_3 P_{\mathrm{O}_2}{}^3 + 4\beta_4 P_{\mathrm{O}_2}{}^4}{4(1 + \beta_1 P_{\mathrm{O}_2} + \beta_2 P_{\mathrm{O}_2}{}^2 + \beta_3 P_{\mathrm{O}_2}{}^3 + \beta_4 P_{\mathrm{O}_2}{}^4)} \qquad (7)$$

The Adair equation (Eqn 7) might seem complex, but it excellently fits all oxygen-binding data so far reported. An extensive characterisation of the behaviour of human haemoglobin in terms of this equation has been carried out by several authors (Imai, 1973; Imai & Yonetani, 1975), and shows that a description of the cooperative properties of haemoglobin necessarily implies that $K_1 < K_4$. The Adair equation is a general thermodynamic formulation, and any model capable of describing the cooperative behaviour of haemoglobin must also be capable of generating four 'apparent' Adair constants. However, it is not a model of cooperative behaviour in itself, because it does not explain why the value of K_1 is lower than that of K_4; or, put in simpler terms, it tells us that there are high- and low-affinity oxygen-binding sites, but does not tell where the high-affinity sites become hidden at low ligand concentration.

Sequential models

In 1935 Pauling proposed the first modern chemical model of haemoglobin, in which oxygen binding to one subunit modifies its structure and transmits information, via the intersubunit contacts, to all or some of the neighbouring subunits (Pauling, 1935). This model is sequential because the ligand affinity of each binding site in any single haemoglobin molecule varies as a function of the number of liganded sites (see the two-state model, below; and Fig. 5 for comparison). Pauling's model is worth a detailed description, not only for historical reasons, but also because it excellently defines what we shall call 'sequential-type' interactions in the following pages, where we will also discuss the intrinsic limits of these models.

The basic hypotheses underlying the model are: (i) haemoglobin contains four haems, each interacting with two (the 'square' case) or three others (the 'tetrahedral' case); and (ii) the four haems are chemically equivalent in their interactions with oxygen and with each other. Only the simplest case, that of tetrahedral haemoglobin, will be discussed; it corresponds to scheme 4, where the dots connecting the haems (H) represent the possible cooperative interactions; HL denotes a liganded subunit and a line an actual interaction.

H H   HL—H   HL—HL   HL—HL   HL—HL
H H   H H   H H   HL—H   HL—HL   (scheme 4)

The first ligand binds with the intrinsic affinity constant K; each of the following bind with an increased affinity constant because of the contribution of the interaction term i, raised to the relevant power:

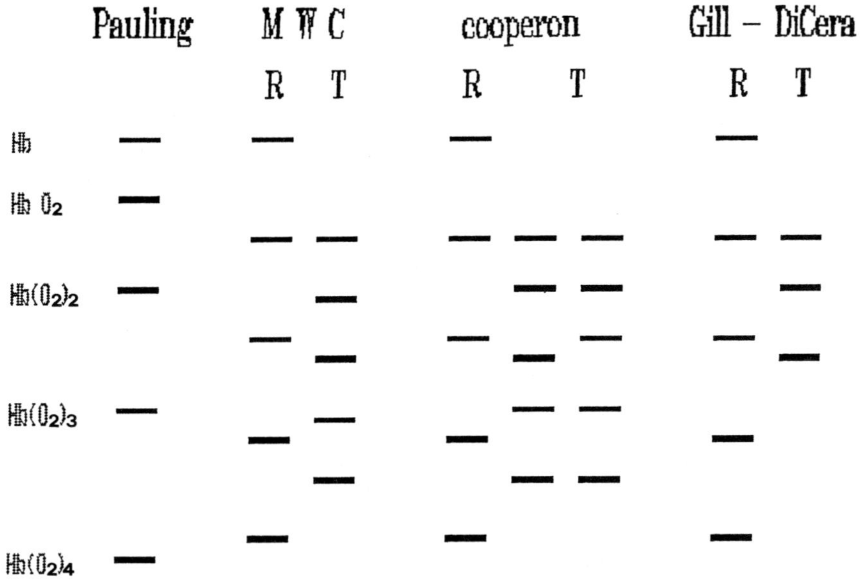

Pauling MWC cooperon Gill – DiCera

R T R T R T

Hb

Hb O_2

$Hb(O_2)_2$

$Hb(O_2)_3$

$Hb(O_2)_4$

Fig. 5. Energy levels of haemoglobin according to the models of Pauling (1935), Monod *et al.* (1965), Brunori *et al.* (1986) and Di Cera *et al.* (1987). The abscissa represents the free energy for each liganded intermediate on an arbitrary scale. Sequential and symmetrical models (e.g. Pauling's and MWC) have only one major path for ligand binding, whereas a combinatorial model has more (see the cooperon model of Brunori *et al.*, which distinguishes two parallel paths of ligand binding to T-state haemoglobin).

$$HbO_2 = K\mathrm{Hb}P_{O_2} \tag{8.1}$$

$$Hb(O_2)_2 = iK\mathrm{HbO_2}P_{O_2} = iK^2{P_{O_2}}^2 \tag{8.2}$$

$$Hb(O_2)_3 = i^2K\mathrm{Hb(O_2)_2}P_{O_2} = i^3K^3{P_{O_2}}^3 \tag{8.3}$$

$$Hb(O_2)_4 = i^3K\mathrm{Hb(O_2)_3}P_{O_2} = i^6K^4{P_{O_2}}^4 \tag{8.4}$$

A statistical (or symmetry) term has to be introduced, since the model uses intrinsic, instead of apparent, constants and the species Hb has four haems apt to react with oxygen, whereas $Hb(O_2)_3$ has only one. The number of species for each saturation intermediate is: 1 for Hb; 4 for HbO_2; 6 for $Hb(O_2)_2$; 4 for $Hb(O_2)_3$ and 1 for $Hb(O_2)_4$; therefore the binding polynomial is:

$$BP = 1 + 4KP_{O_2} + 6iK^2{P_{O_2}}^2 + 4i^3K^3{P_{O_2}}^3 + i^6K^4{P_{O_2}}^4 \tag{9}$$

Consequently, the fractional saturation is:

$$Y = \frac{KP_{O_2} + 3iK^2P_{O_2}{}^2 + 3i^3K^3P_{O_2}{}^3 + i^6K^4P_{O_2}{}^4}{1 + 4KP_{O_2} + 6iK^2P_{O_2}{}^2 + 4i^3K^3P_{O_2}{}^3 + i^6K^4P_{O_2}{}^4} \qquad (10)$$

Figure 5 reports the energy levels of each saturation intermediate of haemoglobin according to Pauling's model.

Sequential models have the great advantage of a simple and intuitive formulation; indeed, the interaction parameter *i* has an immediate physical meaning: '. . . the free energy [of ligand binding] is decreased by $RT \ln i$ for each interaction . . .' (Pauling, 1935). Their weak point is that, even in very symmetrical proteins such as haemoglobin, the interactions between subunits are mediated by different contact regions, presumably characterised by different interaction terms. Because of the lack of information about the stereochemistry of haemoglobin, the sequential models devised after Pauling's used further interaction terms, and allowed several other functional geometries, which assumed the subunits in the tetramer to be arranged as chains, squares, rectangles or tetrahedrons (reviewed by Antonini & Brunori, 1971).

In 1966 Koshland and co-workers published a systematic reappraisal of sequential models, in which all reasonable cases of cooperative interactions were dealt with (Koshland *et al.*, 1966); a scheme was given for building this kind of model, from which almost any sequential model can be derived. The concerted model of Monod, Wyman & Changeux (see below), published in 1965, can also be derived from the scheme of Koshland *et al.*, albeit with some difficulty. The essential features of Koshland's scheme are as follows: (i) cooperative proteins are composed of multiple subunits, but only two structural conformations are freely accessible to each subunit, and only one structure is capable of ligand binding; (ii) the interactions amongst subunits in the allosteric macromolecule can modify the probability of each subunit to be found in the active conformation. Koshland's scheme offers a sound algebraic approach to build sequential models, provides a fundamental framework to order the published sequential models and has stopped their otherwise limitless proliferation. However, its application to haemoglobin has been relatively modest, owing to the intrinsic complexity, the poor agreement with X-ray crystallography and the arbitrary formulation of the model itself, which allows square, linear or tetrahedral structures to be constructed.

The two-state model of Monod, Wyman and Changeux

A real breakthrough in the thermodynamic modelling of haemoglobin function came with the publication of the so-called two-state (MWC) model by Monod, Wyman & Changeux (1965). It is now widely accepted that some of the basic features of this model do indeed describe real events taking place at the molecular level. This model postulates that cooperative proteins are composed of symmetrically arranged, equivalent subunits and that only two allosteric conformations (states) of the polymeric cooperative macromolecule need exist, characterised by different ligand affinity and capable of free interconversion independently of the presence of the ligand (see Figs 1 and 5).

Therefore, according to the two-state model, the reaction of haemoglobin with oxygen can be described by three chemical equilibria (treated in the original paper as dissociation reactions; but here, for the sake of consistency with all other models, converted into association reactions):

$$^{R}Hb(O_2)_{n-1} + O_2 \rightleftharpoons {}^{R}Hb(O_2)_n \qquad \text{(scheme 5.1)}$$

$$^{T}Hb(O_2)_{n-1} + O_2 \rightleftharpoons {}^{T}Hb(O_2)_n \qquad \text{(scheme 5.2)}$$

$$^{R}Hb(O_2)_n \rightleftharpoons {}^{T}Hb(O_2)_n \qquad \text{(scheme 5.3)}$$

where n ranges from 1 to 4 and the superscripts T and R, standing for 'tense' and 'relaxed', respectively, denote the two allosteric conformations. It is important to point out that the oxygen-binding constants of the two states differ (the R state displaying higher ligand affinity), but are insensitive to the number of ligand molecules bound (i.e. the model is not sequential). The transition between R- and T-state haemoglobin is an all-or none- (concerted) phenomenon governed by the allosteric constant L, defined as:

$$L_n = [^{T}Hb(O_2)_n]/[^{R}Hb(O_2)_n] \qquad (11)$$

A solution of haemoglobin contains T- and R-state molecules and the relative weight of the two populations at equilibrium depends on the value of the allosteric constant L and on the concentration of oxygen; the larger the fractional weight of the R-state population, the higher the overall apparent ligand affinity of the system. In the absence of ligand the T state must be strongly favoured with respect to the R state if a cooperative behaviour is observed (L_0 typically ranges between 1000 and 1 000 000).

It is easy to demonstrate that the values of the allosteric constants L_1 to

L_4 can be derived from the value of K_T, K_R and L_0; in fact, considering only the first oxygenation step, one has the reaction scheme:

$$\begin{array}{ccc} {}^{R}Hb + O_2 & \overset{K_R}{\rightleftharpoons} & {}^{R}HbO_2 \\ L_0 \updownarrow & & \updownarrow L_1 \\ {}^{T}Hb + O_2 & \overset{K_T}{\rightleftharpoons} & {}^{T}HbO_2 \end{array} \qquad \text{(scheme 6)}$$

The following relation applies:

$$L_1 = L_0(K_T/K_R) \qquad (12)$$

and, in general:

$$L_n = L_0(K_T{}^n/K_R{}^n) \qquad (13)$$

The binding polynomial for this model (taking as a reference the concentration of R-state unliganded haemoglobin) is:

$$BP = L_0(1 + K_T P_{O_2})^4 + (1 + K_R P_{O_2})^4 \qquad (14)$$

The fractional saturation of haemoglobin is predicted to be:

$$Y = \frac{L_0 K_T P_{O_2}(1 + K_T P_{O_2})^3 + K_R P_{O_2}(1 + K_R P_{O_2})^3}{L_0(1 + K_T P_{O_2})^4 + (1 + K_R P_{O_2})^4} \qquad (15)$$

The two-state model withstood 25 years of haemoglobin research and proved able to offer a conceptual framework for interpreting a wide range of experiments, from nanosecond ligand-rebinding kinetics (see Hofrichter *et al.*, 1983 and references therein) to X-ray crystallography (Perutz, 1970*a*). Its main limit lies in the inadequate description of the effect of pH on the oxygen affinity (Imai, 1973, 1983) and several other concerted models have been proposed to bypass this difficulty. We will discuss this aspect after considering information now available on the molecular structure.

The search for a stereochemical model and its thermodynamic implications

The subjects discussed in this section constitute a homogeneous body of information obtained from analysis of the crystallographic structures of several haemoglobin derivatives, which we shall split, somewhat arbitrarily, into three parts. An additional part will describe the thermodynamic interpretation of the crystallographic data.

The allosteric transition

The two-state model satisfactorily describes most of the experiments carried out on haemoglobin, but its most impressive success is probably linked to the strict agreement with the structural information gathered by X-ray crystallography (almost at the same time) and computer modelling.

Perutz and co-workers (Perutz, 1970*a*,*b*, 1990; Perutz *et al.*, 1987) demonstrated that oxy- and deoxyhaemoglobin have different crystallographic structures (R and T, respectively); furthermore, he and others have shown that the structure of partially liganded haemoglobin derivatives is either R-like or T-like and no intermediate structures were ever found. It is important to realise that the global deoxy (T), or oxy (R), allosteric conformation imposes the structure of all the subunits, whether they are or are not liganded, so that the liganded T state and unliganded R state are well characterised from the crystallographic and computer-modelling analysis of partially liganded intermediates. The structures of oxy- and deoxyhaemoglobin are different enough to prevent a direct comparison by superimposition, but the structure of the $\alpha_1\beta_1$ dimer in the two states changes surprisingly little, and most of the overall change is to be attributed to rotation and shifting of each dimer with respect to the other. Thus, we shall first describe the allosteric changes intervening upon haemoglobin oxygenation with reference to the $\alpha_1\beta_1$ dimer (Baldwin & Chothia, 1979).

The unliganded porphyrin is domed and its five-coordinated, paramagnetic haem iron is pulled towards helix F by its bonding to proximal His (F8); when a ligand is bound, the haem tends to flatten and the six-coordinated diamagnetic iron atom is brought towards the porphyrin plane. As a consequence of the movement of the iron, helix F and the FG corner of the α and β subunits move, while the other helices stand almost still. To summarise the structural changes occurring in the dimer upon ligand binding, we see that (i) the haem iron moves by, at most, 0.6 Å from unliganded T state to liganded R state; (ii) the structural change of each subunit is small, aside from the region of helix F which is tilted and shifted by more than 1 Å; (iii) the $\alpha_1\beta_1$ interface, contributed by helices B, G and H, is almost insensitive to the presence of the ligand, but (iv) the $\alpha_1\beta_2$ interface, which is contributed by helices F and C, undergoes a major rearrangement.

The flattening of the haem and the development of this structural transition in the tetrameric molecule may be inhibited by the constraint imposed by the other T-state unliganded subunits. But, if some ligand is already bound, and the dimer fully rearranges, a cataclysmic change takes

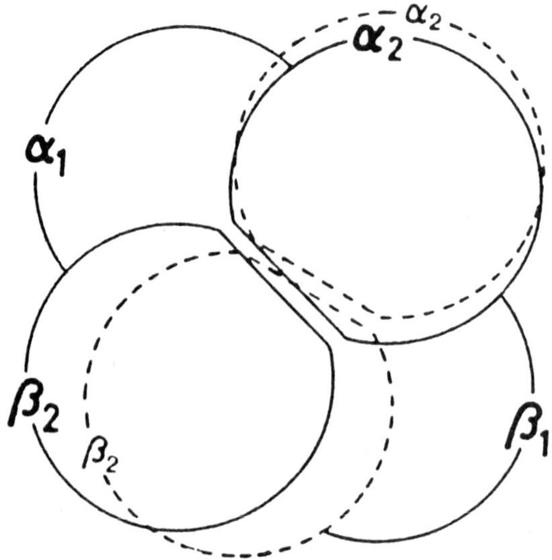

Fig. 6. Rearrangement of the two $\alpha\beta$ dimers in the quaternary transition (modified after Baldwin & Chothia, 1979). The position of the $\alpha_1\beta_1$ dimer is used as reference; the broken and continuous lines indicate the position of $\alpha_2\beta_2$ in the R and T states, respectively.

place, switching the quaternary conformation of haemoglobin from T to R.

A more refined description of the events which accompany the allosteric transition requires that the contacts α_1FG-β_2FG, α_1FG-β_2C and α_1C-β_2FG, which constitute the $\alpha_1\beta_2$ interface, are separately examined; the reader is referred to Baldwin & Chothia (1979) for a more detailed description. Helix F and the FG corner of the α_1 chain essentially slide over helix C of the β_2 chain and the same amino acid residues form different contacts in the T and R structures. The contacts between helices F and G of both chains behave similarly, and slightly slide on each other, replacing the H-bond between Asn αG4(97) and Asp βG1(99) of T-state haemoglobin with the symmetric one between Asp αG1(94) and Asn βG4(102). Helix C of the α_1 chain and helix F and FG corner of the β_2 chain form a special unit and do not slide on each other; the packing of bulky His βFG4(97) is so tight that it must jump from the T position, between Thr αC6(41) and Pro αCD2(44), and settle between Thr αC6(41) and Thr αC3(38) in the R state. Moreover, the transition from T to R breaks the H-bond between Asp βG1(99) and Tyr αC7(42). The peculiar stability properties of this contact partially explain the all- or none-character of the allosteric transition. Perutz (1970*a*,*b*, 1990) has shown the quaternary transition to result in a rotation of 12–15° and a translation of 0.8 Å of the $\alpha_1\beta_1$ dimer relative to the other (Fig. 6), and to cause the breaking of four salt bridges; in fact, the constraint imposed by the quaternary structure on the chains is mainly due to salt bridges formed by the C- and N-termini of the subunits.

The Bohr effect and binding of organic phosphates

According to Perutz's stereochemical model, the Bohr effect is mainly linked to the terminal residues His βHC3(146) and Val αNA1(1) which, when positively charged, form two additional salt bridges and further stabilise the T state. In deoxyhaemoglobin the imidazole group of His βHC3 forms a salt bridge with Asp βFG1(94), which breaks in oxyhaemoglobin; the removal or substitution of His βHC3 substantially reduces the Bohr effect. Val αNA1 is a Cl^--dependent Bohr group (Cl^- binds between the NH_3^+ group of Val αNA1 and the hydroxyl group of Ser αH14(131)) (Fantl *et al.*, 1987). In addition to these two, other residues are likely to be involved in the Bohr effect (Bucci & Fronticelli, 1985), which may also be influenced by the interaction of buffer ions with either oxy- or deoxyhaemoglobin.

The model also aims to explain the Root effect on a molecular basis, by ascribing a key role to the weak interactions established by His βHC3(146). As described above, this residue is, in human haemoglobin, the pivotal point of a network of salt bridges involving Asp βFG1(94) in ^{T}Hb and Lys βHC1(144) in ^{R}Hb. In trout Hb IV and carp haemoglobin, which display the Root effect, Cys βF9(93) of mammalian haemoglobin is substituted by Ser. In the deoxy T structure, Ser F9 donates a H-bond to the unbound oxygen atom of His βHC3 and accepts one from the peptide –NH– of the same residue, stabilising the C-terminal salt bridges. Trout Hb IV and carp haemoglobin also have Gln at position βHC1 (and not Lys, like mammals), which is unable to form the salt bridge established by Lys and characteristic of the R state. Thus, in Root-effect haemoglobins, T is stabilised by substitutions at positions βF9 and βHC1. Another important consequence of the oxygen-linked breakage of the salt bridge between Asp βFG1(94) and His βHC3(146) is the disordered structure of the C-terminal part of the β chain in oxyhaemoglobin and the reduced affinity of organic phosphates, whose binding site is contributed by Val NA1(1), His NA2(2), Lys EF6(82) and His H21(143) on each β chain (Perutz, 1970*b*; Arnone, 1972).

In teleost fishes the physiological allosteric effectors are ATP and, to a lesser extent, GTP; their binding site is very similar to that of human haemoglobin described above. In trout Hb IV, Lys EF6 is conserved and the two other residues are Asp NA2 and Arg H21. The carboxylate of Asp accepts an H-bond from the N-6 $-NH_2$ of adenine; additional bonds neutralise the four negative charges of ATP. If the ligand is GTP, an additional bond can be established between Val β2NA1 and the 2′-O of ribose. The number of amino acid sequences of fish haemoglobin is steadily increasing, and notably those of polar teleosts. These species live in extreme environments which have determined physiological adaptations. As will be mentioned below, work is in progress to correlate the functional properties of these haemoglobins to the molecular structure. Crystals have also become available for X-ray analysis.

Crystallographic study of partially oxygenated haemoglobin and its analogues

As in any strongly cooperative system, unliganded and fully liganded haemoglobin molecules are more stable than any of the saturation intermediates; thus, it is impossible directly to crystallise haemoglobin molecules bearing one or two molecules of oxygen. Moreover, crystals of oxy- and deoxyhaemoglobin are not isomorphous and crystals of deoxyhaemoglobin usually crack when exposed to air, making it almost impossible to prepare crystals of saturation intermediates from crystals of deoxyhaemoglobin. Nevertheless, it has been possible to obtain crystallographic information about partially saturated haemoglobin derivatives, exploiting three different experimental techniques, and we will briefly discuss their structure.

Partial titration of deoxyhaemoglobin crystals with air yields crystals of 50% oxygen-saturated haemoglobin; these maintain the T-crystallographic structure (thus do not break) and bear the ligand only on the α chains (Perutz *et al.*, 1987). Chemical modification of some haemoglobins, while maintaining the cooperative ligand-binding properties, may prevent the crystals from breaking when the ligand is added or removed. For example, crystals of R-state horse methaemoglobin, reacted with bis(N-maleimidomethyl) ether, can be reduced with dithionite and yield crystals of T-state deoxyhaemoglobin, whose structure is indistinguishable from that of normal horse deoxyhaemoglobin (reviewed by Perutz *et al.*, 1987). Finally, substitution of the iron in the haem with Ni, Mn or Co, in either type of chain, and reconstitution with the native complementary chain yields a haemoglobin derivative still capable of cooperative ligand binding (usually only to the native subunit: Ikeda-Saito *et al.*, 1977; Ackers & Smith, 1987). These haemoglobins have been shown by X-ray crystallography to assume either the T or R structure, never an intermediate one (Fermi *et al.*, 1982; Luisi & Shibayama, 1989; see also the review of Perutz *et al.*, 1987).

It is well demonstrated that haemoglobin can be crystallised only in two basic structures, corresponding to the T and R states. Does this finding prove the validity of the two-state model? A word of caution is necessary. Aside from the (remote) possibility of crystallising a third conformation of haemoglobin (see note added in proof, p. 134), it was pointed out by Szabo & Karplus (1972) that thermodynamically different states need not be greatly different in structure.

The Szabo–Karplus thermodynamical analysis of the stereochemical model

An interesting attempt to clarify the relationships between the structural data and the thermodynamic properties of haemoglobin was that of Szabo & Karplus (1972). Although a complete description cannot be given here, it is important to draw at least an approximate picture of their theory. The Szabo–Karplus (SK) model implies two quaternary structures and two major thermodynamic states, but suggests that once the intrinsic ligand affinity of one of the two is determined, along with allosteric constant L_0, it is possible to derive the other from the known structural changes, which essentially involve breaking or formation of salt bridges and other weak interactions. Thus, the authors set as essential parameters the free energy changes associated with the ligand binding to the α and β subunits in R-state haemoglobin, the allosteric R–T transition, the formation (in the R–T transition) of four equivalent and of two pH-dependent, inequivalent salt bridges.

In this model two ligands, oxygen and hydrogen ions (which influence two of the six salt bridges) are considered, and six free-energy terms (or equilibrium constants) are needed, which makes the analysis of oxygen-binding curves quite difficult. This model explains the Bohr effect; all its parameters (except the free energy of the allosteric transition) can be either directly measured (the free energy of ligand binding to the isolated, R-state subunits) or constrained within the limits considered reasonable for the reaction involved (the free energies of salt bridges).

The SK model represents a fundamental attempt to analyse the thermodynamic implications of structural information, and its significance must not be compared with that of the MWC model: it is a sort of theoretical experiment in itself, and helps to compare quantitatively pieces of information that would otherwise make an array of qualitatively consistent data. Moreover, the model gives the crucial suggestion that three or more thermodynamic states may coexist with two crystallographic structures (in fact the free energy of the T state is not fixed, but pH-dependent). As a theoretical experiment, the SK model allows extensions and repetitions, thus including the contribution of other weak interactions or that attributable to the tetramer–dimer equilibrium (Johnson & Ackers, 1982).

Experimental data, other than X-ray crystallography, interpreted in terms of the two-state model

Evidence in support of the model

All thermodynamic models describe the sigmoid oxygen-binding curve of haemoglobin; but the two-state model, as stated above, also perfectly fits the crystallographic data. We shall now refer to further experimental evidence strongly supporting this model.

The kinetics of ligand binding to, and release from, haemoglobin has been extensively studied by means of stopped flow or flash and laser photolysis. The principle of the photolytic technique is simple: derivatives of ferrous haemoglobin are photosensitive and light absorption has a defined probability of removing the bound ligand (quantum yield). Therefore, if a solution of ligand-saturated haemoglobin is transiently exposed to intense light, part of the ligand is removed and, after switching the light off, its rebinding can be studied. Depending on the intensity and duration of the light pulse, the amount of ligand removed ranges from 100% to the smallest fraction compatible with the sensitivity of the detecting apparatus. The population of T and R intermediates at zero time is statistical (at least to a first approximation) and a biphasic (or multiphasic) time course is obtained, each phase corresponding to one (or some) intermediate(s).

Using photographic flashes as light sources, it has been possible to record the time course of binding of carbon monoxide to 0 to 90%-saturated haemoglobin, and it has been demonstrated that ligand rebinding is described by only two phases, the faster predominating at low level of photolysis (Gibson, 1959; reviewed by Brunori & Giacometti, 1981). The use of short laser pulses (usually in the nanosecond time scale), revealed some more kinetic phases (Sawicki & Gibson, 1976, 1977; Hofrichter *et al.*, 1983) which may be summarised as: (i) two monomolecular, very fast rebinding reactions, ascribed to ligand molecules which have not diffused away from the protein matrix (geminate pairs); (ii) one or two monomolecular phases whose spectral properties do not correspond to those of ligand rebinding, attributed to tertiary and quaternary rearrangements; and (iii) two bimolecular phases attributed to ligand rebinding from the bulk. Almost all the available photolysis data can be and have been described using the MWC model; indeed, the result which would directly contradict this model, i.e. a single trace with three bimolecular phases, has never been reported.

Evidence not supporting the model

Hydrogen ions, carbon dioxide and organic phosphates bind to specific sites of the haemoglobin molecule and, while inducing very small changes in its basic T or R structures, significantly affect the oxygen-binding properties of at least the T state (Imai & Yonetani, 1975; Imai, 1983). This contradicts the implicit assumption of the two-state model that any substance capable of interacting with either or both conformations of haemoglobin may modify the value of the allosteric constant L_0, and change the overall ligand affinity, but must not change the intrinsic equilibrium constants for the ligand K_T and K_R (otherwise it would be possible to titrate a transition from, say, the T state to some T* state and a third conformation would be introduced).

For the following discussion it is unnecessary to describe the interaction of haemoglobin with each effector; only protons will be considered, with the assumption that the conclusions will qualitatively apply to all other allosteric effectors. The oxygen affinity of the R state of human haemoglobin is almost pH-independent but that of the T state greatly changes with pH (Imai, 1973); this effect is coupled with the increase of the allosteric constant L_0. The Root effect of fish haemoglobins depends on the cooperation of three factors (Brunori *et al.*, 1987): (i) the larger increase of the allosteric constant which stabilises the T state throughout the titration with the ligand; (ii) the intrinsically low affinity of the T state at low pH (i.e. the Bohr effect of T-haemoglobin); and (iii) the functional inequivalence of the two chains in T-state haemoglobin, which may be so relevant that only half of the oxygen-binding sites may be saturated with oxygen at a pressure of several atmospheres. The last two factors are not fully compatible with the two-state model.

Other experimental evidence, whose interpretation along the lines of the MWC model is relatively unsatisfactory, is as follows. (i) The analysis of the tetramer–dimer equilibrium, which has been used to evaluate the free energies for ligand binding of liganded intermediates of haemoglobin, such as those formed by mixing cyano-met or metal-substituted β chains with deoxy α chains (Smith & Ackers, 1985). One of these hybrids was demonstrated not to have a free energy of ligand binding compatible with the two-state model, though it deviates only slightly from the predicted behaviour. Moreover, it was shown that intermediates with the same ligation state (e.g. diliganded tetramers), but with different intramolecular distribution of the ligand (e.g. $\alpha_2{}^{CO}\beta_2$ and $\alpha^{CO}\alpha\beta^{CO}\beta$), may not be equivalent. (ii) The population and distribution of partially liganded intermediates measured by isoelectrophoresis (Perrella *et al.*, 1986; Samaja *et al.*, 1987). The relative amount of intermediates is small (slightly over 10% of the total pigment), and therefore the analysis is run under unfavourable conditions. This experiment demonstrates that the distribution of the same MWC intermediate (e.g. the diliganded species) strongly favours specific intramolecular patterns (e.g. the symmetric diliganded species $\alpha_2{}^{CO}\beta_2$ or $\alpha_2\beta_2{}^{CO}$ are conspicuously absent). Since in the MWC model the molecule is symmetric and the chains are equivalent, all distribution patterns of the ligand in the tetramer should be equivalent. (iii) The NMR spectroscopy of haemoglobin. Some resonances have been attributed to the quaternary structure (e.g. hydrogen atoms involved in state-specific H-bonds); it was shown that, in valency hybrid haemoglobins, some of these markers signal T state while others suggest R state, implying that intermediate states may exist (Miura *et al.*, 1982).

Recent thermodynamic models of haemoglobin cooperativity

How can a two-state model be extended to comply with the Bohr effect, or even more, with the Root effect of fish haemoglobin? Several attempts have been made, which can be divided into two groups, i.e. that of the three-state model (Minton & Imai, 1974) and that of the models which introduce sequential effects inside either or both the major T and R crystallographic structures.

The **three-state model** of Minton & Imai (1974): MI, which is the only other pure 'concerted' model, is an extension of the two-state model. A third conformational state, S, is introduced, characterised by the ligand affinity constant K_S; its conformational equilibrium, with respect to the R state, is defined by the additional allosteric constant M_0:

$$M_0 = {}^{S}\mathrm{Hb}/{}^{R}\mathrm{Hb} \quad (16)$$

The binding polynomial is analogous to that of the MWC model (Eqn 14):

$$\mathrm{BP} = M_0(1 + K_S P_{O_2})^4 + L_0(1 + K_T P_{O_2})^4 + (1 + K_R P_{O_2})^4 \quad (17)$$

This model has been very scarcely used as a tool for describing haemoglobin cooperativity, owing to the large number of freely varying parameters. It predicts that, under selected experimental conditions, it should be possible to generate a mixture of states in which the time course of ligand rebinding is described by three exponentials: a prediction which, however, has never been fulfilled. Support for the presence of three conformations of haemoglobin comes from the work of Ackers and co-workers (Smith & Ackers, 1985; Ackers & Smith, 1987) and Perrella and co-workers (Perrella *et al.*, 1986; Samaja *et al.*, 1987) who found that three energy states, and not two, describe their experimental data (see p. 124 above). None of these authors, however, explicitly used the MI model, because their results require that the subunits be non-equivalent in their ability to promote the structural transition, or (better) that different arrangements of the same number of liganded subunits be non-equivalent (e.g. when considering the two MWC and MI equivalent di-liganded species $\alpha^{CO}\alpha\beta^{CO}\beta$ and $\alpha_2{}^{CO}\beta_2$; the latter, according to Ackers & Smith (1987), is in the same state as fully liganded haemoglobin, whereas the former is in the same state as monoliganded derivatives). Thus, these authors suggest that three energy levels (states) are accessible to haemoglobin and that the switch between them depends on the number and position of the bound ligand molecules (i.e. the simple relation described by Eqn 13 does not apply).

We shall now describe two recent models, which incorporate some features of older sequential models into the two-state model and successfully describe many haemoglobins, including Root-effect haemoglobins, and also other cooperative oxygen carriers. The **'cooperon' model** (Brunori *et al.*, 1986; Gill *et al.*, 1986) suggests that structural changes attributable to ligand binding to one subunit can be transmitted to other subunits belonging to a defined group, without changes in the overall T or R structure of the whole multimeric molecule; the group of interacting sites is called the cooperative unit or the cooperon. Each liganded subunit of a given cooperon influences the ligand affinity of the other subunits of the same cooperon to a fixed extent, and does not exert any effect on subunits belonging to a different cooperon; therefore this model allows T- or R-state molecules to bind the ligand cooperatively and to be influenced by allosteric effectors. In its original formulation, this model suggests that two dimeric cooperons (the $\alpha_1\beta_1$ dimers) are assembled in the T state, whereas no sequential interactions (therefore no cooperons) are present in the R state. The total distribution of haemoglobin species for the T state at any given ligand activity is given by:

$$\mathrm{BP_T} = (1 + 2K_\mathrm{T}P_{\mathrm{O_2}} + iK_\mathrm{T}{}^2P_{\mathrm{O_2}}{}^2)^2 \tag{18}$$

where the term i describes the interaction of the two subunits of the cooperon in the T state; $i > 1$ implies positive cooperativity and $i < 1$ negative cooperativity (compare with Pauling's model, Eqn 9). When $i = 1$, this model is identical to the MWC model. The binding polynomial describing the R-state saturation intermediates corresponds to that of the MWC model:

$$BP_R = (1 + K_R P_{O_2})^4 \quad (19)$$

Expanding and multiplying by the proper cofactors, the function describing the fractional saturation of human haemoglobin is obtained:

$$Y = \frac{L_0 K_T P_{O_2}[1+(2+i)K_T P_{O_2} + 3iK_T{}^2 P_{O_2}{}^2 + i^2 K_T{}^3 P_{O_2}{}^3] + K_R P_{O_2}(1 + K_R P_{O_2})^3}{L_0(1 + 2K_T P_{O_2} + iK_T{}^2 P_{O_2}{}^2)^2 + (1 + K_R P_{O_2})^4} \quad (20)$$

The cooperon model satisfactorily describes the oxygen-equilibrium curves of human haemoglobin in the presence and absence of allosteric effectors (see Vandegriff *et al.*, 1989) and has been used to fit ligand binding to several giant oxygen carriers (Coletta *et al.*, 1986). It easily describes the Root-effect haemoglobins, imposing a very large value of L_0 and a very small value of i (that is, a markedly anti-cooperative behaviour of the two subunits composing the T-state cooperon).

An interesting feature of this model lies in its generalised formulation, which allows other structural arrangements to be taken into account (e.g. the trimeric units nested in hexameric haemocyanins from Arthropoda); a drawback, shared with Pauling's and Koshland's models, is that the model may be arbitrarily modified, simply by changing the dimensions of the cooperon. It will be noticed that, as shown in Fig. 5, this model predicts that the effect of ligands with respect to the allosteric transition is 'combinatorial', i.e. the energy level of a diliganded tetramer is sensitive to the ligand distribution and a fully saturated cooperon will more easily induce the quaternary transition than two half-saturated cooperons. It is also worth noting that this model predicts a three-exponential time course for ligand rebinding may be found, but also implies that deviations from the two-exponential behaviour would be small, since the population of molecular species one would detect in a photolysis experiment are R, T and *i*T (where *i*T is the T structure whose cooperons contain liganded subunits). Ligand binding to just one site would convert each T cooperon into *i*T, whereas two or three ligands have to be bound before T converts to R.

A somewhat similar allosteric model, devised to incorporate all the known structural features of human haemoglobin, was proposed by Gill and co-workers (Di Cera *et al.*, 1987; Gill *et al.*, 1987). The basic experimental observation prompting this proposal was the very low population of the triply liganded species $Hb(O_2)_3$, lower than that predicted by the two-state model (Eqn 13). Indeed, in the original experiments the amount of triply liganded haemoglobin was so low, that fitting according to the Adair equation (Eqn 7) invariably gave a value of zero for the equilibrium constant of the third ligation step. This paradoxical result, which has been widely questioned, finds a confirmatory observation in the X-ray crystallographic map of partially liganded haemoglobin. Indeed, as stated above, when oxygen was diffused through crystals of deoxy-haemoglobin, a non-statistical distribution was found: 50% of the haems (only those of the α chains) bound the ligand. Thus, the model of Di Cera *et al.* suggests that in T-state haemoglobin only the α chains are active in ligand binding and, in order to obtain a ligand saturation higher than 50%, the molecule must switch to the R state. Fitting of the experimental data requires, however, that the binding of the first two molecules of oxygen is cooperative, i.e. that some 'sequential' cooperativity is present inside the T-state haemoglobin. The binding polynomial for this model (taking the R-state unliganded haemoglobin as unity) is:

$$\mathrm{BP} = L_0(1 + 2K_\mathrm{T}\alpha P_{\mathrm{O}_2} + iK_\mathrm{T}\alpha^2 P_{\mathrm{O}_2}{}^2) + (1 + K_\mathrm{R}P_{\mathrm{O}_2})^4 \qquad (21)$$

The similarity with the binding polynomial corresponding to the cooperon model (sum of Eqns 18 and 19) is obvious, but it is important to stress that in this model a cooperon formed by the two α chains of T-state haemoglobin is coupled with the unit formed by the two 'inactive' β chains: T-state haemoglobin can be envisaged as an asymmetric unit formed by the α_2 cooperon and the very low-affinity β_2, with properties inaccessible to investigation. The fractional ligand saturation predicted by this model is:

$$Y = \frac{2L_0K_\mathrm{T}\alpha P_{\mathrm{O}_2}(1 + iK_\mathrm{T}\alpha P_{\mathrm{O}_2}) + K_\mathrm{R}P_{\mathrm{O}_2}(1 + K_\mathrm{R}P_{\mathrm{O}_2})^3}{L_0(1 + 2K_\mathrm{T}\alpha P_{\mathrm{O}_2} + iK_\mathrm{T}\alpha^2 P_{\mathrm{O}_2}{}^2) + (1 + K_\mathrm{R}P_{\mathrm{O}_2})^4} \qquad (22)$$

This model offers a complete description of the reaction of human haemoglobin with haem ligands and allosteric effectors (see, for example, Doyle *et al.*, 1987), and can be used to fit most data on other haemoglobins, most notably Root-effect haemoglobins, which suffer from severe chain inequivalence in the T state. However, there have been several questions about the validity of its basic assumptions: in fact (i) the chain heterogeneity in T-state human haemoglobin, measured with different techniques, is relatively small, and (ii) the concentration of triply liganded species, though very small, might be higher than that suggested by this model (Perrella *et al.*, 1986; Samaja *et al.*, 1987).

Conclusions

As already emphasised, the study of haemoglobin chemistry encompasses almost a century of biochemistry, and this short overview of modelling of haemoglobin function omits much more information than it describes. Only those models which may be considered as the prototype of a way of thinking about haemoglobin have been selected for discussion, and many details have been disregarded for the sake of simplicity. Three-dimensional modelling of haemoglobin function is highly advanced, and molecular simulations and energy calculations allow us reasonably to predict or explain the interactions with allosteric effectors and drugs (Lalezari *et al.*, 1990), or to infer the consequences of site-directed mutagenesis, the powerful technique of molecular biology which allows the introduction of selected mutations into cloned genes (Nagai *et al.*, 1985; Springer & Sligar, 1987). This continuously growing body of knowledge on the structure–function relationship in haemoglobin produces, and will keep producing, important practical achievements (e.g. in the treatment of haemoglobinopathies) and will be of benefit for the general understanding of protein function.

However, all new information raises further questions. For example, and even if we limit our attention to the interpretation of the action of allosteric effectors, the increasing number of sequences of Antarctic fish haemoglobins reveals the difficulty of establishing clear-cut correlations with the molecular structure. Going into some detail, D'Avino *et al.* (1991), Caruso *et al.* (1991) and di Prisco *et al.* (1991) show that (i) some of the β-chain residues proposed to be essential for the Bohr and Root effects and for the binding of ATP (e.g. in positions NA2, F9, FG1, HC1, HC3, H21) have also been found in haemoglobins not displaying these effects; (ii) conversely, some Root-effect haemoglobins do not have some of the essential residues for the ATP (GTP) site in NA2, EF6, H21. Thus, the residues of the Antarctic Hb sequences that agree with the model are the invariant Gln HC1, Lys H21 (the only one in the ATP site) and Glu (Asp) FG1, found also in haemoglobins, without Bohr and Root effects; (iii) at least one Root-effect haemoglobin has Cys in F9 (like mammals), confirming the conclusion arising from site-directed mutagenesis experiments (Nagai *et al.*, 1985): the Cys–Ser replacement, once thought to be the prime cause of the Root effect (Perutz & Brunori, 1982), is indeed insufficient to give the H-bond which would overstabilise the T state, and other substitutions may be required in order to place Ser βF9 and His βHC3(146) in close enough contact.

Thermodynamic modelling of haemoglobin function appears to have several drawbacks too. Thus, we will merely draw some very general conclusions, based on (almost) unchallenged experimental data: (i) discrete conformational structures of haemoglobin exist, corresponding to at least two (or maybe more) thermodynamic states; (ii) the MWC model is a minimal one, because it includes only two states. Nevertheless, it is almost always adequate to describe the experimental data and, even when they are inadequate, they offer a 'language' to define haemoglobin function; (iii) some experiments demonstrate that the MWC model only provides an approximate description of haemoglobin, that three energy states are necessary and that the switching between states may be controlled by complex rules; but (iv) nobody at present is able to give hints as to what this third state is likely to be, whether a third crystallographic structure or a slight change in one of the two states known so far.

Haemoglobin has been a model for structure–function relationships in proteins and is going to maintain this special position in the foreseeable future. Although no other protein is known in such great detail, its understanding is yet far from being complete. For example, ascribing a functional feature to only a single, or a few, amino acid residues appears more and more difficult, since refinements (dictated by the structural effect of replacements or other modifications in other domains of the molecule) will soon become necessary. In conclusion, the history of modelling of haemoglobin stresses that the advancements of knowledge are indeed associated with the privilege of addressing more unsolved questions.

Acknowledgements

The stimulating discussion and helpful suggestions of Professors M. Brunori, G. Amiconi, M. Coletta and L. Camardella are gratefully acknowledged. This work was partially supported by the Italian National Programme of Antarctic Research.

References

Ackers, G.K. & Smith, F.R. (1987). The hemoglobin tetramer: a three-state molecular switch for control of ligand affinity. *Annual Review of Biophysics and Biophysical Chemistry* **16**, 583–609.

Adair, G.S. (1925). The hemoglobin system. VI. The oxygen dissociation curve of hemoglobin. *Journal of Biological Chemistry* **63**, 529–45.

Antonini, E. & Brunori, M. (1971). Hemoglobin and myoglobin in their reactions with ligands. In *Frontiers of Biology*, Vol. 21, ed. A. Neuberger & E.L. Tatum. Amsterdam: North Holland.

Antonini, E., Wyman, J., Bellelli, L., Rumen, N. & Siniscalco, M. (1964). The oxygen equilibrium of some lamprey hemoglobins. *Archives of Biochemistry and Biophysics* **105**, 404–8.

Arnone, A. (1972). X-ray diffraction study of binding of 2,3-diphosphoglycerate to human deoxyhaemoglobin. *Nature* **237**, 146–9.

Baldwin, J. & Chothia, C. (1979). Haemoglobin: the structural changes related to ligand binding and its allosteric mechanism. *Journal of Molecular Biology* **129**, 175–220.

Benesch, R. & Benesch, R.E. (1967). The effect of organic phosphates from the human erythrocyte on the allosteric properties of hemoglobin. *Biochemical and Biophysical Research Communications* **26**, 162–7.

Brittain, T. (1987). The Root effect. *Comparative Biochemistry and Physiology* **86B**, 473–81.

Brunori, M. (1975). Molecular adaptation to physiological requirements: the hemoglobin systems of trout. *Current Topics in Cellular Regulation* **9**, 1–39.

Brunori, M., Bellelli, A., Giardina, B., Condò, S.G. & Perutz, M.F. (1987). Is there a Root effect in *Xenopus* hemoglobin? *FEBS Letters* **221**, 161–6.

Brunori, M., Coletta, M. & Di Cera, E. (1986). A cooperative model for ligand binding to biological macromolecules as applied to oxygen carriers. *Biophysical Chemistry* **23**, 215–22.

Brunori, M. & Giacometti, G.M. (1981). Photochemistry of hemoproteins. *Methods in Enzymology* **76**, 582–95.

Bucci, E. & Fronticelli, C. (1985). Anion Bohr effect of human hemoglobin. *Biochemistry* **24**, 371–6.

Caruso, C., Rutigliano, B., Romano, M. & di Prisco, G. (1992). The hemoglobins of the cold-adapted Antarctic teleost *Cygnodraco mawsoni*. *Biochimica et Biophysica Acta* (in press).

Coletta, M., Di Cera, E. & Brunori, M. (1986). A cooperative model for ligand binding to oxygen carriers. In *Invertebrate Oxygen Carriers*, ed. B. Linzen, pp. 375–81. New York: Springer-Verlag.

D'Avino, R., Caruso, C., Camardella, L., Schininà, M.E., Rutigliano, B., Romano, M., Carratore, V., Barra, D. & di Prisco, G. (1991). An overview of the molecular structure and functional properties of the hemoglobins of a cold-adapted Antarctic teleost. In *Life Under Extreme Conditions. Biochemical Adaptation*, ed. G. di Prisco, pp. 15–33. New York: Springer-Verlag.

Di Cera, E., Robert, C.H. & Gill, S.J. (1987). Allosteric interpretation of the oxygen-binding reaction of human hemoglobin tetramers. *Biochemistry* **26**, 4003–8.

di Prisco, G., D'Avino, R., Caruso, C., Tamburrini, M., Camardella, L., Rutigliano, B., Carratore, V. & Romano, M. (1991). The biochemistry of oxygen transport in red-blooded Antarctic fishes. In *Biology of Antarctic Fish*, ed. G. di Prisco, B. Maresca & B. Tota. New York: Springer-Verlag (in press).

Doyle, M.L., Di Cera, E., Robert, C.H. & Gill, S.J. (1987). Carbon dioxide and oxygen linkage in human hemoglobin tetramers. *Journal of Molecular Biology* **196**, 927–34.

Eastman, J.T. (1988). Ocular morphology in Antarctic notothenioid fishes. *Journal of Morphology* **196**, 283–306.

Edelstein, S.J., Rehmar, M.J., Olson, J.S. & Gibson, Q.H. (1970).

Functional aspects of the subunit association–dissociation equilibria of hemoglobin. *Journal of Biological Chemistry* **245**, 4372–81.

Edsall, J.T. (1980). Hemoglobin and the origins of the concept of allosterism. *Federation Proceedings* **39**, 226–35.

Fantl, W.J., Donato, A.D., Manning, J.M., Rogers, P.H. & Arnone, A. (1987). Hemoglobin specifically carboxymethylated at the α- and β-chain NH_2-terminal residues as an analogue of carbamino hemoglobin: X-rays and solution studies. *Journal of Biological Chemistry* **262**, 12700–13.

Fermi, G., Perutz, M.F., Dickinson, L.C. & Chien, J.C.W. (1982). Structure of human deoxy cobalt hemoglobin. *Journal of Molecular Biology* **155**, 495–505.

Gibson, Q.H. (1959). The photochemical formation of a quickly reacting form of haemoglobin. *Biochemical Journal* **71**, 293–303.

Gill, S.J., Di Cera, E., Doyle, M.L., Bishop, G.A. & Robert, C.H. (1987). Oxygen binding constants for human hemoglobin tetramers. *Biochemistry* **26**, 3995–4002.

Gill, S.J., Robert, C.H., Coletta, M., Di Cera, E. & Brunori, M. (1986). Cooperative free energies for nested allosteric models as applied to human hemoglobin. *Biophysical Journal* **50**, 747–52.

Gillen, R.G. & Riggs, A. (1972). Structure and function of the hemoglobins of the carp, *Cyprinus carpio*. *Journal of Biological Chemistry* **247**, 6039–46.

Hill, A.V. (1910). The possible effects of the aggregation of the molecules of hemoglobin on the dissociation curve. *Journal of Physiology (London)* **40**, 190–205.

Hofrichter, J., Sommer, J.H., Henry, E.R. & Eaton, W.A. (1983). Nanosecond absorption spectroscopy of hemoglobin: elementary processes in kinetic cooperativity. *Proceedings of the National Academy of Sciences, USA* **80**, 2235–9.

Ikeda-Saito, M., Yamamoto, H. & Yonetani, T. (1977). Studies on cobalt myoglobins and hemoglobins. Electron paramagnetic resonance investigation on iron–cobalt hybrid hemoglobins and its implications for the heme–heme interaction and for the alkaline Bohr effect. *Journal of Biological Chemistry* **252**, 8639–44.

Imai, K. (1973). Analyses of oxygen equilibria of native and chemically modified human adult hemoglobin on the basis of Adair's stepwise oxygenation theory and the allosteric model of Monod, Wyman, and Changeux. *Biochemistry* **12**, 798–807.

Imai, K. (1983). The Monod–Wyman–Changeux allosteric model describes haemoglobin oxygenation with one adjustable parameter. *Journal of Molecular Biology* **167**, 741–9.

Imai, K. & Yonetani, T. (1975). pH dependence of the Adair constants of human hemoglobin. Nonuniform contribution of successive oxygen bindings to the alkaline Bohr effect. *Journal of Biological Chemistry* **250**, 2227–31.

Johnson, M.L. & Ackers, G.K. (1982). Thermodynamic analysis of human hemoglobins in terms of the Perutz mechanism: extension of the Szabo–Karplus model to include subunit assembly. *Biochemistry* **21**, 201–11.

Koshland, D.E., Némethy, G. & Filmer, D. (1966). Comparison of experimental binding data and theoretical models in proteins containing subunits. *Biochemistry* **5**, 365–85.

Lalezari, I., Lalezari, P., Poyart, C., Marden, M., Kister, J., Bohn, B., Fermi, G. & Perutz, M.F. (1990). New effectors of human hemoglobin: structure and function. *Biochemistry* **29**, 1515–23.

Luisi, B. & Shibayama, N. (1989). Structure of haemoglobin in the deoxy quaternary state with ligand bound at the α haems. *Journal of Molecular Biology* **206**, 723–36.

Minton, A.P. & Imai, K. (1974). The three-state model: a minimal allosteric description of homotropic and heterotropic effects in the binding of ligands to hemoglobin. *Proceedings of the National Academy of Sciences, USA* **71**, 1418–21.

Miura, S., Ikeda-Saito, M., Yonetani, T. & Ho, C. (1982). Oxygen equilibrium studies of cross-linked asymmetrical cyanomet valency hybrid hemoglobins: models for partially oxygenated species. *Biochemistry* **26**, 2149–55.

Monod, J., Wyman, J. & Changeux, J.P. (1965). On the nature of allosteric transitions: a plausible model. *Journal of Molecular Biology* **12**, 88–118.

Nagai, K., Perutz, M.F. & Poyart, C. (1985). Oxygen binding properties of human mutant hemoglobins, synthesized in *Escherichia coli*. *Proceedings of the National Academy of Sciences, USA* **82**, 7252–5.

Pauling, L. (1935). The oxygen equilibrium of hemoglobin and its structural interpretation. *Proceedings of the National Academy of Sciences, USA* **21**, 186–91.

Perrella, M., Sabbioneda, L., Samaja, M. & Rossi-Bernardi, L. (1986). The intermediate compounds between human hemoglobin and carbon monoxide at equilibrium and during approach to equilibrium. *Journal of Biological Chemistry* **261**, 8391–6.

Perutz, M.F. (1970*a*). Stereochemistry of cooperative effects in haemoglobin. Haem–haem interaction and the problem of allostery. *Nature* **228**, 726–34.

Perutz, M.F. (1970*b*). Stereochemistry of cooperative effects in haemoglobin. The Bohr effect and combination with organic phosphates. *Nature* **228**, 734–9.

Perutz, M.F. (1987). Species adaptation of a protein molecule. *Molecular and Biological Evolution* **1**, 1–28.

Perutz, M.F. (1990). Mechanisms regulating the reactions of human hemoglobin with oxygen and carbon monoxide. *Annual Review of Physiology* **52**, 1–25.

Perutz, M.F. & Brunori, M. (1982). Stereochemistry of cooperative effects in fish and amphibian haemoglobins. *Nature* **299**, 421–6.

Perutz, M.F., Fermi, G., Luisi, B., Shaanan, B. & Liddington, R.C. (1987). Stereochemistry of cooperative mechanisms in hemoglobin. *Accounts of Chemical Research* **20**, 309–21.

Perutz, M.F., Kilmartin, J.V., Nishikura, K., Fogg, J.H., Butler, P.J.G. & Rollema, H.S. (1980). Identification of residues contributing to the Bohr effect of human haemoglobin. *Journal of Molecular Biology* **138**, 649–70.

Phillips, S.E.V. & Schoenborn, B.P. (1981). Neutron diffraction reveals oxygen–histidine hydrogen bond in oxymyoglobin. *Nature* **292**, 81–2.

Riggs, A. (1988). The Bohr effect. *Annual Review of Physiology* **50**, 181–204.

Root, R.W. (1931). The respiratory function of the blood of marine fishes. *Biological Bulletin* **61**, 427–56.

Root, R.W. & Irving, L. (1941). The equilibrium between hemoglobin and oxygen in whole and hemolyzed blood of the tautog, with a theory of the Haldane effect. *Biological Bulletin* **81**, 307–23.

Samaja, M., Rovida, E., Niggeler, M., Perrella, M. & Rossi-Bernardi, L. (1987). The dissociation of carbon monoxide from hemoglobin intermediates. *Journal of Biological Chemistry* **262**, 4528–33.

Sawicki, C.A. & Gibson, Q.H. (1976). Quaternary conformational changes in human hemoglobin studied by laser photolysis of carboxyhemoglobin. *Journal of Biological Chemistry* **251**, 1533–42.

Sawicki, C.A. & Gibson, Q.H. (1977). Quaternary conformational changes in human oxyhemoglobin studied by laser photolysis. *Journal of Biological Chemistry* **252**, 5783–8.

Scholander, P.F. & Van Dam, L. (1954). Secretion of gases against high pressures in the swimbladder of deep-sea fishes. I. Oxygen dissociation in blood. *Biological Bulletin* **107**, 247–59.

Smith, F.R. & Ackers, G.K. (1985). Experimental resolution of cooperative free energies for the ten ligation states of human hemoglobin. *Proceedings of the National Academy of Sciences, USA* **82**, 5347–51.

Springer, B.A. & Sligar, J.G. (1987). High-level expression of sperm whale myoglobin in *Escherichia coli. Proceedings of the National Academy of Sciences, USA* **84**, 8961–5.

Szabo, A. & Karplus, M. (1972). A mathematical model for structure–function relations in hemoglobin. *Journal of Molecular Biology* **72**, 163–97.

Tan, A.L., De Young, A. & Noble, R.W. (1972). The pH dependence of the affinity, kinetics and cooperativity of ligand binding to carp hemoglobin, *Cyprinus carpio. Journal of Biological Chemistry* **247**, 2493–8.

Vandegriff, K.D., Medina, F., Marini, M.A. & Winslow, R. M. (1989). Equilibrium oxygen binding to human hemoglobin cross-linked between the α chains by bis(3,5-dibromosalicyl) fumarate. *Journal of Biological Chemistry* **264**, 17824–33.

Wells, R.M.G., Tetens, V. & Brittain, T. (1983). Absence of cooperative haemoglobin–oxygen binding in *Sphenodon*, a reptilian relict from the Triassic. *Nature* **306**, 500–2.

Wittemberg, J.B. & Wittemberg, D.K. (1974). The choroid rete mirabile. I. Oxygen secretion and structure: comparison with the swimbladder rete mirabile. *Biological Bulletin* **146**, 116–36.

Wyman, J. (1964). Linked functions and reciprocal effects in hemoglobin: a second look. *Advances in Protein Chemistry* **19**, 223–86.

Note added in proof

A third crystallographic conformation has recently been described for the pathologic Hb Ypsilanti (*see* Smith F.R., Lattman E.E. & Carter C.W. (1991). The mutation of β 99 Asp-Tyr stabilises Y, a new composite quaternary state of human hemoglobin. *Proteins* **10**, 81–90). The relevance of this finding is still unassessed.

JOHANNES H.G.M. VAN BEEK

Fractal models of heterogeneity in organ blood flow

Introduction

Measurements of local blood flow in an organ such as the heart or lung are subject to limited spatial resolution. The situation is schematically depicted in Fig. 1. Here the large square represents the organ in which local flow is measured, e.g. by injecting radiolabelled microspheres which are trapped in proportion to local capillary flow. The small sub-squares represent the spatial resolution of the local flow measurement. In the case of microsphere measurements the sub-squares correspond to the pieces the organ is cut into, which must each contain sufficient microspheres so that enough radioactive disintegrations can be counted. When spatial resolution is improved and one looks inside the sub-squares, usually heterogeneity is found also at the smaller scale. Indeed, where any one of the sub-squares is itself divided in exactly the same number of sub-squares as the organ was divided into, the relative deviations from the average amongst sub-squares of an organ or amongst sub-sub-squares of a sub-square may look the same. This process of zooming in, and revealing more heterogeneity at smaller spatial scales may be repeated a number of times.

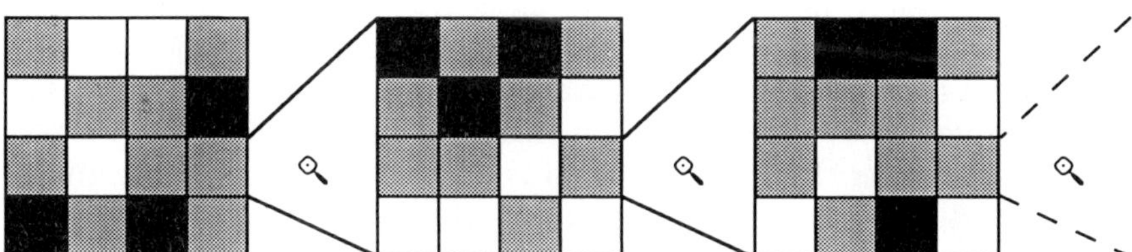

Fig. 1. Repetitive improvement of measurement resolution of local blood flow within an organ. The large square on the left indicates an organ. The small sub-squares indicate the measurement resolution. The grey scale in the sub-squares gives the local blood flow measured in the sub-square. When measurement resolution is improved, symbolised by the magnifying glass, one finds also heterogeneity within a sub-square.

Society for Experimental Biology Seminar Series 51: *Oxygen Transport in Biological Systems*, ed. S. Egginton & H.F. Ross.

Such a repetitive pattern of heterogeneity has been found for blood flow in the heart muscle for at least a two orders of magnitude change of the measurement volume (Bassingthwaighte *et al.*, 1989). For lung perfusion this pattern of flow heterogeneity has even been found over at least three orders of magnitude of the measurement's volume resolution (Glenny & Robertson, 1990). Since microspheres are trapped in capillaries and small arterioles it may be expected that tissue blocks containing a large fraction of large vessel luminal volume trap fewer microspheres. Although this effect plays a role it was shown in tracer studies that it does not account for much of the observed flow heterogeneity (Gonzalez & Bassingthwaighte, 1990). Therefore, the capillary flow distribution is determined by the distribution via the arterial tree to the capillary network.

Although there is flow heterogeneity we will see that the flow distribution is not merely random, but that there is a strong correlation between flows in adjacent regions. This correlation decreases relatively slowly when the distance between regions increases. We will also see that the pattern of flow distribution may be described by a model based on a simple rule that is recursively repeated to generate flow heterogeneity at progressively finer scales. This model is fractal in nature.

What are fractals?

We saw above that the blood flow distribution within an organ shows the property of self-similarity: after magnification a part resembles the whole (see Fig. 1). This property is the hallmark of a class of irregular geometrical structures (or sets) that have been studied by mathematicians, and have been named **fractals** by Benoit B. Mandelbrot (1983). Unlike the straight lines, squares and polygons of classical geometry, fractals resemble natural objects such as clouds, coastlines, botanical and vascular trees in their irregularity. We shall see that fractals provide a very useful tool for the mathematical modelling of biological objects. A good mathematical introduction on fractal geometry at an intermediate level has been written by Falconer (1990), while Feder (1988) gives a good introduction to applications in the physical sciences.

Figure 2 gives an example of one particular type of fractal set at three levels of magnification. The type of set shown in this example was devised by and named after the French mathematician Gaston Julia who defined the Julia set by a very simple mathematical rule (see legend of Fig. 2 for details). The essential feature is that the rule is recursively applied, e.g. in a computer program, whereby detail is progressively added at finer scales and a wonderfully complex geometrical figure emerges. The recursive application of relatively simple replication rules resulting in complex irregular structures is a very common tool in the study of fractals.

Upon magnification of an arbitrarily small part, a Julia set shows much the same pattern as seen at larger scales (see Fig. 2): the set has the property of self-similarity, just as we saw for blood flow in the heart and lung. The coastlines of islands or lakes are further examples of natural fractals that show self-similarity (Mandelbrot, 1983).

Network models

A strategy one may adopt to explain flow heterogeneity in an organ is to document accurately the anatomic structure of the vascular tree and then apply haemodynamic laws to predict the flow distribution in the organ. Following this strategy Pries *et al.* (1990) obtained precise anatomic data on the vascular system in a small part of the mesentery. The vascular structure was taken as the basis of a haemodynamic computer model. Various haemodynamic effects that play important roles in the microvasculature were incorporated, e.g. the dependence of haematocrit and viscosity on vessel diameter and shear rate, and the disproportional distribution of total flow and red blood cells at vascular bifurcations. Pries *et al.* (1990) found that various physiological parameters measured in individual vessels, such as haematocrit and even the direction of flow, were not reproduced at all by the model. The failure to predict correctly the flow distribution by Pries's elaborate haemodynamic model illustrates the extreme difficulty of precise predictions in models that have a complex structure and are governed by non-linear laws. The precision of the data needed for precise predictions may well exceed what is obtainable. Such difficulties that are already apparent in Pries's models, containing 400–1000 vessel segments, will be insurmountable for large organs such as the heart, containing on the order of 10^9 vessel segments. As long as models containing precise anatomical data and haemodynamic laws do not explain haemodynamic behaviour, the fractal model described below, although it is based on simple, to a large extent unrealistic anatomic and haemodynamic laws may be helpful for the description of gross flow heterogeneity, and may provide insight into underlying control mechanisms and the effect of altered parameters. Thus, the power is illustrated of fractal modelling based on recursive application of simple rules to generate a self-similar structure.

A

B

C

Fractal network models

It has been noted that many vascular trees appear to be self-similar (Mandelbrot, 1983; Bassingthwaighte, 1988): branching patterns in small vessels resemble those in larger vessels. Fractal concepts and renormalisation theory have been applied to the structure of the bronchial tree (West *et al.*, 1986). We went beyond structure and investigated the flow distribution in a fractal self-similar network. The first modelling attempt explicitly considered the radii and lengths of the vessels in such a network (Bassingthwaighte & van Beek, 1988). This attempt to model flow distribution in a fractal network followed the successful application of a fractal power law to describe blood flow heterogeneity (Bassingthwaighte, 1988; see below). The fractal vascular network is of a very simple nature and does not resemble the anatomical structure of the coronary vascular tree in detail. Nevertheless, the simple model will prove very useful for the description of blood flow distribution.

The network consisted of repetitive vascular bifurcations. The daughter branches at each bifurcation were shorter and narrower than the parent vessel, and different from each other (see Fig. 3*A*). The self-similarity of the fractal network required that the radii and lengths changed by the same factors at each bifurcation, irrespective of the size of the vessel or the branching generation number. An arterial tree constructed following these principles was connected with a similar venous tree

---

Fig. 2. A filled-in Julia set generated on an Apple Macintosh computer using the computer program MandelZot 3.0.2 which was developed by Dave Platt. The definition of this set is based on the relation $Z_{n+1} = Z^2_n + C$, where Z and C are complex numbers. In this example the real part of the complex number C is 0.35, and the imaginary part is also 0.35. The relation is iteratively applied, and n indicates the iteration number. The iteration starts from Z_0. The values of Z_0 for which Z_n remains bounded form the filled-in Julia set with parameter C. The Julia set itself is the boundary separating the black figure, i.e. the filled-in Julia set, from its white surroundings. In panel *A* a picture of the whole set is seen. In panel *B* we see the upper-left part of panel *A* 44 times magnified. In panel *C* the upper-left part of panel *B* is in turn 18 times magnified. Note that a magnified small part of the set in panel *C* resembles the less magnified, much bigger part of the set in panel *B*: the set has the property of self-similarity. The process of zooming-in and revealing self-similar structures may in principle be repeated an infinite number of times, but will in practice be limited by computer hardware limitations.

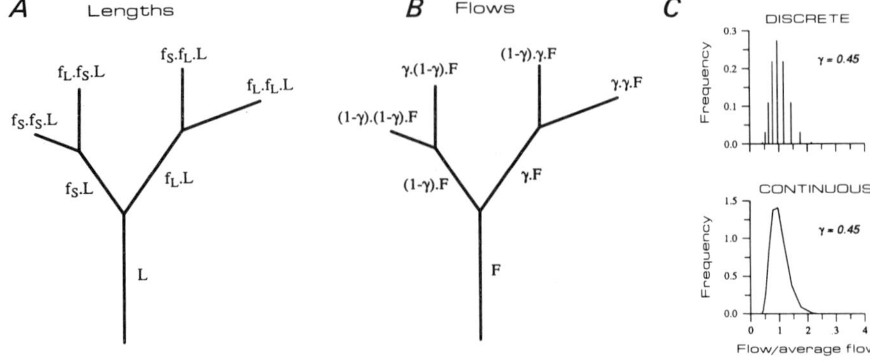

Fig. 3. A recursively bifurcating fractal vascular network. *A*, The lengths at a vascular bifurcation change according to a simple rule: the daughter branches have lengths f_S, <1, and f_L, <1, times the length of the mother branch L. This pattern is repeated at each bifurcation. Only two of arbitrarily many generations of bifurcation are shown. A similar rule applies to the radii of the vessels. *B*, A fraction γ of the flow F entering a bifurcation enters one daughter branch, the remaining fraction $1-\gamma$ enters the other daughter branch. Only two generations of bifurcation are shown. *C*, The histogram of flow in the vessels after eight generations of bifurcation. The distribution is discrete but an approximate continuous distribution has been drawn as well.

via a capillary bed. The veins could differ in diameter from the corresponding arteries. There could also be two venules to each arteriole in the model, as is often seen in the heart (Bassingthwaighte *et al.*, 1974). We worked out the distribution of flow in this fractal network based on the Poiseuille law for laminar flow of a Newtonian fluid. Of course, real microvascular haemodynamics is more complicated than we assumed here and depends on vessel diameter. In spite of this there are two reasons for preferring simple Poiseuille haemodynamics. First, it does not make sense to combine a simple geometrical description for a vessel network with details of advanced haemodynamics. Secondly, application of more complicated haemodynamics will interfere with the fractal concept of a simple self-similar pattern whose usefulness we are trying to investigate. It should be noted that the distribution of geometrical characteristics in the network does not merely reflect the passive anatomical properties but takes the vasoregulatory state of vascular smooth muscle into account.

Analysis of the flow distribution in the fractal vessel model described above showed that a simple law describes flow distribution at each bifurcation (Bassingthwaighte & van Beek, 1988): irrespective of size and position of the vascular bifurcation a fraction γ of the flow that entered the bifurcation enters the daughter vessel with the higher resistance and the remaining fraction $1-\gamma$ enters the other daughter vessel, with $0<\gamma<0.5$ (see Fig. 3*B*). It should be noted that the unequal distribution of flow at a bifurcation is not merely determined by the resistance of the first daughter vessels but by the total resistance of the dependent part of the arterial and venous networks. As a result of the self-similar structure the flow distribution is the same at each bifurcation. Thus, the flow distribution shows a self-similar fractal pattern. Self-similarity is found not only in the structure of the vessel network, but also in the flow distribution in the network. The probability density function (or histogram) of flow in the $N=2^n$ vessels after n bifurcations was shown to be right skewed (see Fig. 3*C*; van Beek *et al.*, 1989).

Since after more generations of vascular bifurcation the same total flow goes through more vessels the average flow per vessel decreases. For the same reason the standard deviation of the flow distribution decreases with an increase in the branch generation. However, the relative spread of the flow histogram with respect to the mean increases. This can be seen by dividing the standard deviation by the average flow at a certain branching generation level, a quantity called relative dispersion, RD, by Bassingthwaighte (1988). The relative dispersion for the flow distribution was found to depend on the number of vessels $N=2^n$ after n generations of bifurcation (van Beek *et al.*, 1989):

$$\mathrm{RD} = \left[N[\gamma^2 + (1-\gamma)^2]^{\log N/\log 2} - 1\right]^{\frac{1}{2}} \tag{1}$$

This model may be suited to describe flow distributions in branching vessel networks, but has not, to my knowledge, been applied for this purpose so far.

We went one step further and made a simple assumption to be able to apply the vascular network model to the distribution of local capillary flow in tissue as determined with microspheres. We assumed that each vessel at a certain bifurcation level supplied an equal volume of tissue in the organ. Thus, the blood flow distribution in an organ divided in N subvolumes was modelled. This assumption is probably very unrealistic and is discussed in the section on Multifractal Distributions (p. 144) where we suggested another, more abstract interpretation of the bifurcation model that does not refer to a vascular network but yields exactly the same mathematical result and is presumably more appropriate for the description of blood flow measurements using microspheres. The assumption of an exact correspondence between vascular supply volume and the sampling volume for microsphere measurement is also unrealistic and will be discussed below.

The experimentally determined relative dispersion of the histogram of myocardial flow for various levels of subdivision of the heart into N pieces, shown in Fig. 4, was fitted with Eqn 1 (van Beek *et al.*, 1989). The fit is shown in Fig. 4*A* as the curve for model I. It was surprising that the relation between RD and N was rather well fitted by Eqn 1, which contained only one parameter, since the relation between RD and the number of sample pieces was strongly curved (see Fig. 4). To obtain this fit it was assumed that the first bifurcation in the vessel network is completely symmetric in terms of flow distribution, with $\gamma = 0.5$. The reason for this is explained in the next paragraph. Parameter γ, which describes the asymmetry of flow distribution, was found to be 0.45–0.47 in sheep and baboon myocardium. This value of γ differs relatively little from 0.5, the value for a completely homogeneous flow distribution. The marked heterogeneity of flow in the fractal vascular tree is thus brought about by repetitive slightly asymmetric distribution of flow at successive bifurcations.

It is very unlikely that in the experiment the sample pieces are cut exactly along the boundaries of the regions supplied by the blood vessels. Hence, the effect of mismatch between sample piece and the region supplied by the blood vessel was investigated by computer simulation (van Beek *et al.*, 1989). Flow was supplied to tissue in the simulation by the vascular network with the first bifurcation having the same value for γ as the rest of the bifurcations. Then mismatch between the vessel's supply region and the sample piece was simulated and the resulting flow histograms for the sample pieces were analysed with Eqn 1. This led to overestimation of the value of γ and thus to underestimation of the

---

Fig. 4. The relative dispersion (RD), i.e. the standard deviation divided by the mean, of the probability density function of local flow. The RD measured in baboon myocardium is given by the triangles. The flow was measured by the local deposition of microspheres. The myocardium was divided in a varying number of pieces N. Note that the plot is double logarithmic. Data are averages of 10 baboons (from van Beek *et al.*, 1989, by permission of the American Physiological Society). *A*, The measured relative dispersions were fitted, using a non-linear least-squares procedure, by Eqn 1, resulting from model I, and by Eqn 3, resulting from model II. *B*, The measured relative dispersions (RDs) of the probability density function of local flow fitted by model III, which assumes a random distribution of flow in the vascular bifurcation. Model IV is an approximative analytical formula for model III.

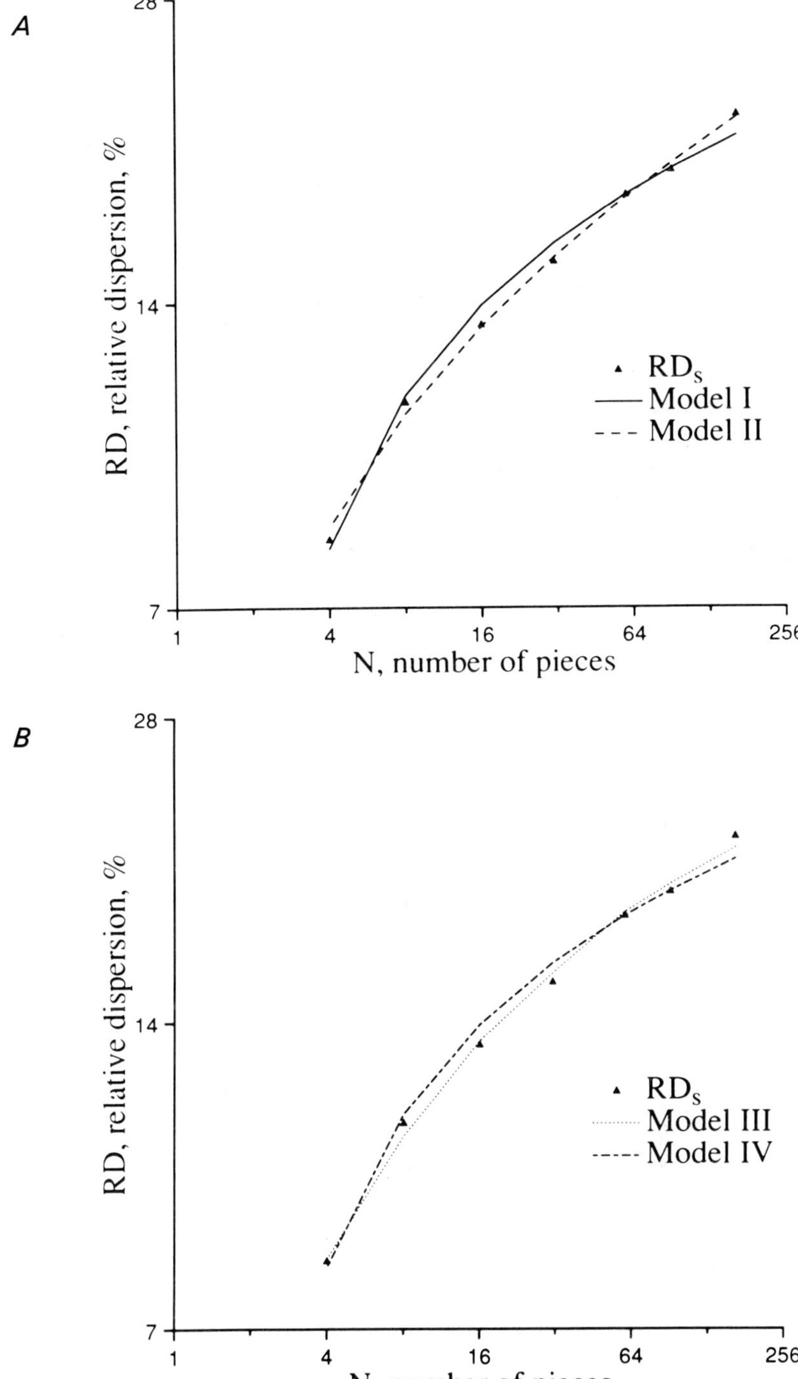
A
28
14
7
RD, relative dispersion, %
1
4
16
64
256
N, number of pieces
RDs
Model I
Model II
B
28
14
7
RD, relative dispersion, %
1
4
16
64
256
N, number of pieces
RDs
Model III
Model IV

degree of asymmetry. However, when it was assumed for the analysis that the first bifurcation was completely symmetric, $\gamma = 0.5$, the estimated values of γ were now almost exactly equal to the values of γ used in the simulation. Since we found that the fractal vascular tree model also fitted the experimental data on myocardial flow distribution well, with the additional assumption that the first bifurcation in the fractal vascular tree is symmetric, the assumption of symmetric flow distribution in the first bifurcation may be regarded as merely a correction for the mismatch between sample piece and flow field.

We now arrive at the remarkable conclusion that the fractal vascular network model, in spite of very simple and even unrealistic assumptions, describes the macroscopic flow distribution measured with microspheres very well. In contrast, a very detailed model of vascular flow distribution on the microscopic scale (Pries *et al.*, 1990), based on extensive anatomical data and incorporating detailed microvascular haemodynamic effects, was not very successful in explaining the flow distribution in a network. This suggests that the fractal recursive self-similar modelling technique is very powerful for describing and modelling very complex vessel networks that are governed by complex non-linear fluid-dynamical laws.

On the one hand, it must be realised that a number of unrealistic assumptions had to be made in the fractal bifurcation model. On the other hand, the bifurcation model fits the data very well. Therefore, a more abstract interpretation of the fractal bifurcation model is suggested in the next section.

Multifractal distributions

We derived the flow distribution in a fractal vascular network and used it to describe the flow distribution in tissue measured by the deposition of microspheres. The unrealistic assumption had to be made that each vessel supplies an equal volume of tissue. In order to avoid this assumption we have suggested another interpretation of the mathematics of the fractal vascular network model (van Beek *et al.*, 1989). Imagine that the tissue in which flow is distributed is divided in half. Upon measurement a fraction γ (<0.5) of the flow is found in one half, the remaining fraction $1-\gamma$ is found in the other half, even when measurement error is negligible. Then each half is in turn divided into two sub-halves, and the total flow entering the original half is divided over the sub-halves according to the same asymmetry parameter γ. This process of asymmetric subdivision is repeated over and over again. The mathematical derivation for this repetitive subdivision interpretation is exactly the same as for the vascular network model. Irrespective of the interpretation chosen for the bifurcation model, the vascular network interpretation or the repetitive subdivision interpretation, the model gives a good description of the measured data with one parameter.

It turned out that the idea of a fractal distribution in space by repetitive asymmetric distribution in halves had already been found by De Wijs (1951; cf. Mandelbrot, 1983), and was thus merely rediscovered by us. De Wijs, a mining engineer, developed this idea in an attempt to find a statistical description of the spatial distribution of ore. A similar distribution of mass in 1-dimensional space (imagine a thread of variable diameter) has also been studied by mathematicians and is called the Besicovitch process (cf. Mandelbrot, 1983; Feder, 1988). It contains a range of fractal subsets with a range of values for the fractal dimension (which is explained below) and has therefore been called a multifractal (cf. Feder, 1988). Mandelbrot (1983) considers De Wijs's equivalent of our asymmetry parameter γ an unsuitable measure because it applies to only one model, in contrast with the fractal dimension D which is generally applicable. However, the meaning of γ is easy to grasp for non-mathematicians working in the life sciences (see Fig. 3*B*), in contrast with the meaning of the fractal dimension D, and I therefore prefer to use γ. The multifractal distribution model may be useful as a means to describe heterogeneous spatial distributions in biology such as population densities and capillary density in tissue. Since an extensive range of methods is used to characterise spatial distributions (Egginton, 1990), it remains to be investigated whether multifractal distributions have something to add to established methods. It may be that the recognition of spatial correlation, implicit in multifractal distributions, is an important contribution of multifractal distributions to spatial statistical methods.

Deterministic fractal network model with asymmetry increasing with decreasing scale

In an attempt to improve further the fit of the fractal bifurcation model I of Fig. 3 to the data the model was slightly modified. We now assumed that the asymmetry in the fractal network model increased with the number of bifurcations, i, the vascular tree has gone through (van Beek *et al.*, 1989). This is expressed by the asymmetry coefficient γ which changes according to:

$$\gamma_i = \gamma_1 f^{i-1} \tag{2}$$

where γ_1 is the value of γ at the first branch point with $0 < \gamma_1 < 0.5$, and f is constant with $0 < f < 1$. Since $f < 1$ it follows that γ_i deviates further from 0.5, and thus from symmetric flow distribution, the higher the generation number i. This version of the bifurcation model was called model II, and the model with constant γ we discussed above was called model I. How the relative dispersion of the flow distribution increased with the generation number n of the bifurcation was derived (van Beek *et al.*, 1989):

$$\mathrm{RD} = \left[2^n \prod_{i=1}^{n} (2f^{2i-2}\gamma_1{}^2 - 2f^{i-1}\gamma_1 + 1) - 1\right]^{\frac{1}{2}} \tag{3}$$

Figure 4*A* shows the fit of model II compared with the fit of model I to the myocardial flow data. It should be noted that model II in general tended to fit the data better. The value for γ_1 was around 0.46–0.47 and the value for f was around 0.995. Thus, a slight modification of model I led to an improved description of the data at the expense of introducing one more parameter. As is the case for model I, model II may be interpreted as a fractal vascular network or as an abstract repetitive distribution of matter in space.

Stochastic fractal network models

Apart from deterministic fractal models, there are also stochastic fractals in which the recursive replication rules are modulated with a random influence (Mandelbrot, 1983). Since modulation of basic deterministic growth rules by random external influences may be of importance in biological objects, we decided to construct a fractal vascular network in which flow is distributed randomly at each bifurcation (van Beek *et al.*, 1989). At a branch point a fraction $\gamma_i = 0.5 + \delta$ of the total flow goes into one daughter branch, while a fraction $1 - \gamma_i = 0.5 - \delta$ enters the other daughter branch. Here δ is a random variable drawn from the normal distribution with mean zero and standard deviation σ. This model was called model III. The flow distribution was calculated on the computer in this case, using values of δ generated with a random number generating routine. For each value of σ the simulation was repeated a number of times, to reduce the effect of random fluctuations. The first bifurcation was treated deterministically with a fixed γ_i, since it was found that random fluctuations otherwise had too much effect on the flow distribution. The stochastic model fitted the data about as well as model II with deterministic, increasingly asymmetric branching (see Fig. 4*B*); the value for σ was found to be 0.043–0.047. This model again shows that relatively small deviations of γ from 0.5 explain the marked heterogeneity of flow found in myocardium attributable to repetitive slightly asymmetric distribution of flow. Although this has been derived from macroscopical flow distribution it is remarkable that a similar degree of asymmetry has been found by microscopic observation in muscle capillaries. Tyml & Ellis (1989) measured red blood cell velocity in adjacent pairs of capillaries in frog sartorius muscle. They found that the ratio of the lowest to the highest velocity in the capillary pair was 0.74 during hyperaemia, corresponding to a value for γ of 0.425, or for σ of 0.075. The asymmetry in resting frog muscle was more marked, with γ about 0.37. In both cases there was marked scatter in the asymmetry ratio, suggesting that the stochastic fractal model is realistic.

Fractal flow distribution and oxygen transport

Heterogeneity of perfusion may have important consequences for oxygen transport. The quantity of direct importance is the local oxygen consumption to perfusion ratio rather than perfusion *per se*. When this ratio does not differ we expect to find the same local venous and tissue oxygen tensions everywhere and there is no net diffusion amongst adjacent capillary supply regions. However, when the local oxygen consumption to perfusion ratio differs, oxygen tensions will be lower in regions with a higher oxygen consumption to perfusion ratio. Diffusion over small distances between adjacent regions will then counteract the effect of oxygen consumption to perfusion heterogeneity, while diffusion will not be able to counteract uneven perfusion in larger regions with greater diffusion distances. Thus, it is to be expected that flow heterogeneity within small regions is of little relevance for the probability density function of oxygen tension. We (J. H. G. M. van Beek, J. P. Barends and F. Spit, unpublished data) investigated the effect of diffusion amongst adjacent capillary exchange regions on the local oxygen tension in heterogeneously perfused tissue with a computer model of O_2 transport in the myocardial wall to define the spatial detail of flow heterogeneity that should be known to calculate an accurate probability density function of oxygen tension.

The calculations were first done for a slab of tissue representing the myocardial wall and perfused with saline solution, and were then extended to the blood-perfused situation. We model the unphysiological situation of perfusion with saline solution since the applicability of the transport model has been shown for this situation (van Beek *et al.*, 1991) by testing it with experimental data on oxygen diffusion across the surface of isolated, perfused guinea pig heart (Loiselle, 1989). The calculations are relatively simple because of the linear relation between oxygen tension and concentration in the saline perfusate. The oxygen tensions in the modelled situation are high so that desaturation of oxymyoglobin is unlikely and facilitated diffusion by myoglobin therefore plays no role. In the model the myocardial wall is represented by a slab of tissue. The 1-dimensional oxygen transport equation is:

$$\alpha\lambda D\frac{d^2P}{dx^2} - 2\varrho F\alpha(P - P_a) = \varrho\dot{V}_{O_2} \tag{4}$$

where P is the average O_2 tension in a capillary exchange region, P_a the arterial O_2 tension and x is the 1-dimensional spatial coordinate. Further, α is O_2 solubility, λ is the ratio of O_2 solubility in tissue to the O_2 solubility in the perfusate, D is the diffusion coefficient, ϱ is tissue density, F is flow per unit tissue mass and $\dot{V}_{O_2}$ is the O_2 consumption per unit tissue mass. A similar equation has been used by Popel & Gross (1979) to model oxygen diffusion in tissue with capillary perfusion. Flow heterogeneity is simulated using the fractal model I, Fig. 3*B*, which was described above. The lower panel of Fig. 5 shows an example of a fractal distribution with asymmetry parameter γ equal to 0.45. For comparison also a homogeneous flow pattern is drawn. The O_2 consumption was taken to be homogeneous, so that there was marked oxygen consumption to perfusion heterogeneity. There is oxygen diffusion across the epicardial and endocardial surface. The profile of the average O_2 tension in the capillary region, calculated on the computer applying a finite difference approach, is shown in the upper panel of Fig. 5. In this example, based on experiments of Loiselle (1989), medium with an O_2 tension of about 680 mm Hg was in contact with both surfaces of the tissue. The arterial O_2 tension was also about 680 mm Hg. We see a steep oxygen tension gradient extending a few hundred μm into tissue. This gradient drives the oxygen diffusion flux into tissue that has been measured by Loiselle (1989) for isolated guinea pig heart suspended in 95% O_2-containing gas. In homogeneously perfused tissue there is virtually no diffusion gradient in the centre of the slab of tissue. In the heterogeneously perfused case the O_2 tension profile undulates. It should be noted that sharp transitions in flow level are smoothed out in the oxygen tension profile because of O_2 diffusion.

We proceeded to investigate the effect of flow heterogeneity at different scales. This was done by calculating the average value and the standard deviation of the O_2 tension histogram for different degrees of flow heterogeneity. We subtracted the O_2 tension profile for homogeneous flow from the profile for heterogeneous flow so that the boundary effect attributable to oxygen diffusion across the surfaces on the oxygen tension histogram was cancelled out. We generated flow heterogeneity down to high spatial resolutions. The average O_2 tension for heterogeneous flow on a scale of 20 μm, which is of the order of the intercapillary distance in the heart, was only about 1.1 mm Hg lower than

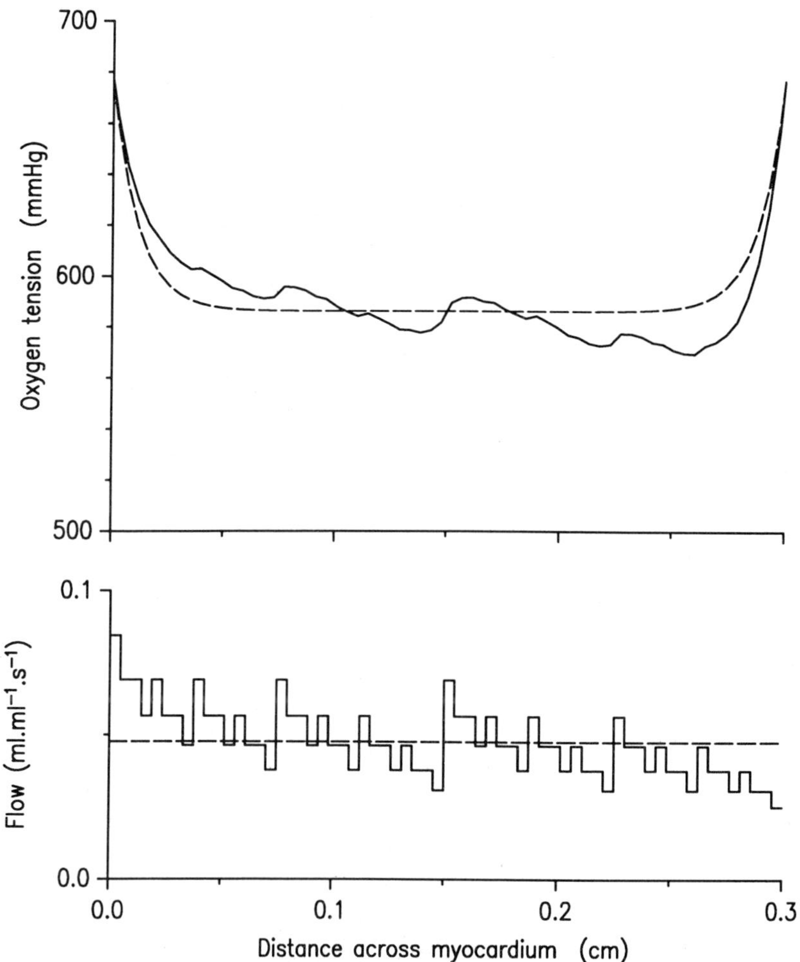

Fig. 5. In the lower panel a profile of a heterogeneous flow distribution across a slab of tissue is drawn (continuous line), and for comparison also a homogeneous flow distribution (dashed line). The slab represents the left ventricular wall of saline-perfused arrested guinea pig heart to model experiments by D. S. Loiselle (see text). The heterogeneous flow distribution is generated by the fractal vascular network, model I, to a spatial resolution of about 50 μm, i.e. internal flow heterogeneity within a region of this size is neglected. In the upper panel the oxygen tension profile is drawn calculated according to our model. At both surfaces of the slab, at 0 and 0.3 cm, there is diffusion between tissue and surroundings. The oxygen tension at both surfaces and the arterial oxygen tension are all 677 mm Hg under these experimental circumstances.

for homogeneous flow for a normal O_2 diffusion coefficient in tissue, $D = 1.5 \times 10^{-5}\,cm^2\,s^{-1}$, and with an oxygen-carrying capacity equal to that of blood. The standard deviation characterising the P_{O_2} differences between the profiles for homogeneous and heterogeneous flow was about 3.3 mm Hg. Next, we simulated that flow heterogeneity was known with a lower spatial resolution by taking adjacent regions together and replacing the heterogeneous flow in each part of this aggregate with the average flow in the whole aggregate. The average and the standard deviation of the difference in O_2 tension between homogeneous and heterogeneous flow was again calculated and was plotted as a function of the spatial resolution, i.e. the dimension of the region in which the fractal heterogeneous flow was averaged. In the case of Fig. 6 this was done for an oxygen-carrying capacity similar to that of blood. This was also done for lower and higher values of D. The average value of the O_2 tension in tissue is always lower for heterogeneous flow than for homogeneous flow. When there is no diffusion to counteract the effect of flow heterogeneity the average O_2 tension keeps decreasing when more flow heterogeneity at finer scales is added. However, when there is diffusion amongst adjacent microvascular exchange regions the average difference and the standard deviation do not increase any further when flow heterogeneity at smaller spatial scales is added. This means that for the calculation of the oxygen tension histogram it suffices to know flow heterogeneity with a limited spatial resolution. For conditions in blood-perfused working heart the standard deviation of the O_2 tension histogram is predicted to be in error by less than 1% when the spatial resolution is better than 100 μm. For the normal O_2 diffusion coefficient and for the conditions in the saline-perfused heart the standard deviation is known with an error of less than 1% when the spatial resolution is better than 500 μm. This value of the spatial resolution thus defines a unit size for flow measurement: internal flow heterogeneity within such a unit adds very little to the spread of the oxygen tension distribution. The unit size was found to depend on the flow rate, on the diffusion coefficient and on the oxygen-carrying capacity of the perfusate.

Although we made use of a model that omits details of oxygen tension gradients at the capillary level, the model analysis allowed us to investigate the interaction between flow heterogeneity at large scales and O_2 diffusion. We conclude that spatial heterogeneity on small scales is sometimes of little relevance for transport since diffusion amongst adjacent regions tends to counteract the effect of flow heterogeneity. The cutoff distance which determines this lower limit of flow heterogeneity relevant for transport is smaller when the diffusion coefficient of the considered substance is lower, and when the effective solubility of the substance is higher.

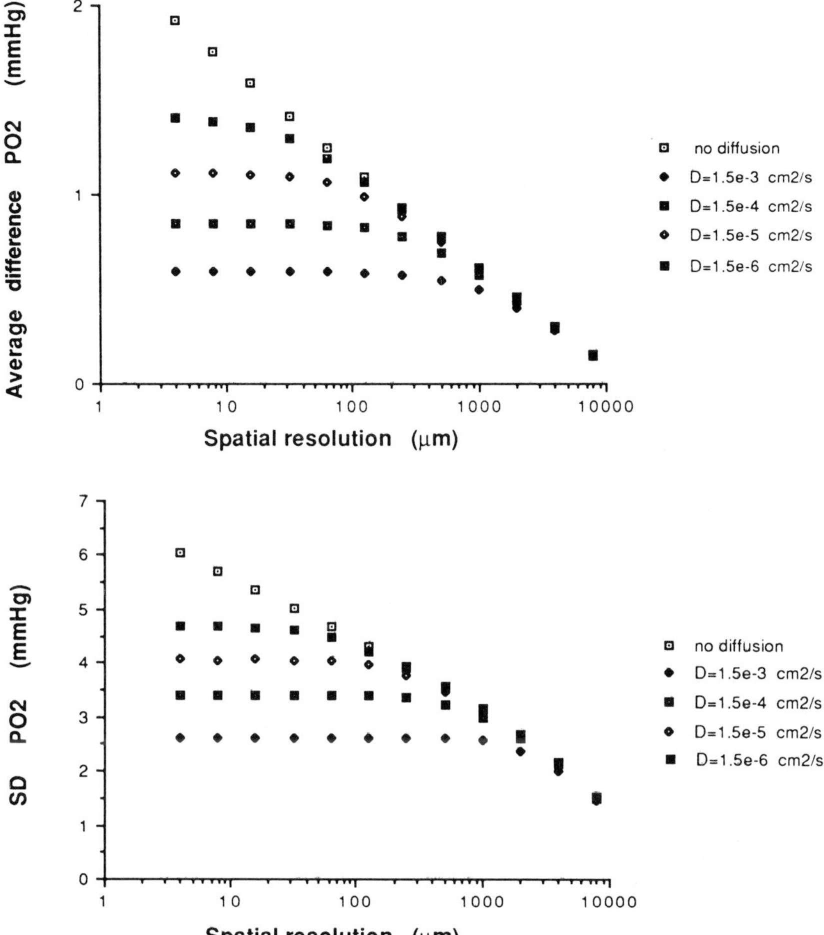

Fig. 6. The average difference and the standard deviation of the difference between the oxygen tension profiles in a slab of tissue, as in Fig. 5, calculated for homogeneous flow and heterogeneous flow. In this case the oxygen tensions were calculated for perfusate with an oxygen-carrying capacity similar to that of blood. The spatial resolution is the size of the regions in which the flow heterogeneity generated using the fractal vascular network model I is averaged, so that all internal fractal flow heterogeneity within these regions was neglected. The calculation was done with no diffusion amongst adjacent vascular exchange regions and for diffusion with four values for the diffusion coefficient D. The normal value of the oxygen diffusion coefficient is $D = 1.5 \times 10^{-5}\,\mathrm{cm}^2\,\mathrm{s}^{-1}$. The calculated standard deviation and average deviation of the oxygen tension histogram remain the same below a certain threshold for the spatial resolution. This threshold becomes higher when the diffusion coefficient becomes higher.

Fractal power law analysis

Apart from the fractal bifurcation models discussed above a quite different type of analysis based on fractal concepts of self-similarity has been used to describe flow heterogeneity: fractal power law analysis of the relative dispersion of the flow histogram. Power law analysis was the first type of analysis to be applied (Bassingthwaighte, 1988), but is treated as second in this chapter since it is conceptually more difficult and may be understood better at this point. Although fractal power law analysis is completely different from the fractal network approach presented above, it is also based on the fundamental fractal concepts of self-similarity and repetitive recursion.

To give an example of fractal power law analysis we return to the Julia set of Fig. 2. It is known that the length L of the coastline of such a fractal 'island' increases when the linear dimension ε decreases, and the extra spatial detail taken into account when L is measured. In Fig. 2 the Julia set is plotted with a certain spatial detail: small bays and inlets for a certain stretch of the coastline have been omitted and replaced by straight lines. Imagine that one walks a divider with opening ε between the legs along the map so that the length L of the coastline obtained is the number of steps made with the divider times ε. When smaller bays are now also taken into account by decreasing ε the length L of the coastline increases. The increase is found to be given by a power law:

$$L = L_0(\varepsilon/\varepsilon_0)^{1-D} \tag{5}$$

where L_0 is the length for an arbitrary value ε_0 and where D has a value between 1 and 2 (Mandelbrot, 1983). The exponent $1 - D$ has been given as 'one minus the parameter D' since D has an important interpretation: D has been shown to have the mathematical properties of a dimension and is therefore called the fractal dimension (Mandelbrot, 1983). For a straight line or a smooth curve, which we would normally consider as 1-dimensional, D indeed has the value 1, and for a smooth plane D has the value 2. However, for the complicated shape of a Julia set the fractal dimension D has a non-integer value between 1 and 2. On the one hand, the coastline of the Julia set is a line (its topological dimension is one); on the other hand, its length is infinite since each arbitrarily small part contains a detailed wiggling coastline that resembles the whole. In a sense the coastline of the Julia set to some extent fills part of the plane in which it is drawn with its infinitely long, infinitely complex perimeter so that at less than infinite magnification one sees a ribbon rather than a sharp line. The fractal power law for the length of a geometrical object applies both to mathematical fractal sets and to natural fractals such as the coast of

Britain (Mandelbrot, 1983). The fractal dimension D can be determined from the relation in Eqn 5, but also in various other ways, and is considered as the most fundamental characteristic of the complexity and heterogeneity of fractal sets. We will now apply an equation analogous to Eqn 5 to characterise the self-similar fractal flow distribution.

Power law analysis of flow heterogeneity indicates a non-random distribution

The heterogeneous distribution of flow can be rendered by a histogram of local flow. The spread of the histogram can be characterised by the standard deviation. For flow histograms the ratio of the standard deviation of flow divided by the mean flow, which in statistics is known as the coefficient of variation, has been called the relative dispersion (Bassingthwaighte *et al.*, 1989). Bassingthwaighte conjectured that the relative dispersion of the local myocardial flow increased with a power function of the inverse of the mass of the pieces into which the myocardium had been divided, in analogy with the fractal power law for the length of the coastline in Eqn 5. Therefore:

$$\mathrm{RD} = \mathrm{RD}_0(m/m_0)^{1-D} \tag{6}$$

where m is the mass of a piece of myocardium which is inversely related to the spatial resolution, and where $1 < D < 2$. The subscript 0 indicates an arbitrary reference mass, e.g. of 1 g. The analogy between length (Eqn 5) and relative dispersion of flow (Eqn 6) is not easy to see and will be discussed below. The mass of the sample pieces m is the total myocardial mass M divided by the number of pieces N into which the myocardium is divided so that:

$$\mathrm{RD} = \mathrm{RD}_0(N/N_0)^{D-1} \tag{7}$$

We will illustrate the fit of Eqn 7 rather than Eqn 6 to a data set to facilitate comparison with Fig. 4. The fractal power law was tested using the data on microsphere deposition in the myocardium discussed in the Introduction to this chapter. The myocardium had been divided in the smallest pieces which allowed reasonably accurate flow measurement. The effect of smaller spatial resolution was simulated by aggregating a group of adjacent pieces and summing the flows measured in the pieces. The histogram of the flow per gram of tissue was determined for the aggregates, which were of about the same mass. In Fig. 7 a fit of Eqn 7 to data on the RD of flow vs the number of aggregates is shown in a double-logarithmic plot. The power law of Eqn 7 gives a straight line on this plot. The data points are reasonably well described by the fitted line, but one

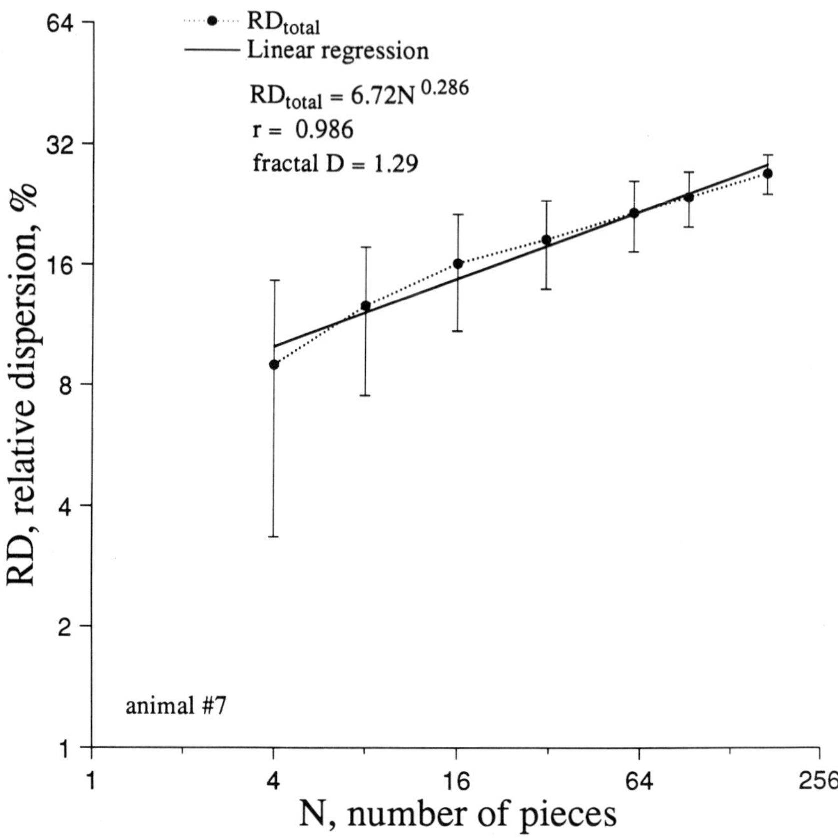

Fig. 7. The relative dispersion of the flow distribution in baboon myocardium as a function of the number of pieces the myocardium has been divided in. Data are from one experiment. Note that the plot is double logarithmic. The relative dispersion was fitted with the fractal power law of Eqn 7.

usually sees a systematic deviation: for large aggregates, i.e. small numbers of pieces N, the data points usually lie under the fitted line for the power law. However, it may be seen from the figure that when the small piece numbers are omitted, the power law gives a good description of the measured data points.

It was found that Eqn 6 describes the experimentally found relation between the aggregate mass of the pieces and the relative dispersion for that piece size well when the piece's mass is between approximately 0.1 and 10 g (Bassingthwaighte *et al.*, 1989). However, for larger piece sizes the relative dispersion falls consistently below the value predicted by extrapolating Eqn 6 from the range 0.1–10 g to values > 10 g. We will see below that this lack of fit of Eqns 6 and 7 reflects the limited size of the organ, in contrast with the infinite size of fractal objects in mathematics.

The fractal dimension D found from the fit of local myocardial flow data to Eqns 6 and 7 was about 1.15 (Bassingthwaighte *et al.*, 1989) and for the blood flow in the lung it was about 1.1 (Glenny & Robertson, 1990). From classical statistics it is well known that Eqns 6 and 7 are valid with $D = 1.5$: the standard deviation of the average value of a sample of size n is the standard deviation of the population divided by $n^{0.5}$, assuming that the samples are *not* correlated. Note that the sample size n equals the total population size n_0 divided by the number of samples N in Eqn 7. The flow data, where D is clearly lower than 1.5, must therefore be correlated. Sample pieces with a high flow tend to have neighbours with high flow, and pieces with low flow tend to have neighbours with low flow. It was shown (van Beek *et al.*, 1989) that Eqn 6 is equivalent to saying that the flow in adjacent pieces is correlated with a nearest-neighbour correlation coefficient r, and r is constant, irrespective of the volume of the aggregates the organ has been divided into. It was further found (van Beek *et al.*, 1989) that:

$$r = 2^{3-2D} - 1 \qquad \text{for } 1 < D < 2 \tag{8}$$

Thus, when $r = 0$ the fractal dimension $D = 1.5$, and when $r = 1$ the fractal dimension $D = 1$.

Since the fractal dimension D from Eqn 2 turns out to have a value of 1.15 for regional myocardial blood flow, the nearest-neighbour correlation coefficient is about 0.63. Since the correlation coefficient has an appreciable value for both large and small aggregates it is clear that a fractal flow distribution in space exhibits correlation over very large distances. However, organs like the heart and lung are of limited size. Glenny & Robertson (1991) calculated the correlation between local flows in small sample volumes of the lung as a function of the distance between the sample volumes. They found that the correlation coefficient was high for nearest neighbours but went to negative values when the distance between the sample volumes approached the size of the lung. Indeed, when flows near a sample volume with high flow are above average, thus reflecting positive correlation between neighbours, flows farther away within an organ must be below average: e.g. when flow in one half of the organ is above the average for the whole organ, the flow in the other half must be below the average so that the correlation coefficient becomes negative. This results in negative correlations at large distances between sample volumes. Owing to this negative correlation at large distances, Eqns 6 and 7 cannot be applicable for large masses or a small number of sample pieces, respectively. This derives from the fact that the power law of Eqns 6 and 7 implies that the correlation is the same, irrespective of the size of the sample volume chosen. This is presumably reflected by the experimental finding that for the lowest spatial resolutions, flow measurements in one half, one quarter or one eighth of the organ, Eqn 7 predicts consistently too high relative dispersions of flow (see Fig. 7).

It should be noted that D in this section has been used as defined by Bassingthwaighte *et al.* (1989). We shall see below that the true fractal dimension of blood flow is presumably exactly 2 higher than D obtained from Eqns 6 or 7 (see section on 'The fractal dimension of myocardial blood flow: embedding in four-dimensional space', p. 157). Thus, the actual fractal dimension should presumably be 3.15 for myocardial flow and 3.1 for lung flow, since analysis with Eqns 6 and 7 does not take embedding in 4-dimensional space into account, but artefactually embeds blood flow in 2-dimensional space.

Spatial correlation and fractional Brownian motion

The length of a fractal boundary increases as a power of the spatial resolution of the measurement, see Eqn 5. This inspired the application of Eqn 6 which indicates that the relative dispersion increases with a power of the spatial resolution (Bassingthwaighte, 1988). It is not clear why relative dispersion can be substituted for length. In this section it is shown that Eqn 6 is indeed valid for an example of a mathematical fractal model, fractional Brownian motion.

Normal Brownian motion can be modelled in mathematics by a random walk. Imagine you are moving on a footpath. A coin is thrown and when heads comes up you take one step forward and when tails comes up you take one step backward. You are now taking a random walk. Mathematical analysis shows that on the average your distance from your starting position increases with the square root of the number of steps you have taken, or written in a peculiar way:

$$\textit{distance proportional with } (\textit{number of steps})^{2-D} \tag{9}$$

where D has a value of 1.5. Mandelbrot (1983) has generalised the random walk, or Brownian motion, by allowing D to have values between 1 and 2. The resulting walk with D unequal to 1.5 was called **fractional Brownian motion** (fBm) and D is the fractal dimension of this fractal process. When $1 < D < 1.5$ you will find that on the average you are moving away from your starting point faster than when $D = 1.5$. The outcomes of the coin throws are not independent any more. To obtain this result one needs a strangely biased coin such that when more heads than tails have come up in the past, it is more likely that a head will come up than a tail at the next throw. Thus, you will tend to walk on in the direction you have been moving. When $1.5 < D < 2$ you are moving away more slowly than for $D = 1.5$, and when more heads than tails have come up in the past it is more likely that a tail will come up and that you will move back towards your starting position. The process of fractional Brownian motion, which is similar to the random walk, has been worked out in strict mathematical terms (Mandelbrot & van Ness, 1968; Feder, 1988).

It has been shown for fractional Brownian motion that the distance moved during the last n steps before a certain point in time is correlated with the distance moved during the first n steps after that time and that the correlation coefficient r is exactly given by Eqn 8 (Feder, 1988) with D the fractal dimension of the fractional Brownian motion. Since Eqn 6, with the number of steps n substituted for m, can be derived from Eqn 8 this means that Eqn 6 also applies to fractional Brownian motion: when during fractional Brownian motion the displacement for n steps is measured repeatedly in subsequent intervals of n steps, the relative dispersion (standard deviation/mean) of the displacements for n steps is given by Eqn 6. It is thus shown that by analysing an example of a fractal signal, fBm, by relating the relative dispersion of the displacements with the number of steps n, using Eqn 6, the fractal dimension D of the fBm can be derived. The fractal power law for length also applies to fBm: when the displacement from the starting position of fBm is plotted against time the length of this graph is given by Eqn 5. Thus, it can be shown that Eqn 5 for the length and Eqn 6 for the relative dispersion both apply to an example of a fractal signal, fractional Brownian motion, and that the fractal dimension determined from both equations is the same.

The fractal dimension of myocardial blood flow: embedding in four-dimensional space

We have seen that a fractal dimension of about 1.15 was reported for myocardial blood flow based on fractal power law analysis (Bassingthwaighte *et al.*, 1989). I will show that for myocardial blood flow distributed in a 3-dimensional organ the true value of the fractal dimension should presumably be 3.15, and that the method of analysis based on Eqns 6 and 7 underestimates the fractal dimension by exactly 2. For an object embedded in 2-dimensional space, e.g. the graph of the displacement of fractional Brownian motion plotted as a function of time discussed in the last section, one expects a fractal dimension D between 1 and 2. For a fractal surface embedded in 3-dimensional space, e.g. the profile of the surface of the earth, where the height is a function of the 2-dimensional coordinates of the position, one finds a fractal dimension of about 2.15 in many instances (Voss, 1988). A quantity distributed in 3-dimensional Euclidian space is embedded in 4-dimensional space: for each 3-dimensional coordinate a fourth coordinate is needed to specify a quantity, such as the water vapour density in a cloud, temperature or blood flow in myocardium (Voss, 1988). [The reader who feels uncomfortable with dimensions higher than 3 may get an imaginative introduction to them in the science fiction novel *Flatland* (Abbott, 1884).] We must therefore expect a fractal dimension between 3 and 4 for the blood flow distribution. The RD analysis based on Eqn 6 has resulted in a reduction in fractal dimension. Equation 6 contains but one variable, the mass of a piece of tissue, and 3-dimensional information is discarded by this analysis. The blood flow distribution is thus analysed in 1-dimensional sections through 3-dimensional space. The fractal dimension of blood flow in 3-dimensional space reduces by two on intersection with a line (Mandelbrot, 1983; Falconer, 1990). I therefore conjecture that the correct fractal dimension of myocardial blood flow is 3.15 instead of the value of 1.15 given by analysis with Eqn 6 and that the fractal dimension is underestimated by exactly 2.

An assumption hidden under the 1-dimensional analyses of fractal distributions in 3-dimensional space is that the structure of blood flow heterogeneity is isotropic: the correlation as a function of distance is the same in all directions. A preliminary report by Schosser *et al.* (1989) suggested that blood flow distribution in myocardium indeed was not completely isotropic. The assumption of isotropy in the blood flow distribution should be examined more thoroughly.

The derivation of Eqn 8 was based on the analysis of Eqn 6. When the real fractal dimension of the blood flow distribution embedded in 4-dimensional space is used the relation between correlation and fractal dimension becomes:

$$r = 2^{7-2D} - 1 \qquad \text{with } 3 < D < 4 \tag{10}$$

Thus, with the true fractal dimension of myocardial blood flow equal to 3.15, the nearest-neighbour correlation coefficient still is 0.63.

Towards integrated modelling of anatomical structure and flow distribution: L-systems?

Although they are useful to model macroscopic flow heterogeneity the fractal network models discussed above are certainly not suitable to model the anatomic structure of the coronary vascular tree. To describe a complex, irregular biological structure like a vascular network mathematically one could store a large number of data: e.g. the 3-dimensional coordinates of the start and end of each vessel segment in the network and its diameter. However, in the mammalian heart there are in the order of 10^9 vessel segments so that gathering such data and storing them would constitute a gigantic task. The same would apply to the mathematical description of the coastline of Britain, certainly if one includes details on scales finer than 1 m.

However, if one wants to generate a picture of a coastline using a computer, an alternative to storing huge amounts of data is available: program the computer with a simple mathematical rule so that the rule is applied recursively to generate spatial detail at ever finer scales (see Fig. 2). Not only will this result in a simulation of a curve which appears to the eye as a coastline (and may share many statistical characteristics with a real coastline) but it also points to the fact that complex, irregular objects may have an underlying simple structural basis. However, a limitation of this fractal approach is that the recursive application of the simple rule leads to geometries which appear similar, but are not in general exactly like the simulated natural object.

For the computer graphical generation of botanical trees, which often resemble vascular trees, promising and elegant results have been obtained with the use of L-systems (Prusinkiewicz & Lindenmayer, 1990). An example of L-system simulations of botanical trees which resemble bronchial trees or vascular trees is given in Fig. 8. L-systems share some characteristics with fractals: they involve recursive applications of simple rules, usually on a computer, and can be used to generate self-similar structures. Indeed, L-systems have been successfully applied to generate graphs of fractals using a digital computer.

The letter L in L-systems is derived from the name of Aristid Lindenmayer who introduced in 1968 a formal system to describe the development of multicellular filamentous organisms, e.g. algae. The L-system consists of a linear connected string of elements which are represented by symbols, e.g. 'aab' is an example of such a string. Each symbol in such a string is replaced by a string of symbols during a growth step according to prescribed rules, e.g. 'a' becomes 'b', 'b' becomes 'aab'. The string 'aab' thus becomes 'bbaab' in one step. In another type of L-system, called context-sensitive, the change of a symbol depends on the neighbouring symbols in the string, e.g. 'a' changes to 'b' only if it does not have a 'b' as neighbour, otherwise it remains 'a'. L-systems have been used to model the growth of algae and the symbols then represented various types of algal cell. L-systems may branch, e.g. a symbol 'c' is replaced by 'c[c]c' where the brackets indicate that the middle symbol is

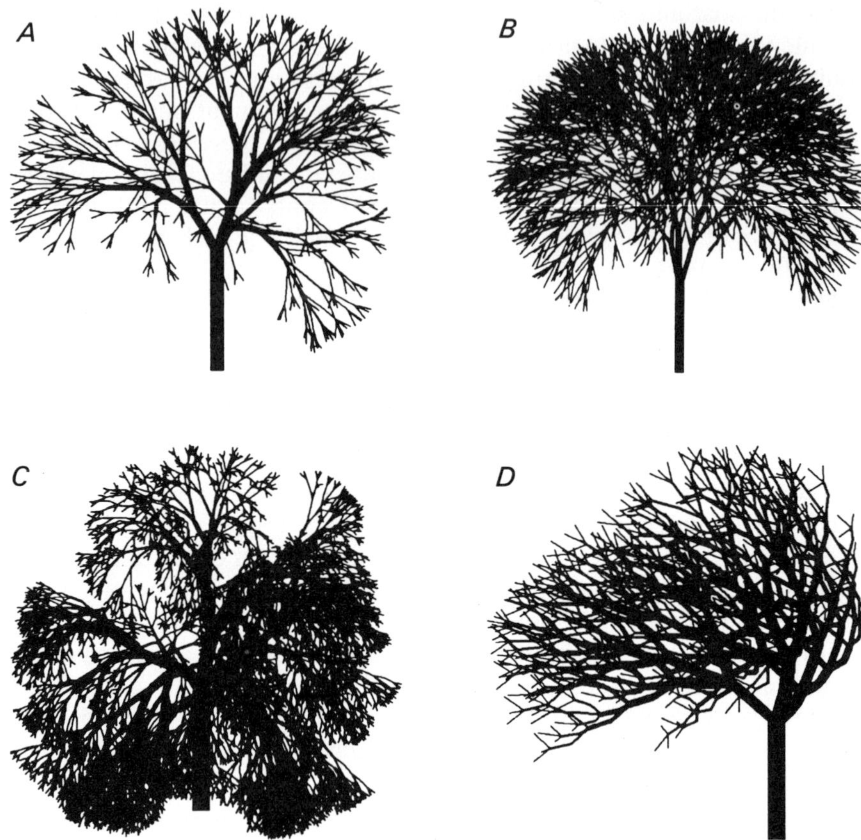

Fig. 8. Trees simulated by computer graphics, using L-systems. (From *The Algorithmic Beauty of Plants*, by Prusinkiewicz & Lindenmayer (1990), by permission from Springer, New York.)

on a side branch of the main branch which in this example contains the first and last 'c'. Also a vector of parameters can be attached to a symbol, for instance representing the length of the plant cell, the angle under which it branches off or the concentration of an important signal molecule in the cell.

L-systems can, by virtue of their ability iteratively to produce self-similar structures, be used to produce fractal geometrical objects. To achieve this the symbols in the string are interpreted as commands for a plotter pen or a LOGO-style turtle graphics device. For instance 'F' may be interpreted as: move forward 1 cm while drawing a line, 'f' may mean:

move forward without drawing a line, '+' may mean: change direction of movement and drawing by a certain angle, etc. Interpreted in this way L-systems can be used to draw fractals in a plane. A 'mole' controlled by an L-system might dig fractal tunnels in 3-dimensional space or blood vessel networks in a model organ, in analogy with the drawing turtle.

It may be expected that by using L-systems very good models of the vascular anatomy can be developed. Since parameters such as the vessel's radius, length and branch angle can be associated with the elements of the L-system's strings it is possible to calculate haemodynamic resistance and the flow distribution in the vascular trees. Like fractals, L-systems allow a very concise description of complex structures, well suited for computer modelling.

Conclusion

We have seen that fractal concepts are well suited for the description of the macroscopic flow heterogeneity which has been found in a number of mammalian organs. One fractal approach is based on a fractal power law relation between the spread of the flow histogram and the spatial resolution of the flow measurement. It implies a constant correlation coefficient of about 0.6–0.7 between flows in neighbouring regions in an organ, irrespective of the region's size. The other fractal approach is based on recurrent self-similar branching structures to model the heterogeneity of perfusion within organs. Fractals based on the latter branching pattern are applicable on all scales, while the constant nearest-neighbour correlation fractal approach (equivalent with the fractal power law) is not applicable for large scales that approach the size of the organ. Although perfusion heterogeneity may be expected to occur on all scales, heterogeneity on small scales is of little importance for oxygen transport because diffusion over small distances counteracts the effects of flow heterogeneity.

L-systems, like fractals, apply simple rules recursively and seem very promising for the modelling of the anatomy of vascular systems within organs, and may allow one to calculate the haemodynamic behaviour and flow distribution in the vascular system.

Acknowledgements

I am very grateful to Dr James B. Bassingthwaighte who inspired much of my work on fractals. Further, I am grateful to Jan Paul Barends and Fred Spit who very skilfully assisted with computer programming and graphics for part of the work reported here, and to Hans Schepers for discussion of the relation between fractional Brownian motion and the fractal power law.

References

Abbott, E.A. (1884). *Flatland*, revised (1952) edn. New York: Dover Publications.

Bassingthwaighte, J.B. (1988). Physiological heterogeneity: fractals link determinism and randomness in structures and functions. *News in Physiological Sciences* **3**, 5–10.

Bassingthwaighte, J.B., King, R.B. & Roger, S.A. (1989). Fractal nature of regional myocardial blood flow heterogeneity. *Circulation Research* **65**, 578–90.

Bassingthwaighte, J.B. & van Beek, J.H.G.M. (1988). Lightning and the heart: fractal behavior in cardiac function. *Proceedings of the IEEE* **76**, 693–9.

Bassingthwaighte, J.B., Yipintsoi, T. & Harvey, R.B. (1974). Microvasculature of the dog left ventricular myocardium. *Microvascular Research* **7**, 229–49.

De Wijs, H.J. (1951). Statistics of ore distribution. *Geologie en Mijnbouw (The Hague)* **13**, 365–75.

Egginton, S. (1990). Morphometric analysis of tissue capillary supply. *Advances in Comparative and Environmental Physiology* **6**, 73–141.

Falconer, K. (1990). *Fractal Geometry. Mathematical Foundations and Applications*. Chichester: John Wiley.

Feder, J. (1988). *Fractals*. New York: Plenum Press.

Glenny, R.W. & Robertson, H.T. (1990). Fractal properties of pulmonary blood flow: characterization of spatial heterogeneity. *Journal of Applied Physiology* **69**, 532–45.

Glenny, R.W. & Robertson, H.T. (1991). Spatial correlation: a corollary of fractal pulmonary perfusion. *FASEB Journal* **5**, A404.

Gonzalez, F. & Bassingthwaighte, J.B. (1990). Heterogeneities in regional volumes of distribution and flows in rabbit heart. *American Journal of Physiology* **258**, H1012–24.

Loiselle, D.S. (1989). Exchange of oxygen across the epicardial surface distorts estimates of myocardial oxygen consumption. *Journal of General Physiology* **94**, 567–90.

Mandelbrot, B.B. (1983). *The Fractal Geometry of Nature*. New York: W.H. Freeman.

Mandelbrot, B.B. & van Ness, J.W. (1968). Fractional Brownian motions, fractional noises and applications. *SIAM Reviews* **10**, 422–37.

Popel, A.S. & Gross, J.F. (1979). Analysis of oxygen diffusion from arteriolar networks. *American Journal of Physiology* **237**, H681–9.

Pries, A.R., Secomb, T.W., Gaehtgens, P. & Gross, J.F. (1990). Blood flow in microvascular networks. Experiments and simulation. *Circulation Research* **67**, 826–34.

Prusinkiewicz, P. & Lindenmayer, A. (1990). *The Algorithmic Beauty of Plants. A Virtual Laboratory*. New York: Springer-Verlag.

Schosser, R., Zwissler, B., Schwickert, Ch., Weiss, M., Iber, V., Spengler, P., Weiss, Ch. & Messmer, K. (1989). Evidence that fractal dimension is a biological constant of blood flow heterogeneity in dogs. *International Journal of Microcirculation: Clinical and Experimental* **8**, S36.

Tyml, K. & Ellis, C.G. (1989). Localized heterogeneity of red cell velocity in skeletal muscle at rest and after contraction. *Advances in Experimental Medicine and Biology* **248**, 735–43.

van Beek, J.H.G.M., Loiselle, D.S. & Westerhof, N. (1991). Calculation of oxygen diffusion across the epicardial surface of isolated perfused heart. *FASEB Journal* **5**, A1468.

van Beek, J.H.G.M., Roger, S.A. & Bassingthwaighte, J.B. (1989). Regional myocardial flow heterogeneity explained with fractal networks. *American Journal of Physiology* **257**, H1670–80.

Voss, R.F. (1988). Fractals in nature: from characterization to simulation. In *The Science of Fractal Images*, ed. H.-O. Peitgen & D. Saupe, pp. 21–70. New York: Springer-Verlag.

West, B.J., Bhargava, V. & Goldberger, A.L. (1986). Beyond the principle of similitude: renormalization in the bronchial tree. *Journal of Applied Physiology* **60**, 1089–97.

S. EGGINTON and H.F. ROSS

Planar analysis of tissue capillary supply

Introduction

This chapter will deal primarily with analysis of the structural elements which facilitate or limit oxygen transport to tissue. The physiological importance of the capillary bed is generally accepted in terms of providing adequate oxygen delivery to working muscle, and facilitating abnormal growth of tumours. It has been assumed since the pioneering work of Krogh (1919) that the capillary bed serves mainly to supply oxygen, and this aspect has been emphasised in most studies of skeletal muscle as well as heart microvasculature. This is not surprising as it has also been known for a long time that red muscles, subsequently characterised by highly aerobic metabolism, have a much more dense capillary network than white muscles with glycolytic metabolism, and that capillary density in the heart is almost an order of magnitude higher than in glycolytic muscles. However, capillaries are also needed for removal of different metabolites, one of the most important in skeletal muscle being lactate, and heat produced during muscle contraction. Nevertheless, it is far from clear to what extent oxygen demand or accumulation of waste products regulate capillary growth. Indeed, if one examines the observed interrelationship between the microvascular supply and tissue phenotype, our inadequate understanding quickly becomes evident (Hudlicka *et al.*, 1992). As direct measurement of oxygen release from capillaries and its consumption by surrounding tissue is fraught with technical problems, much effort has been invested in developing models of tissue oxygen supply. There is, therefore, a need for quantitative (morphometric) analyses which adequately describe the anatomical capillary bed, in order to derive estimates of the functional capillary supply. Skeletal muscle is particularly well suited to morphometric analysis as the predominant constructs are composed of cylinders. This has been known since the earliest histological investigations, and indeed allowed Krogh to elaborate his famous conceptual approach to the study of oxygen transport. Although the basic model has been developed and refined subsequently

Society for Experimental Biology Seminar Series 51: *Oxygen Transport in Biological Systems*, ed. S. Egginton & H.F. Ross.

(see Hoofd, this volume), all models of physiological processes require realistic anatomical data as their primary input. We will therefore examine some approaches that can provide realistic, quantitative estimates of tissue capillary supply which may be used in such models.

Given that capillaries act as the final path in the respiratory cascade transferring oxygen from environment to cell, oxygen transport capacity must be related in some manner to the physical extent of the microcirculation, i.e. blood supply at a cellular level, although whether this reflects an optimal or a compromised design is a matter of some debate. The vascular supply of vertebrate striated muscle is matched with functional demand of the tissue, e.g. capillary density (CD) reflects muscle phenotype and/or level of activity (Hudlicka *et al.*, 1992). Although many indices of capillary supply are used, it is clear that none provides a complete description of either physical dimensions or functional capacity of the capillary bed. This is in part attributable to the fact that many indices, and the models which incorporate such information, greatly simplify the complex arrangement evident *in vivo*. Most are implicitly based on a rather uncritical interpretation of the Krogh model, using the same unrealistic assumptions about tissue structure, and consequently have limited utility. It must be remembered that capillaries perform functions other than oxygen delivery, and in some cases this may be of prime importance, such as delivery of substrates and removal of products from endocrine tissue. Also, where capillaries do act primarily as oxygen conduits there is a hierarchy of physiological adaptations that can modify the efficacy of oxygen delivery (e.g. macrocirculation, microcirculation and mitochondrial distribution). Finally, most indices and models are essentially restricted to use in cases of high CD and geometrically regular capillary networks, whereas most tissues have a relatively low CD and a heterogeneous distribution of capillaries. Clearly, we have to develop morphometric indices of capillary supply which can account for at least some of these factors if we are to provide the mathematical models of oxygen transport with a realistic anatomical basis.

Simple morphometric analyses

It is important to recognise at the outset that morphometric indices of structure can only estimate the maximal capacity of a system. The functional characteristics are difficult to assess without additional physiological data; in the case of the microcirculation we need to know local blood flow, capillary haematocrit, regional variation in tissue oxygen consumption, etc. Global morphometric indices may therefore reflect control exerted at higher levels of biological organisation, but are subject to possible errors of simplification, while analyses having a greater resolution may reveal control at the lower levels of organisation, but fail to account for cooperativity between these different levels of organisation. The eventual aim must be to reconcile analyses describing structure of both the macro- and microcirculation, and integrate morphometric data and mathematical models of diffusion, in order to develop a quantitative theory of blood flow and oxygen transport to tissue.

In any morphometric study it is important to consider the many ways in which erroneous results may be generated; they are numerous! (see Egginton, 1990*a*). For example, adequate tissue preparation is an obvious prerequisite if realistic cellular dimensions are to be retained. Different criteria seem to be adopted by different groups, however. For example, those who maintain that the minimal wall thickness seen in perfusion-fixed capillaries accurately reflects *in vivo* barrier thickness for diffusion, and those who consider these dimensions merely to reflect the inherent compliance of vessels; those who prefer hyperosmotic fixatives for the clarity of electronmicrographs this produces, and those who accept an apparent reduction in quality of organelle structure with isosmotic fixatives in order to preserve spatial relationships within a tissue by avoiding shrinkage, and so on. Equally obvious is the requirement for unambiguous identification of capillaries, although given the wide variety of methods used to visualise the microcirculation it is not surprising that many discrepancies occur between workers on similar tissue. These methods are divided essentially into those using particles or dyes to mark flow through capillaries, with the uncertainty of knowing if all vessels are patent, and histological or histochemical procedures which provide *in situ* localisation of capillaries, but with an accuracy dependent on the degree to which the staining is unique or the enzyme activity differentiated. Unbiased sampling is essential if the data are not to be influenced by the analytical procedure adopted. The most common problems seem to be those of having an inadequate sample size, which can introduce very serious errors, and of ignoring the shape and size of cells when collecting data. The problem can be overcome easily by collecting morphometric data from sub-samples of a population within finite boundaries (quadrats or fields), such that an object can be counted only once with a series of contiguous counting frames (Egginton, 1990*b*). Problems with preparation, identification and/or sampling can easily explain the six-fold range in estimates of capillary density reported for cat gastrocnemius (Plyley & Groom, 1975).

A final requirement for a successful morphometric study of tissue capillary supply is the choice of an appropriate analysis, which clearly depends on the ultimate goal of the study, and the many options are reviewed elsewhere (Egginton, 1990*a*). Analytical problems are often minimised by examining a highly structured, or anisotropic tissue such as skeletal muscle. Our discussion will concentrate on analysis of data from cross-sections of this tissue, but the principles will be broadly applicable. The capillary supply can be quantified using 0-, 1-, 2- or 3-dimensional indices (Egginton & Ross, 1989*a*). By this we mean simple counts of capillary number, their linear separation and the area or volume of tissue associated with each vessel. These approaches are discussed in more detail elsewhere (Egginton, 1990*a*).

0-dimensional

Simple counts are adequate for most studies where gross changes in the capillary supply are expected and such studies have been concerned mainly with global estimates, i.e. mean values for individual muscles. Numerical indices based on counts of capillaries and fibres within a given field are obviously the most direct form of analysis, with the data commonly expressed as the capillary-to-fibre ratio (C:F). This has the advantage that it is relatively insensitive to errors in section orientation and dimensional artefacts resulting from tissue processing, but will be affected by changes in fibre size. As most physiological adaptations that result in an altered capillary supply also show muscle hypertrophy or atrophy, this is a serious limitation. Therefore, it should not be assumed automatically that an altered C:F ratio implies a change in the density of capillaries per unit area of tissue: it might arise from a change in fibre size.

1-dimensional

Given that most capillary beds appear to have a heterogeneous spacing, linear analyses may be used to determine the degree of clustering or variability in intercapillary distances. Spatial point process statistics may be used to model the distribution of capillaries, taken to be point sources. This allows a test for aggregations or repulsions, equivalent to zones of angiogenesis or growth inhibition. Unfortunately, the form of the distribution may be misleading, as the benchmark 'random' distribution is impossible for capillaries, which are physically excluded from 80 to 90% of the tissue by the muscle fibres.

Linear indices of separation of capillaries have been popular with modellers of transport processes, particularly where the Krogh-type approach has been adopted as, on average, intercapillary distance (ICD) is twice the radius of the Krogh tissue cylinder, R_K. Mean spacing of capillaries is often assumed to be a limiting factor in oxygen delivery, and hence to co-vary with muscle performance, although evidence is accumulating which brings this relationship into question (Hudlicka *et al.*, 1988). ICDs can be estimated by specific application of point process statistics to derive the probability distribution, such as closest-individual or nearest-neighbour analyses (Loats *et al.*, 1978; Kayar *et al.*, 1982*a*,*b*), but the possibility of erroneous interpretation in equating calculated and observed patterns of capillary distribution mentioned above needs to be remembered. The mean ICD is often calculated from capillary density, assuming a regular geometric array of vessels in tissue cross-section. The limitations in this approach are self-evident, as such distributions are suggested from mean values, rather than derived from individual distances. Direct mensuration of adjacent capillary separations has been attempted by construction of a unique set of non-overlapping triangles connecting all points in a plane, where half the mean side length equates with R_K (Renkin *et al.*, 1981*a*), although this is both tedious and subject to some small error in sampling bias. We have adopted an objective, unbiased sampling procedure based on our preferred index of spatial distribution (the capillary domain) which addresses these limitations (see p. 174).

2-dimensional

One of the most common indices of capillary supply to tissue is that of capillary density (CD; mm^{-2}), based on the assumption that the density of point sources of oxygen in a section through any tissue will be a limiting factor for, or accurately reflect, the level of oxidative metabolism. Although it is true that more aerobic muscles (e.g. from wild vs domestic species or from trained vs sedentary individuals) tend to have a higher CD, there are sufficient exceptions to this rule to caution against its uncritical usage (Hudlicka *et al.*, 1988, 1992). In general, for the more aerobic tissue with a regular capillary distribution, CD reflects the level of oxidative metabolism (Romanul, 1965; Gray & Renkin, 1978), while in tissue with significant glycolytic potential and often irregular capillary network the low CD reflects more the capacity for removal of metabolites (Egginton & Johnston, 1983; Hudlicka *et al.*, 1987). CD is, however, more sensitive to variations in preparative technique (often seen as tissue shrinkage, leading to erroneously high values) than C:F. Further limitations in the value of this index follow from its inability to allow for local variations in fibre composition or size, and that it ignores the heterogeneity of capillary spacing, which has been shown to affect both mean tissue P_{O_2} and proportion of poorly oxygenated tissue (Turek & Rakušan, 1981; Rakušan *et al.*, 1984). As theoretical studies become more realistic it is increasingly evident that oxygen transport to tissue (particularly the diffusion-limited component) requires a true planar analysis in order to account for interactions between neighbouring capillaries, in addition to those problems mentioned above. The development of such an analytical approach, and its applications, will form the bulk of this chapter.

3-dimensional

Tissues act as an integrated whole in 3 dimensions, a fact that is sometimes forgotten when the subtleties of the simpler indices are examined in great detail. The mean volume of tissue supplied by individual capillaries may be derived from global estimates of CD or ICD, by integrating the area of tissue around an average capillary over its path length, although this is strictly valid for linear networks of a regular geometric arrangement. Identification of the area around individual capillaries (see 'Domains', p. 172) may provide a method of overcoming these restrictions for irregular networks.

By far the most successful approach so far has been the application of stereological techniques (see chapters by Mayhew and Perry, this volume; also Hoppeler, 1984). This is the powerful application of geometric probability theory that permits extrapolation of structural information from n to $n+1$ dimensions. It can describe the maximal capacity of a system, providing variations in orientation that underlie the apparent order of tissue components are also assessed. It is clear that capillaries do not simply run along muscle fibres, unbranching and straight as required by the Krogh model, but that they form a sinuous path with an averaged orientation parallel to the fibre axis. In addition, there are cross-connections, or anastomoses, between adjacent capillaries. Whilst both these factors are clearly important when defining physical limits to the capillary network and its connectivity, incorporation into indices of capillary supply is problematical (Egginton, 1990*a*). The functional implication of this tortuosity is clearly to increase the effective capillary length and surface area per unit volume of tissue over that estimated from simple counts in transverse sections. The length density (length of capillaries per unit volume of muscle, J_V) equates with the product of CD and a constant of proportionality, which varies according to the angle of sectioning relative to the longitudinal axis of muscle fibres. Using an assumption about the distribution of capillary orientation around the fibre axis, a rather simple method has been devised to overcome this problem whereby a dimensionless concentration parameter, K, may be used to denote the degree to which capillaries follow a preferred axis of orientation. The ratio of CD from two sections at 90° to each other (usually transverse section, TS, and longitudinal section, LS) may be used to calculate this parameter. A correction factor, $c(K,0)$, can then be used to convert the CD from TS (0 radians) directly to the length density (Mathieu *et al.*, 1983):

$$J_V = c(K,0) \cdot CD$$

Although other methods are available, this appears to be quite accurate and efficient. In skeletal muscle fixed at resting length the coefficient of anisotropy, $c(K,0)$, can range from 1.016 (i.e. capillary orientation deviates from perfect alignment with the fibre axis by 1.6%) to as much as 1.42 depending on the muscle and method of fixation (Egginton & Johnston, 1983; Mermod *et al.*, 1988). Capillary length density is probably the fundamental parameter which best describes the extent of capillary supply, but it can also be used to derive two indices of more obvious value, capillary volume (V_V) and surface (S_V) densities, which provide an estimate of the instantaneous volume of blood in the microcirculation, and the surface area available for gaseous exchange, respectively. Any errors involved in estimating J_V may therefore seriously affect the validity of conclusions regarding adaptation of the capillary bed. Perhaps most obvious is the change in tortuosity that occurs when a fibre contracts, either because of the length at which it is held during fixation or because of the method of fixation. If one imagines pieces of string loosely attached along the length of a collapsible tube the effect is obvious, decreasing fibre length (increasing girth) must increase capillary tortuosity. This has been shown experimentally using vascular casts and led some authors to emphasise the importance of determining sarcomere length before reported values of J_V can be correctly interpreted (Mathieu-Costello, 1987). In practice, however, this does not often pose a serious problem. In samples of muscles fixed at resting length from different species ranging from fishes to mammals, we have routinely obtained values for sarcomere length (SL) of fixed material which only range between 2.2 and 2.4 μm for different muscles, a range over which the effect of SL on $c(K,0)$ is quite insensitive (see Hudlicka *et al.*, 1992).

The use of stereology to overcome the pitfalls inherent in indices based on single counts clearly has many problems of its own. It can, however, provide a valid alternative to popular indices such as C:F if the 3-dimensional relationships are quantified. In the heart, where myocytes are not as ordered as skeletal muscle fibres, Mattfeldt (1987) interpreted capillary proliferation to have occurred when each unit length of myocyte is associated with a greater length of capillary, using a novel 3-dimensional C:F ratio of J_V(capillaries):J_V(myocytes). There are, however, relatively few papers (most published in the last 10 years) which utilise stereological techniques in the quantitative evaluation of capillary supply.

Capillary domains

Although simple counts are adequate for most studies where gross changes in capillary supply are expected, diffusion theory suggests that oxygen transport to tissue is critically dependent on the heterogeneity of capillary spacing. However, simply measuring (or calculating) the linear separation of capillaries in a cross-section of muscle has limited descriptive potential, as diffusion does not occur only along a line joining adjacent capillaries but throughout the surrounding tissue. In other words, functional resistance to intramuscular diffusion has to reflect the integration of all linear separations between an individual capillary and its neighbours. It follows that a true planar analysis, based on the area of tissue associated with each capillary, is probably a more useful approach as it can take account of the interactions between neighbouring capillaries, and provide a 2-dimensional index of capillary separation sensitive to the local environment. Analysis of the 'domains of influence' for a population of capillaries should then provide a more realistic estimate of the heterogeneity underlying the capillary distribution in the plane of sectioning.

Construction

Given a section of tissue, which can be viewed as part of a plane, on which the capillaries are marked points, it is possible to surround each of those points with the convex polygon consisting of that part of the plane closer to the included capillary than to any other capillary. The collection of polygons thus formed covers the part of the plane under consideration without gaps or overlaps. The construction which divides up the plane in this way is called a Voronoi tessellation (Egginton & Ross, 1989*b*). The polygon associated with each capillary is called its domain, and its boundary is formed by the perpendicular bisectors of lines joining a capillary with all adjacent capillaries. An illustration of the procedure is displayed in Fig. 1, which shows both the polygonal domain and lines joining the capillary to adjacent capillaries. Note that the neighbourhood of a given capillary defined in this objective manner can be quite different from that defined from a single parameter, such as nearest-neighbour analysis or calculated R_K. In particular, it can accommodate sub-adjacent vessels, which is important when analysing sparse capillary networks where some fibres lack direct capillary contact, such as those found in fish skeletal muscle (Egginton *et al.*, 1988).

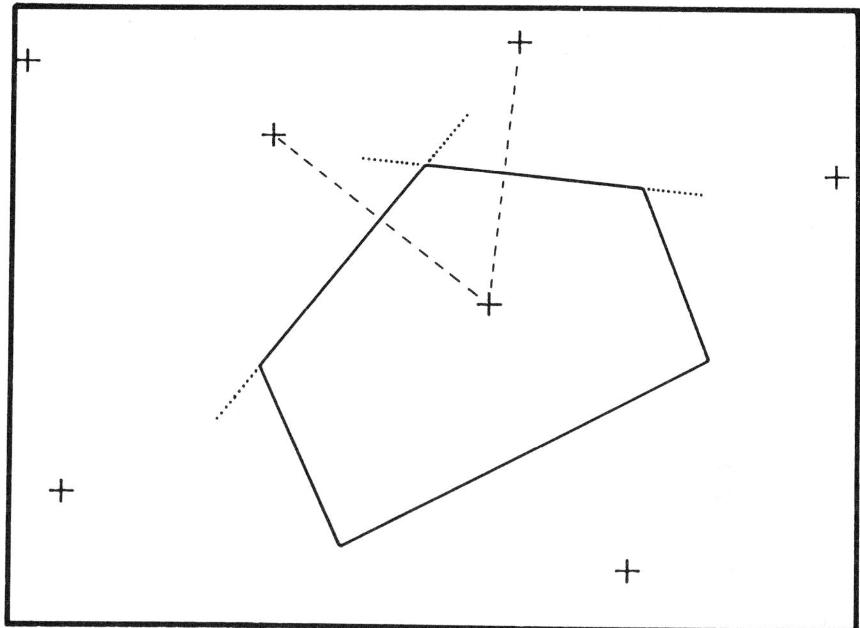

Fig. 1. Method of construction of the capillary domains. Crosses indicate the positions of capillaries; the broken lines connecting capillaries are bisected, and these dotted lines joined to form a convex polygon (Egginton, 1990*a*).

Distribution of domain areas

Figure 2 shows the cumulative distribution of domain areas from a sample of rat soleus muscle plotted on linear and log scales. In both cases the best fitting integrated normal distribution curve is superimposed. The figure shows that the distribution of domain areas is represented better by a log–normal distribution than by a normal distribution, a result which appears to be repeated in the distribution of domain areas from a wide variety of other muscles. The spread of this distribution may be used as a measure of heterogeneity in capillary spacing, most usually in the form of the standard deviation of the logarithms of the areas (unfortunately first referred to in this context by the rather confusing abbreviation of 'logSD').

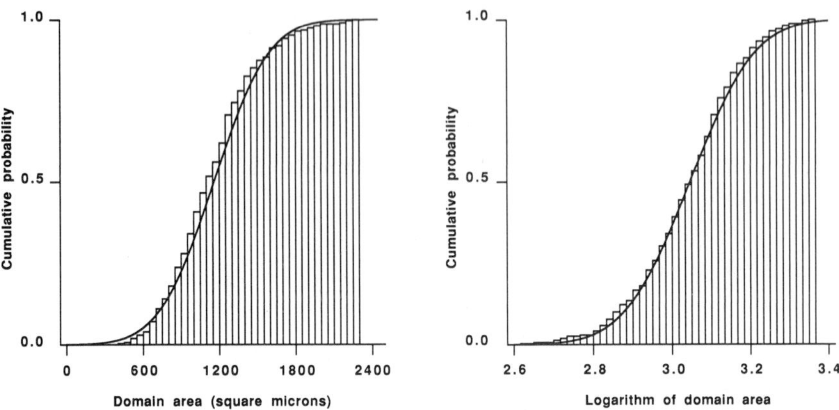

Fig. 2. Cumulative distributions of domain area and logarithm of domain area with the best-fitting cumulative normal distribution in each case. The fit is better in the right-hand panel than in the left, indicating a log–normal distribution of domain area. Data from adult rat soleus muscle, 12 μm frozen section.

Derived linear separation of capillaries

Although direct measurement of ICDs is preferable to derivation from a global quantity, such as CD, there are some practical difficulties to overcome. Use of either the simple estimates described earlier or point process statistics for this purpose has, in addition to achieving an unbiased sample, to overcome ambiguity in the definition of which capillary is a 'neighbour'. Nearest neighbours are strictly valid only for estimating minimum diffusion distances, and an objective scheme is required to account for non-nearest neighbours. The procedure for calculating the capillary domains, which was described above, may be extended to provide an objective method for defining neighbouring capillaries; capillaries are neighbours if their domains share a common edge (Egginton & Ross, 1989*b*). The process is illustrated in Fig. 3, which shows the full construction of a set of domains in a sample of tissue from rat hind limb (extensor digitorum longus) muscle, and the set of intercapillary distances derived from them. The network of lines which connect adjacent capillaries form a set of triangles which cover the plane and is called the Delaunay tessellation. This method has the advantages that distances are measured not derived, that the rules for determining which capillaries are to be deemed adjacent are built into the algorithm in an objective way, and that the analysis is not limited to nearest or *n*-nearest neighbours. The resulting

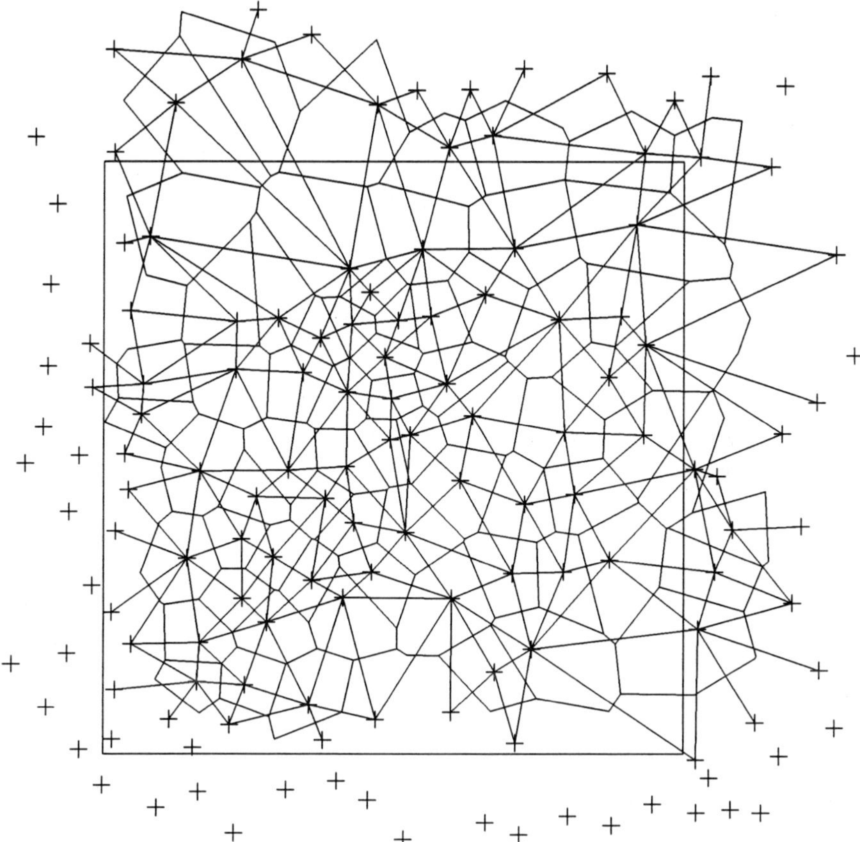

Fig. 3. Showing the construction of a set of capillary domains and the associated set of intercapillary distances. Crosses indicate positions of the capillaries and the domains are the polygons surrounding them (Voronoi tessellation). The lines connecting capillaries are the ones used to build the sample of intercapillary distances (the complementary, Delaunay tessellation). The rectangle is the sampling frame used for unbiased sampling of domains according to their number, rather than the usual area, distribution. 12 μm section, rat EDL.

network of triangles is reminiscent of the triangulation method of Renkin *et al.* (1981*a*), although the form of construction is not imposed. Being derived from an area-based construction means that an unbiased sampling rule may be used to sample fibres or domains in proportion to their numerical, rather than areal, distribution (Egginton, 1990*b*). Analysis of ICD distributions in striated muscle by the methods of concentric circles

Table 1. *Comparison of ICD estimates measured directly, or derived from domain area*

	Mean	S.D	S.E.	*N*
Diaphragm ICD	36.70	14.04	0.4656	909
Diaphragm diameter	35.84	7.19	0.4349	273
Soleus ICD	39.36	13.16	0.4153	1004
Soleus diameter	38.01	5.65	0.3205	311
Tibialis anterior ICD	63.27	22.59	0.9396	578
Tibialis anterior diameter	62.02	11.64	0.8827	174

Direct estimates of ICD are from the respective Delaunay tessellation, while indirect estimates ('diameter') are the equivalent domain diameters.

(Turek & Rakušan, 1981), triangulation (Renkin *et al.*, 1981*a*) and domains (Egginton & Ross, 1989*b*) all give rise to similar, log–normal distributions.

A related parameter which has been used to derive a measure of intercapillary spacing (see also following paragraph) is the domain equivalent diameter. It is computed as the square root of the domain area multiplied by $2/\sqrt{\pi}$. This appears to give mean values and standard errors which do not differ greatly from those obtained from the Delaunay tessellation method. Table 1 shows a comparison of the results from the two methods from three different types of muscle.

Planar separation of capillaries

The method of domains was first used by Hoofd *et al.* (1985) to study heterogeneity of capillary spacing in the myocardium; for capillaries relatively far apart (long ICD) the domains will be larger than for those lying closer together, and the distribution of individual domain areas will therefore reflect the heterogeneity of the anatomical capillary supply.

This study used the equivalent radii of circles whose areas are equal to those of domains in the analysis. However, there are a number of advantages in using area *per se*, rather than derived linear quantities. This should be apparent if we consider what is meant when we refer to a heterogeneous capillary spacing. A plot of ICDs in a regular array will have a narrow distribution. By definition variability is minimised in assumed regular arrays, whatever form they take, but would be broader if the separation of capillaries were to vary from a given pattern or became irregular. When modelling intramuscular oxygen tensions it is clear that the longest diffusion distances have the greatest effect, and so *in situ* variability is potentially of great importance. The use of a 2-dimensional index of separation therefore provides more information about the geometry of the capillary bed as it describes not only the planar separation of individual vessels, but also the influence of the non-nearest neighbours. However, because this approach provides a unique value for each capillary, unlike ICD which provides a value for each of *n* neighbours, it is quite sensitive to sampling error and clearly demonstrates an inverse relationship between the degree of bias and data variance (Egginton, 1990*a*).

One rather simple example will suffice to illustrate the utility of this approach. Using domains to examine the patterns of capillary distribution in skeletal muscles of widely differing metabolic capacity revealed previously unnoticed differences in heterogeneity of supply that are hidden by global indices (Egginton *et al.*, 1988). In muscles from the hind limb of rat there was a significantly lower heterogeneity of capillary spacing in the more oxidative soleus than the mixed fast EDL muscle, although the difference in CD was modest. In contrast, slow and fast muscles of fish show a maximal difference in metabolic capacity, which is reflected in a great difference in CD (35-fold), but showed no difference in heterogeneity of capillary spacing. This suggests that there is a functionally homologous spatial distribution of capillaries irrespective of oxidative capacity; that differences in the capillary bed are quantitative rather than qualitative in nature. Given the local distribution of capillaries relative to individual fibres, it also appears that under conditions of maximal demand, oxygen supply is not restricted to contiguous capillaries, but also involves those remote from the fibre surface.

There is regional variation in structure, such as mitochondrial content, even in muscles with a homogeneous fibre-type composition. This must result in local variations in oxygen consumption and, likely, tissue oxygen tension. How this additional level of heterogeneity may be incorporated into analysis of local capillary supply will be dealt with later, but even using domains to address such a simple comparative problem leads to some very interesting conclusions. In the case of an homologous spatial distribution of capillaries between muscles the qualitative nature of the capillary supply must be independent of the oxidative capacity of the tissue it serves. This is quite different from the textbook story; the local distribution of capillaries shows no apparent correlation with fibre area or mitochondrial content in fish muscle, although the intramuscular heterogeneity is low. In mixed mammalian muscles, however, there appears to be an optimal capillary supply regulated at both tissue and cellular levels.

Calculated tissue oxygen tensions

While many authors have documented the limitations to Krogh's cylinder model it still forms the basis for many analyses of oxygen supply, and may actually provide a reasonable approximation of global oxygen delivery to aerobic tissue (see Hoofd, this volume). For use at the local level, however, it requires a space-filling, non-overlapping representation of the zone of capillary influence. Capillary domains fulfil these requirements and therefore provide a realisation of the Kroghian tissue cylinder in more realistic terms, that of a space-filling polygonal column, for individual vessels. Indeed, it goes further and provides an objective procedure which allocates in an unambiguous way an area of tissue to each capillary. An estimate of mean local diffusion distance, $R\text{K}$, may still be derived, and is then given as the radius of a circle with equivalent area to the domain:

$$R\text{K} = (\text{domain area}/\pi)^{1/2}$$

The more realistic interpretation of tissue supply area given by planar analysis than with mean $R\text{K}$ is shown diagrammatically in Fig. 4, where a field of muscle having a heterogeneous distribution of capillaries is overlaid with overlapping circles depicting the mean Krogh cylinder cross-section, centred on each capillary, and the non-overlapping polygons of the capillary domains.

Several models have been developed which allow the distribution of oxygen tension within skeletal or cardiac muscle to be calculated, and give histograms that are similar to those obtained from surface or needle P_{O_2} electrodes. Given the *in situ* variability of capillary spacing, it is impossible to use a space-filling set of Krogh cylinders with equal $R\text{K}$, and some index of heterogeneity is therefore required. The models of Turek and co-workers originally derived estimates of the mean radius of the Kroghian tissue cylinders, and an estimate of the heterogeneity of capillary spacing, from nearest-neighbour statistics (Rakušan *et al.*, 1984). In later refinements, however, they used capillary domains to obtain an index of spatial heterogeneity (expressed as the standard deviation of the distribution of logarithms of the domain areas, 'logSD') which yields almost identical values for $R\text{K}$, but with a smaller logSD (Turek *et al.*, 1987). The most recent models consist of a series of parallel tissue cylinders, having variable radii which simulate the experimentally determined spatial distribution of capillaries from domains (Turek *et al.*, 1989). Oxygen and CO_2 contents of capillary blood at the beginning of each cylinder are calculated from arterial values of blood gases and acid–base status, and the relevant dissociation relationships. Each tissue

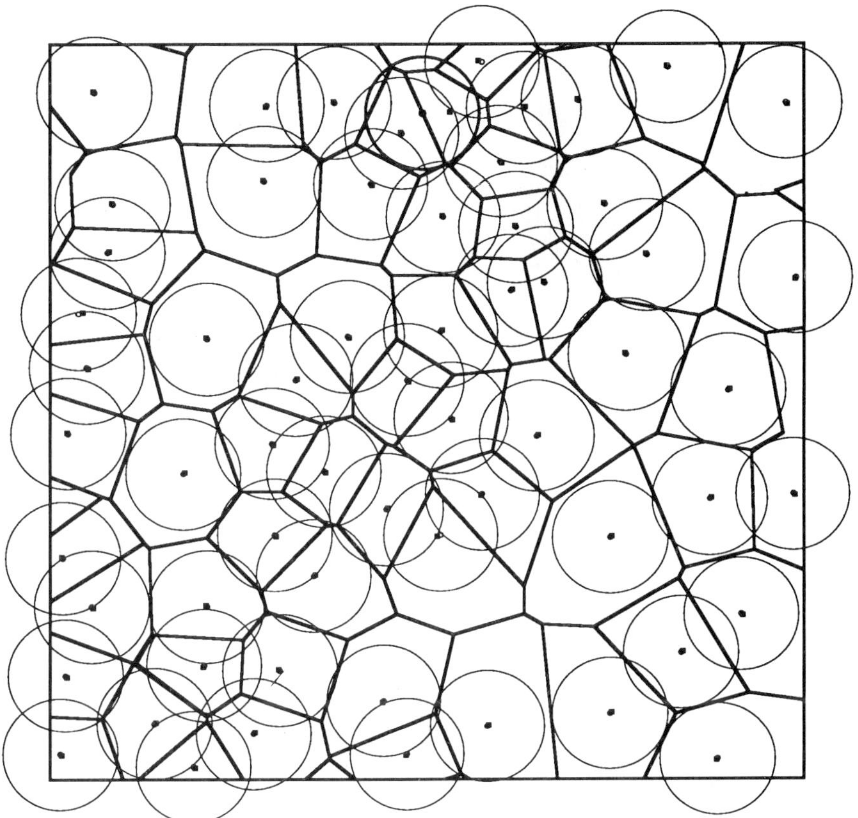

Fig. 4. A sample of capillaries (dots) and their associated domains (heavy polygons) and Krogh cylinders (light circles). The total area of the circles is equal to the total area of the polygons. Note that some areas of the tissue are not covered by circles, while some are covered several times. 12 μm frozen section, rat EDL.

cylinder is then divided longitudinally into a number of tissue slabs, and the P_{O_2} distribution calculated using an iterative procedure (Fig. 5). This Multicylindrical model can then be used to investigate the relative effects of heterogeneity of capillary spacing (expressed as logSD), facilitated diffusion (by myoglobin), differential resistances to oxygen efflux from capillaries (the so-called 'capillary block'), variable distribution of capillary blood flow (similar perfusion per capillary or per unit volume of tissue cylinder) and the kinetics of tissue oxygen consumption (zero-order, or O_2-dependent O_2 consumption) (Turek *et al.*, 1991). Additional

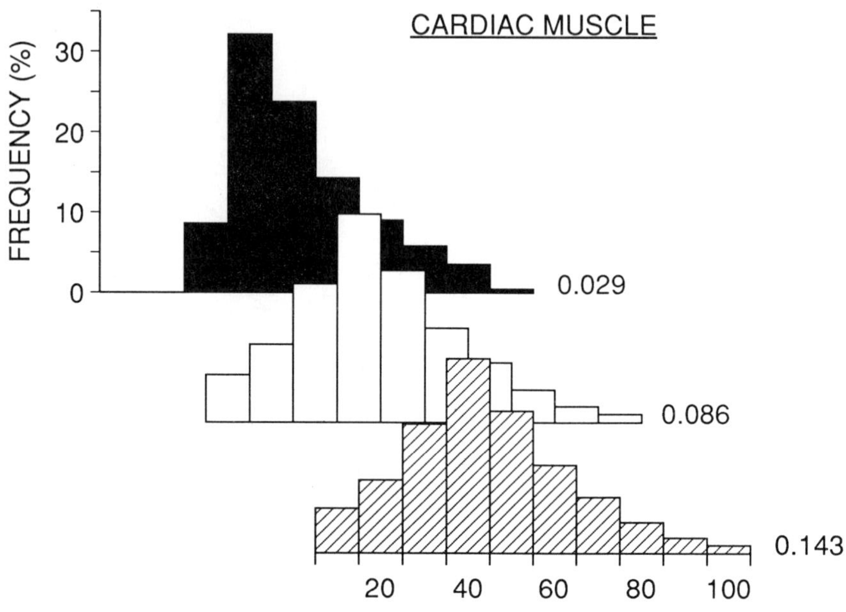
CARDIAC MUSCLE
FREQUENCY (%)
30
20
10
0
0.029
0.086
0.143
20
40
60
80
100

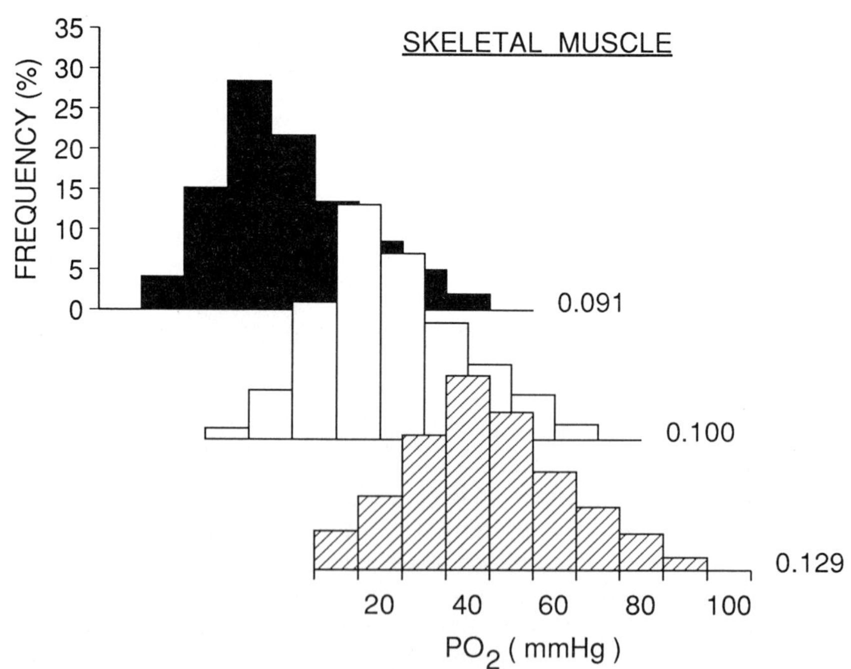
SKELETAL MUSCLE
FREQUENCY (%)
35
30
25
20
15
10
5
0
0.091
0.100
0.129
20
40
60
80
100
PO2 (mmHg)

refinements are possible to accommodate flow heterogeneity at the capillary level (see chapters by Hoofd and Groebe, this volume). However, no matter how well developed the theory of oxygen diffusion is, quantitative results from modelling studies need to be compared with experimental observations. Differences may then reveal deficiencies in either approach, but a realistic anatomical basis for these models is essential before meaningful conclusions can be reached, particularly with respect to the heterogeneity of capillary spacing which is known to play a crucial role in determining tissue P_{O_2} under most circumstances (Rakušan *et al.*, 1984).

Unfortunately, this type of approach is likely to reflect *in situ* levels of oxygenation only in relatively homogeneous tissue, where all capillaries are surrounded by a region of uniform oxygen demand and, at its simplest, each capillary has the same capacity to supply oxygen. These conditions are most likely to be found in highly oxidative tissue, such as the vertebrate myocardium. In an attempt to establish the capillary domain as being not only an objective, but a realistic division of tissue Hoofd *et al.* (1989) calculated the oxygen flux from myocardial capillaries of equal (high) flow (see also Groebe, this volume). In the Multicapillary model the starting point is again a realistic estimate of spatial capillary distribution but, rather than using Krogh cylinders as a basis for the calculations, uses an extension of the Krogh–Erlang formula itself (see Hoofd, this volume). It can be readily seen that the lines of zero flux approximately coincide with the domain boundaries, suggesting that indeed this division of space may have a physiological correlate (Fig. 6*A*). This result is to be expected as, recalling that morphometric (structural) data reflect only the maximal capacity of a system, the influence of neighbouring capillaries on diffusion gradients will cancel midway between them if they have similar transport capacity, and the level of demand of the intervening tissue is also similar. In most striated muscle, however, there is evidence for a considerable degree of temporal intermittency in perfusion of the capillary bed (Renkin *et al.*, 1981*b*; Vetterlein *et al.*, 1982), which would tend

---

Fig. 5. Calculated P_{O_2} distributions using a modified Krogh analysis, with actual domain areas used to examine the effect of varying capillary spacing, and hence *R*k. The values at the edge of the ordinate refer to the logSD; as heterogeneity of spacing decreases (smaller logSD), P_{O_2} distribution becomes narrower and mean P_{O_2} increases. The model incorporates identical capillary perfusion per unit area of domain, the effects of myoglobin-facilitated O_2 diffusion and Michaelis–Menten kinetics for tissue O_2 consumption. Examples taken from rat myocardium (Turek *et al.*, 1991) and rat tibialis anterior (Turek *et al.*, 1989).

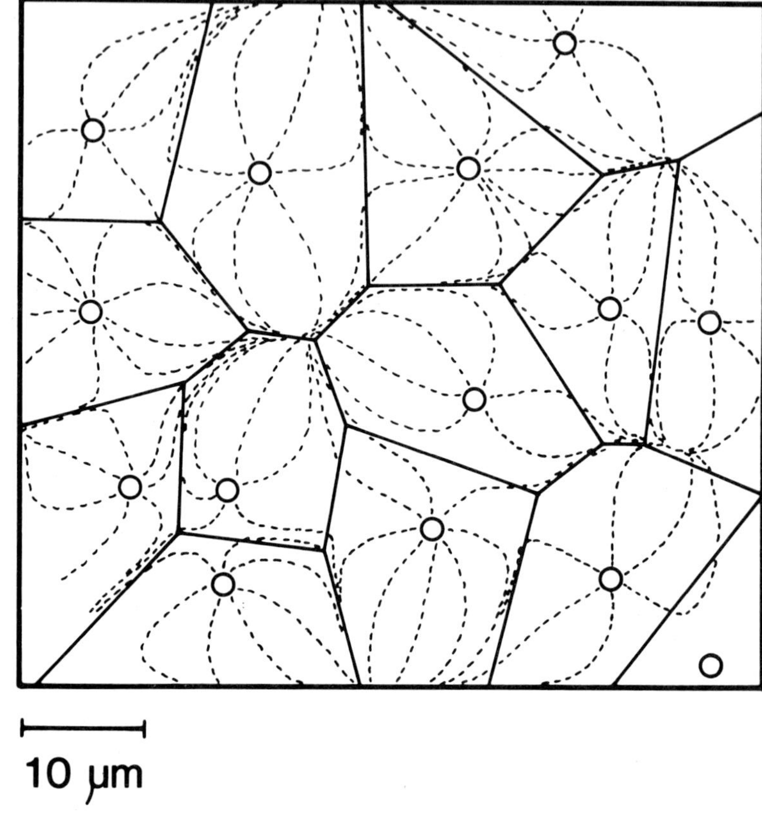

Fig. 6. Structural and mathematical modelling of capillary O_2 supply. Here capillaries (open circles) are assumed to have identical transport capacity, and supply tissue with homogeneous O_2 consumption. The calculated O_2 flux lines (broken lines) coalesce at the point of minimal P_{O_2}, which closely matches the domain boundaries (solid lines) for rat myocardium (*A*), although the fit is not so good for skeletal muscle (EDL; *B*). For estimation of flux lines see Hoofd (this volume). Original data from Egginton *et al.* (1988) and Hoofd *et al.* (1990).

to offset the balance point. Similarly, in skeletal muscle the potential consumption of oxygen is not uniform across the section, and again this mismatch in local supply and demand may be expected to produce a discrepancy between the physiology and anatomy (Fig. 6*B*). We therefore need to modify our basic analysis of the anatomical capillary supply to account for these other variations in tissue composition.

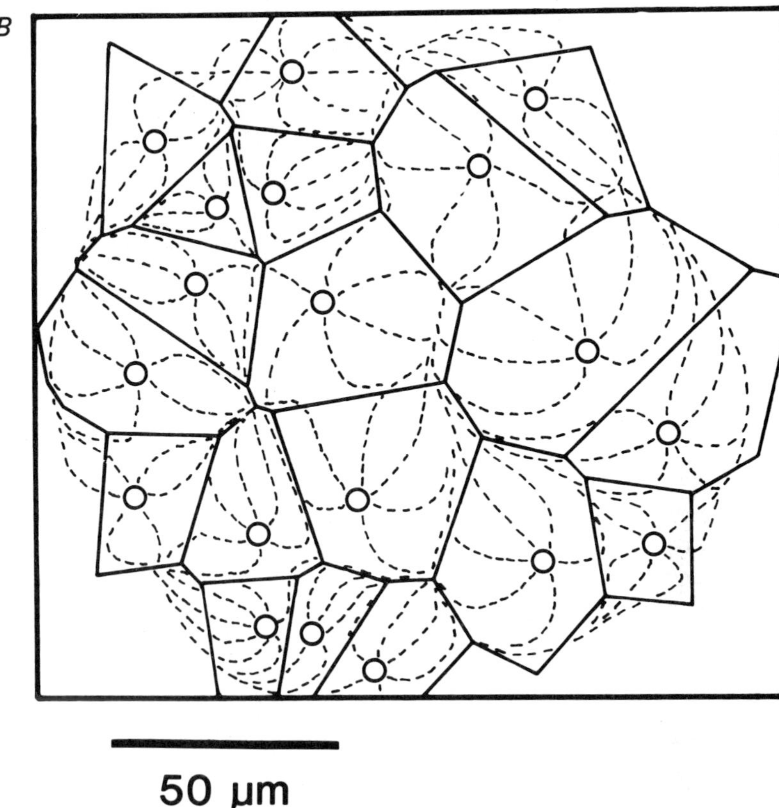

Planar analysis of the capillary–fibre relationship

Effect of fibre size

The most obvious heterogeneity apart from capillary spacing seen *in vivo* is that of cell size. From histological data it is evident that capillaries are found predominantly, though by no means exclusively, in the interstices formed by the close packing of the mainly cylindrical muscle fibres. Capillary location is thus heavily influenced by the size and shape of the intervening cells. As a first approximation, therefore, the underlying growth process which results in a given spread of fibre size will also determine the subsequent pattern of capillary distribution, as the network of interstices will offer a pathway of least resistance for growth of new vessels during development or adaptation to increased oxidative demand, such as endurance training. This is evident in the preferential orientation of new capillaries along the axis of muscle fibres in mouse EDL following experimental hypertrophy (James, 1981).

In an attempt to incorporate these, and other, heterogeneous features of biological tissue into a coherent analysis of tissue capillary supply, new indices of capillary supply related to individual fibres have been developed (Hoofd *et al.*, 1985; Egginton & Ross, 1989*b*). Many advantages accrue from such a local-based approach owing to the additional information available from distribution analyses of domain area and the derived non-integer-based indices (Egginton & Turek, 1990). As well as providing a morphological framework within which to develop mathematical models of oxygen diffusion, the flexibility offered by planar analyses permits the interaction of capillaries and surrounding cells to be examined; primarily this quantifies the overlap between domains and cell profiles. Four main indices can be calculated:

(a) *Fibre:domain ratio* (FDR). This is the number of whole or partial muscle fibre cross-sectional profiles that overlap (intersect) each domain, and represents the number of muscle fibres in the vicinity of (probably supplied by) that capillary. Unlike the integer-based Sharing Factor (SF, number of fibres adjacent to a capillary: Plyley & Groom, 1975) this accounts for non-contiguous capillaries. In most mammalian muscle FDR approximates SF, but will be greater for sparse capillary networks such as in fish skeletal muscle (Egginton *et al.*, 1988).

(b) *Domain:fibre ratio* (DFR). This is the number of whole or partial domains overlapping each muscle fibre profile. This only equates with the number of capillaries around a fibre (CAF: Plyley & Groom, 1975) where CD is high, such as in mammalian slow skeletal or cardiac muscle, and again will be greater (often considerably so) where the capillary supply is sparse or very heterogeneous (Egginton *et al.*, 1988).

These indices, based on area rather than number distribution, and using natural boundaries give a different, but perhaps clearer picture of the anatomical relationship between capillaries and fibres than the more traditional indices. FDR may be considered more appropriate if one considers the capacity for oxygen transport to be limited by eccentric diffusion from capillaries to the surrounding fibres (Krogh, 1919); DFR in the case of concentric diffusion of oxygen from a more homogeneous interstitial level surrounding the fibre periphery (Hill, 1928). In all cases the distribution of the area-based index is skewed to the right and has a greater spread than the numerical index (Egginton & Turek, 1990), being therefore more informative and at least potentially more sensitive to adaptive changes in capillary supply. Greater sensitivity in assessing the role of capillary growth is certainly required, given that many adaptations which are purported to induce angiogenesis only result in a modest change in capillary supply, often of a similar magnitude to methodological resolution (Hudlicka *et al.*, 1992).

(c) *Local capillary:fibre ratio* (LCFR). As a domain usually overlaps more than one fibre, and hence the capillary influence is distributed, the sum of the fractional domain areas overlapping each fibre profile may be calculated. This non-integer index is quite different from the global (overall) capillary:fibre ratio (C:F) which simply reflects the total number of vessels and cells included in the sample. The distribution of LCFR against individual fibre area clearly shows that some fibres interact with only a fraction of a complete zone of influence of any capillary, while for most other fibres the relative supply is calculated to be one or more capillaries' worth. A fibre will, of course, interact with fractions of the supply areas from a number of different capillaries as a domain will always overlap (is in contact with) two or more fibres, as indicated by the FDR. The mean cumulative fraction of domains overlapping a fibre, the LCFR, is approximated by the ratio of mean fibre and domain areas, and represents the transport capacity by diffusion of oxygen and other fuels in terms of 'capillary equivalents' of supply. The method of constructing this index is shown in Fig. 7. For each fibre the contribution from each capillary is computed as the ratio of the domain area which intersects that fibre to the sum of the intersections of the domain with all fibres. The procedure is laborious and time-consuming, both in the digitisation of the sections and the computation of the index. Examination of the distribution of the values of the index suggests that it, too, is described better by a log–normal distribution than by a normal distribution. In view of the method of construction it is hardly surprising that there is quite a strong correlation between LCFR and fibre area. That is shown in Fig. 8*B* in which the data come from rat soleus muscle. Figure 8*A* shows the corresponding picture for a simpler integer-based index, the number of capillaries within 10 μm of a fibre boundary. The rather coarse distribution of points on the vertical scale makes this a less sensitive index of capillary supply.

(d) *Local capillary density* (LCD). The strong correlation between LCFR and fibre size has been noted above. In an attempt to find a size-independent index of capillarisation the effect of cell size can be removed (normalised) by dividing LCFR for each fibre by the fibre area, to give a specific local capillary density. In a few cases we have found that LCD does appear to be independent of fibre size, but in the majority of cases LCD is a decreasing function of fibre size.

The problem of discovering a general relationship between LCFR and fibre area which is invariant across a range of muscle fibre types remains

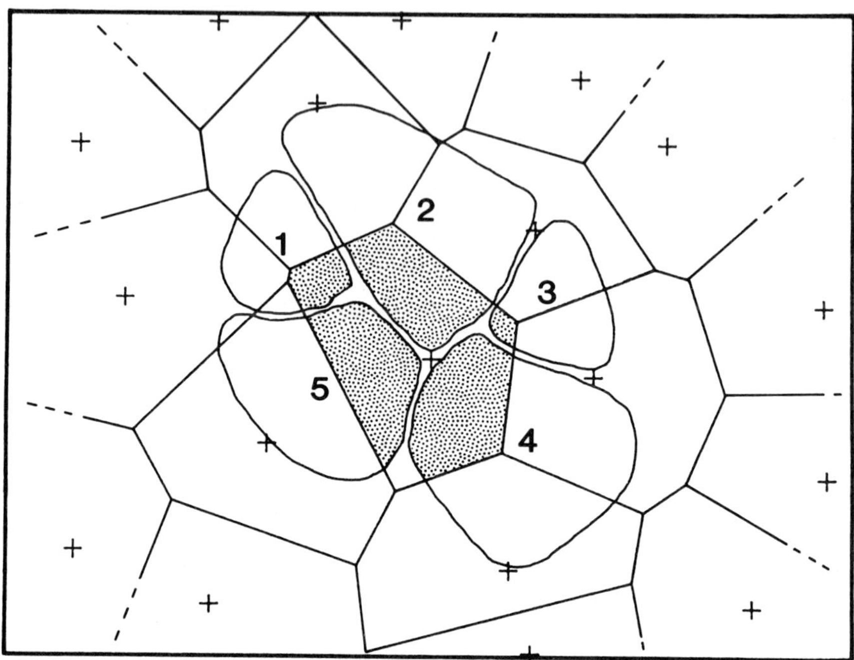

Fig. 7. Interaction of individual capillaries and fibres, illustrating the method of computing the local capillary-to-fibre ratio. The central domain (stipple) overlaps five fibres in proportion to its proximity, and its transport capacity will then be divided among the adjacent fibres in proportion to the degree of overlap if all fibres have equal metabolic demand. Summing the fractions of domains overlapping individual fibres will provide an index of potential local capillary supply (LCFR).

unsolved. Figure 9*A* shows the relationship for a sample of three different fibre types from the same tissue section of rat diaphragm. The single best-fit line shown gives a plausible fit to the whole data set, but closer examination suggests a more complex picture. Fitting separate lines to the individual fibre types suggests that the slopes for the glycolytic fibres may be less than for the two oxidative fibre types, though the difference for this data set is barely significant and the possibility remains that some other function such as a fractional power law could provide a single relationship which could span all fibre types. A further feature shown in Fig. 9*A* is that the residual variability about the line increases with fibre area. It was noted earlier that the fibre areas themselves also tend to have a log–normal distribution. Consequently, a double-log transformation was applied to the data and the analysis repeated. The result is shown in

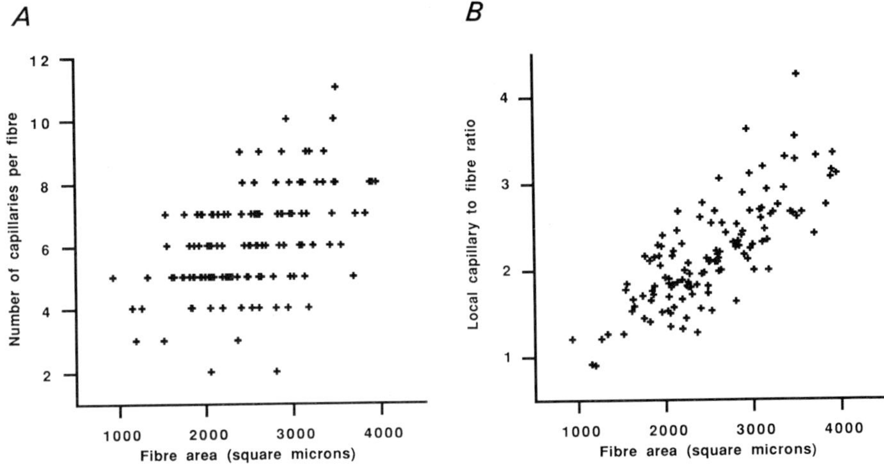

Fig. 8. Two indices of capillary supply to individual fibres plotted against fibre area. *A* shows the number of capillaries per fibre (within 10 μm of the fibre boundary); *B*, the local capillary-to-fibre ratio computed by the method described in the text. The former is an integer-based index which gives a rather coarse distribution. Data from rat soleus muscle, 12 μm frozen section.

Fig. 9*B*. The residual variance has now been made more uniform and any differences between the slopes of lines of the sub-samples are less apparent. More detailed analysis shows that there is no difference between the slopes of the three subsets which are also not significantly different from unity, while the slope of the line fitted to the total data set is significantly different from 1. Taking all those findings together suggested that within a fibre type LCFR increased linearly with fibre area, and therefore LCD would provide a size-independent measure of capillary supply. The mean values of LCD were less for the glycolytic than for the oxidative fibres, which did not differ from one another, and all had slopes which were not significantly different from zero (Egginton & Ross, 1989*c*). However, the application of similar analyses to a number of other data sets (EDL, TA, soleus) showed that those patterns were not repeated in all cases. A considerable range of slopes on the log–log plot has been found, ranging from around 1 down to 0.3 and, so far, no simple pattern can be discerned.

In the case of the thin sheet of muscles which forms the diaphragm, a continuously active and therefore highly oxidative tissue, normalising for fibre area provides an invariant estimate of local capillary supply in the LCD. This scale-independent form of capillarisation may reflect the

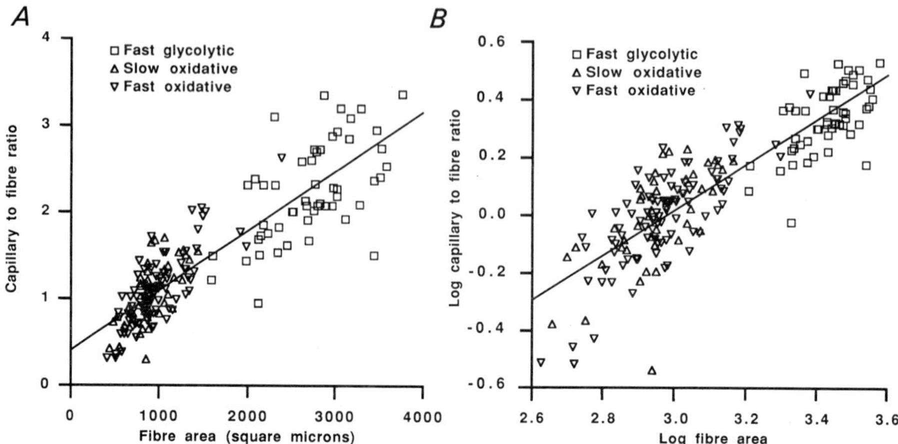

Fig. 9. The relationship between local capillary-to-fibre ratio and fibre area in a heterogeneous muscle (rat diaphragm): *A*, linear plot; *B*, double-log plot. For further details see text.

rather specialised structure, as it suggests that diffusional supply is regulated on an intramuscular (extracellular) level but is independent of cell size, despite the increase in intracellular diffusion distances (normally thought to offer increased resistance to metabolite flux) which occurs during normal growth. Indeed, one might expect the supply network to vary with some linear dimension of the region of consumption, which is the reason for estimating barrier thickness of other laminar structures (see chapters by Mayhew and Perry, this volume). In the case of muscle, therefore, an invariance of local capillary supply with the square root of fibre area might be easier to explain. When the more normal, bulk muscles are examined, where the architecture of the microcirculation may be quite different, one finds something intermediate between these distributions (Table 2). This might be in part a reflection that in these two muscles, chosen to represent the extremes of oxidative capacity in mammalian limb muscles, not only does the anatomy differ from that of the diaphragm, but the composition is also different in that they each have only two metabolic fibre types competing for resources (Type I and IIa in soleus, Type IIa and IIb in the glycolytic region of tibialis anterior). However, the variation in capillary supply cannot be discussed in terms of fibre size alone, as there tends to be an inverse relationship between mitochondrial content (and hence oxygen consumption) and fibre cross-sectional area.

Table 2. *Slopes of the relationship between LCFR and fibre area, for different muscles and fibre types*

Muscle	Fibre type	log–log slope	S.E. of slope	*N*
Tibialis anterior	FG	0.729	0.086	160
Tibialis anterior	FO	0.364	0.334	24
Diaphragm	FG	0.845	0.154	53
Diaphragm	SO	0.956	0.179	42
Diaphragm	FO	1.005	0.082	94
Soleus	SO	0.850	0.062	78
Soleus	FO	0.650	0.142	52

Partitioning of oxygen supply

Perhaps the greatest advantage of an analysis in 2 dimensions is the potential to investigate the interaction or causal relationship between heterogeneity in capillary spacing and other forms of heterogeneity within the tissue. A clear example of this is the need to determine the influence of muscle phenotype on local capillary supply (Plyley & Groom, 1975; Gray & Renkin, 1978). Partitioning of oxygen supply among the (up to four) distinct metabolic fibre types surrounding a capillary in a typical mixed mammalian muscle is clearly beyond the descriptive power of simple counts. While one may infer the relative extent of capillarisation to different fibre types from the variation among muscles, this can only be in rather general terms if information about intramuscular heterogeneity is missing. It is unfortunate that variations in such indices are often assumed to directly reflect oxygen demand, attaching undue importance to such data. The conceptual limitations inherent in these indices are that they ignore the influence of adjacent fibre types, fibre area and proximity of neighbouring capillaries. Perhaps the best attempt at partitioning capillary supply on the basis of muscle composition was that of Gray & Renkin (1978), which attempted to address many of these considerations. This study again analysed individual capillaries, essentially using the sharing factor (SF) to weight supply according to the relative number of a given fibre type around each capillary. The mean fraction of SF then produces a specific capillary : fibre ratio (SCF) which, when normalised for mean fibre area, gives a specific capillary density (SCD) for each fibre type. Although this was a significant improvement over other methods based on capillary counts, it mixes local and global descriptors and is thus unable to scale the microvascular supply according to local variations in ICDs or oxygen consumption. In addition, this is an integer-based index producing a rather coarse distribution of values, with limited descriptive power.

If an area of section is covered by a tessellation of domains and, for each fibre, the contribution of each capillary whose domain intersects the fibre is computed as the proportion of the domain area which overlaps that fibre, the sum of all such contributions from overlapping domains can be taken to represent the effective number of capillaries contributing to that fibre, producing the local capillary: fibre ratio (LCFR). This interaction of individual capillary domains and fibres is really the areal equivalent of Gray & Renkin's numerical fraction of the SF. This continuous distribution of 'capillary equivalents' enables the geometry of the capillary bed to be analysed with respect to both fibre area (above) and, more importantly, oxidative capacity (supply vs demand). LCFR may be normalised by dividing by fibre area to give an index, LCD, which is less dependent on fibre size and may therefore help identify the differential capillary supply to fibres within a mixed muscle. We feel this approach provides the best anatomical data for modelling the functional capacity of the microcirculation in mixed striated muscles to date (Egginton & Ross, 1989*b*).

Other sources of underlying heterogeneity

The concept of capillary domains is proving to be a useful anatomical basis for application of diffusion theory, and has confirmed that the major determinant of muscle capillarity is indeed fibre area, with this basic relationship modified by the oxidative capacity of the individual fibre (Egginton & Ross, 1989*b*). Planar analysis of capillary distribution may also be useful in describing other functional differences (Egginton, 1990*a*). For example, morphometric indices describing the anatomical interaction between capillaries and muscle fibres, whether local or global in resolution, represent a limiting factor only during the relatively infrequent bouts of maximal activity. Domain analysis may provide the geometric requirements for modelling of oxygen transport during the more usual physiological state if used in conjunction with histological sections reflecting temporal differences in capillary perfusion (Egginton & Ross, 1989*a*). In addition, domain area may be weighted in proportion to the relative or absolute level of perfusion of individual vessels, or the complementary differential oxygen demand of the enclosed tissue. Such a scheme may produce domains with straight, convex or concave edges (Boots, 1986), but may go some way to answer Krogh's original plea for the influence of the surrounding tissue to be incorporated into models of tissue oxygenation.

The true limit to distribution of capillaries in any cross-section of tissue may lie in the combined effect of fibre size and radial oxygen demand around individual capillaries. The difficulty, of course, lies in how to scale for differential supply or demand. Given that we can now partition individual capillary O_2 supply among surrounding fibres, it ought to be possible to assess how well this matches tissue O_2 demand. We can classify individual fibres on the basis of their histochemical characteristics, and hence metabolic capacity, and obtain estimates of V_{O_2}max from homogeneous preparations. However, within a given motor unit there is significant heterogeneity in levels of oxidative enzyme activity, suggesting that routine fibre typing may be an inadequate basis to assess the specificity of local capillary supply.

Relative mitochondrial content of domains

An alternate approach is to utilise a structural correlate of oxidative metabolism. In mammals, the specific V_{O_2}max of mitochondria is relatively constant, 2–3 ml $O_2\,g^{-1}$ wet wt h^{-1}, irrespective of the overall aerobic capacity of the muscle from which they were derived (Hoppeler & Lindstedt, 1985). If capillary supply is indeed matched at a local level with maximal tissue O_2 demand, then the mitochondrial density of domains ought to be inversely related to domain areas: small domains (usually overlapping oxidative fibres) having high Vv(mit,f), large domains (primarily overlapping glycolytic fibres) having low Vv(mit,f), with intermediate values found for those domains overlapping a mixture of fibre types. To balance supply and demand, the product of mean relative mitochondrial content and domain area ought to be invariant:

$$V\text{v(mit,f)} \times a(\text{dom}) = V(\text{mit,dom}) = \text{a constant}$$

Such autoregulation of capillary spacing and V_{O_2}max of the intervening tissue is implicit in many studies using 1-dimensional analyses. Examining a rat hind limb muscle of modest oxidative capacity (stimulated EHP), providing relatively small domains and high Vv(mit,f), provides the predicted relationship of V(mit,dom) vs a(dom) having a slope not significantly different from zero (Fig. 10). Some scatter of data points is to be expected, owing both to the heterogeneity of mitochondrial distribution within individual fibres, and the influence of capillary tortuosity on domain position. Nevertheless, it does seem that in this case a reasonable balance exists between O_2 supply and demand.

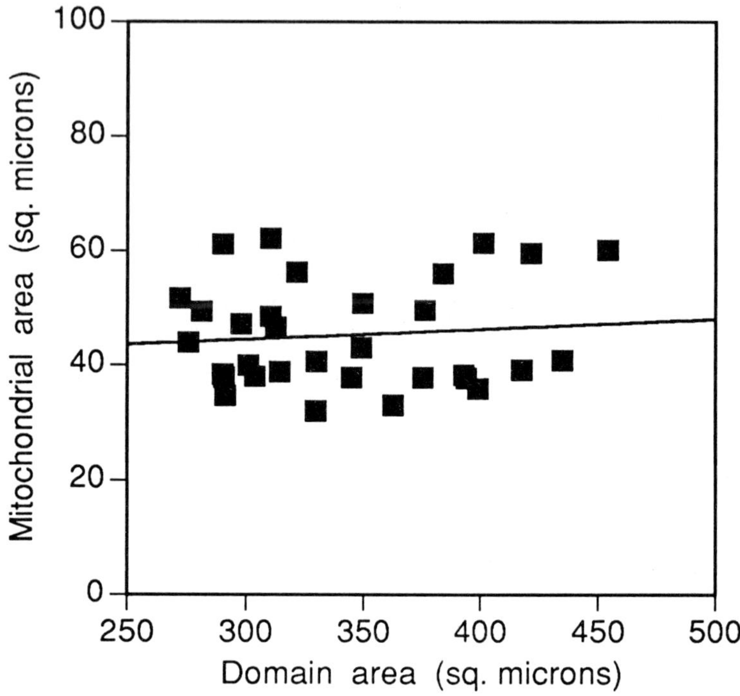

Fig. 10. Matching of O_2 supply and demand. The product of mitochondrial volume density, *V*v(mit,f), for the enclosed muscle fibres and domain area, *a*(dom), provides an estimate of the actual mitochondrial content (i.e. oxygen demand) of each domain. This is seen not to vary in any consistent manner with respect to *a*(dom). The slope of the regression line through all data points is not significantly different from zero. Ultrathin sections, rat extensor hallucis proprius.

Conclusions

In this chapter we have indicated a need for quantitative analyses which adequately describe the anatomical capillary bed, in order to derive estimates of the functional capillary supply. A basic goal for this activity is the provision of a realistic anatomical framework on which mathematical models of oxygen transport might be based. We have emphasised the use of capillary domains because they provide a means of computing improved measures of capillary spatial distribution by assigning a unique oxygen supply neighbourhood to each capillary. This allows the relation to other factors such as fibre size and metabolic type to be incorporated.

We have given some examples of the uses of those techniques in the estimation of P_{O_2} and mitochondrial distribution.

References

Boots, B.N. (1986). *Voronoi (Thiessen) Polygons.* Catmog 45, Geo-Books, Norwich, UK.

Egginton, S. (1990*a*). Morphometric analysis of tissue capillary supply. In *Advances in Comparative and Environmental Physiology*, Vol. 6, *Vertebrate Gas Exchange from Environment to Cell*, ed. R.G. Boutilier, pp. 73–141. Berlin: Springer-Verlag.

Egginton, S. (1990*b*). Numerical and areal density estimates of fibre type composition in a skeletal muscle (rat extensor digitorum longus). *Journal of Anatomy* **168**, 73–80.

Egginton, S. & Johnston, I.A. (1983). An estimate of capillary anisotropy and determination of surface and volume densities of capillaries in skeletal muscles of the conger eel (*Conger conger* L.). *Quarterly Journal of Experimental Physiology* **68**, 603–17.

Egginton, S. & Ross, H.F. (1989*a*). Quantifying capillary distribution in four dimensions. *Advances in Experimental Medicine and Biology* **247**, 271–80.

Egginton, S. & Ross, H.F. (1989*b*). Influence of muscle phenotype on local capillary supply. *Advances in Experimental Medicine and Biology* **247**, 281–91.

Egginton, S. & Ross, H.F. (1989*c*). Planar analysis of tissue capillary supply. *Journal of Physiology* **419**, 49P.

Egginton, S. & Turek, Z. (1990). Comparative distributions of numerical and areal indices of tissue capillarity. *Advances in Experimental Medicine and Biology* **277**, 161–9.

Egginton, S., Turek, Z. & Hoofd, L. (1988). Differing patterns of capillary distribution in fish and mammalian skeletal muscle. *Respiration Physiology* **74**, 383–96.

Gray, S.D. & Renkin, E.M. (1978). Microvascular supply in relation to fibre metabolic type in mixed skeletal muscles of rabbits. *Microvascular Research* **16**, 406–25.

Hill, A.V. (1928). The diffusion of oxygen and lactic acid through tissues. *Proceedings of the Royal Society of London, B* **104**, 39–96.

Hoofd, L., Olders, J. & Turek, Z. (1990). Oxygen pressures calculated in a tissue volume with parallel capillaries. *Advances in Experimental Medicine and Biology* **277**, 21–9.

Hoofd, L., Turek, Z., Kubat, K., Ringnalda, B.E.M. & Kazda, S. (1985). Variability of intercapillary distance estimated on histological sections of rat heart. *Advances in Experimental Medicine and Biology* **191**, 239–47.

Hoofd, L., Turek, Z. & Olders, J. (1989). Calculation of oxygen pressures and fluxes in a flat plane perpendicular to any capillary distribution. *Advances in Experimental Medicine and Biology* **248**, 187–96.

Hoppeler, H. (1984). Morphometry of skeletal muscle capillaries. *Progress in Applied Microcirculation* **5**, 33–43.

Hoppeler, H. & Lindstedt, S.L. (1985). Malleability of skeletal muscle tissue in overcoming limitations: structural elements. *Journal of Experimental Biology* **115**, 355–64.

Hudlicka, O., Brown, M. & Egginton, S. (1992). Angiogenesis in skeletal and cardiac muscle. *Physiological Reviews* **72**, 369–417.

Hudlicka, O., Egginton, S. & Brown, M.D. (1988). Capillary diffusion distances – their importance for cardiac and skeletal muscle performance. *News in Physiological Science* **3**, 134–8.

Hudlicka, O., Hoppeler, H. & Uhlmann, E. (1987). Relationship between the size of the capillary bed and oxidative capacity in various cat muscles. *Pflügers Archiv* **410**, 369–75.

James, N.T. (1981). A stereological analysis of capillaries in normal and hypertrophic muscle. *Journal of Morphology* **168**, 43–9.

Kayar, S.R., Archer, P.G., Lechner, A.J. & Banchero, N. (1982*a*). Evaluation of the concentric-circles method for estimating capillary–tissue diffusion distances. *Microvascular Research* **24**, 342–53.

Kayar, S.R., Archer, P.G., Lechner, A.J. & Banchero, N. (1982*b*). The closest-individual method in the analysis of the distribution of capillaries. *Microvascular Research* **24**, 326–41.

Krogh, A. (1919). The number and distribution of capillaries in muscles with calculations of the oxygen pressure head necessary for supplying the tissue. *Journal of Physiology (London)* **52**, 409–15.

Loats, J.T., Sillau, A.H. & Banchero, N. (1978). How to quantify skeletal muscle capillarity. *Advances in Experimental Medicine and Biology* **94**, 41–8.

Mathieu, O., Cruz-Orive, L.M., Hoppeler, H. & Weibel, E.R. (1983). Estimating length density and quantifying anisotropy in skeletal muscle capillaries. *Journal of Microscopy* **131**, 131–46.

Mathieu-Costello, O. (1987). Capillary tortuosity and degree of contraction and diffusion of skeletal muscle. *Microvascular Research* **33**, 98–117.

Mattfeldt, T. (1987). Estimation of length, surface and number of anisotropic objects: a review of models and design-based approaches. *Acta Stereologica* **6** (Suppl. III), 537–48.

Mermod, L., Hoppeler, H., Kayar, S.R., Straubz, R. & Weibel, E.R. (1988). Validity of fibre size, capillary density and capillary length related to horse muscle fixation. *Acta Anatomica* **133**, 89–95.

Plyley, D. & Groom, A.C. (1975). Geometrical distribution of capillaries in mammalian striated muscle. *American Journal of Physiology* **228**, 1376–83.

Rakušan, K., Hoofd, L. & Turek, Z. (1984). The effect of cell size and capillary spacing on myocardial oxygen supply. *Advances in Experimental Medicine and Biology* **169**, 463–75.

Renkin, E.M., Gray, S.D. & Dodd, L.R. (1981*a*). Filling of the microcirculation in skeletal muscles during timed India ink perfusion. *American Journal of Physiology* **241**, H174–86.

Renkin, E.M., Gray, S.D., Dodd, L.R. & Lia, B.D. (1981*b*). Heterogeneity of capillary distribution and capillary circulation in mammalian skeletal muscles. In *Underwater Physiology*, Vol. 7, ed. A.J. Bachrach & M.M. Matzemm, pp. 465–74. Bethesda, MD: Undersea Medical Society.

Romanul, F.C.A. (1965). Capillary supply and metabolism of muscle fibers. *Archives of Neurology* **12**, 497–509.

Turek, Z., Hoofd, L. & Rakušan, K. (1987). A comparison of the methods for assessment of the heterogeneity of myocardial capillary spacing. *Advances in Experimental Medicine and Biology* **215**, 13–19.

Turek, Z., Olders, J., Hoofd, L., Egginton, S., Kreuzer, F. & Rakušan, K. (1989). P_{O_2} histograms in various models of tissue oxygenation in skeletal muscle. *Advances in Experimental Medicine and Biology* **248**, 227–37.

Turek, Z. & Rakušan, K. (1981). Lognormal distribution of intercapillary distance in normal and hypertrophic rat heart as estimated by the method of concentric circles: its effect on tissue oxygenation. *Pflügers Archiv* **391**, 17–21.

Turek, Z., Rakušan, K., Olders, J., Hoofd, L. & Kreuzer, F. (1991). Computed myocardial P_{O_2} histograms: effects of various geometrical and functional conditions. *Journal of Applied Biology* **70**, 1845–53.

Vetterlein, F., Dal Ri, H. & Schmidt, G. (1982). Capillary density in rat myocardium during timed plasma staining. *American Journal of Physiology* **242**, H133–41.

L. HOOFD

Updating the Krogh model – assumptions and extensions

Introduction

The final stage in delivery of oxygen is transport from the capillaries to sites in the tissue where the oxygen is consumed, i.e. primarily the mitochondria. It has been a long and difficult task to develop theoretical and experimental approaches because of the fine structure of the capillary network. Gas transport by diffusion is only effective over distances in the micrometre range, and hence the capillary network is densely and often whimsically spread between the supplying arterioles and draining venules. The best-known model of tissue oxygenation was developed for tissue with a special, almost parallel arrangement of capillaries, i.e. skeletal muscle. It was the Danish physiologist Krogh (1919) who devised the model of a cylindrical region of tissue supplied with oxygen by one, centrally located, capillary. It is depicted in Fig. 1*A*, with the different variables that can be discerned. His friend, the mathematician Erlang, developed a formula allowing calculation of the oxygen pressure p in any plane perpendicular to the capillary (e.g. the dotted circle in Fig. 1*A*), under the assumption that there is no exchange between the planes. In the notation of Fig. 1, it can be written as:

$$p = p_c - MR^2/(4\mathcal{P})\{\ln(r^2/r_c^2) - (r^2 - r_c^2)/R^2\} \qquad (1)$$

where p_c is the capillary O_2 pressure; for other symbols see the figure legend. The formula is derived by expressing the oxygen flux vector $\vec{J}$ from the p field:

$$\vec{J} = -\mathcal{P}\vec{\nabla}p \qquad (2)$$

and applying mass conservation, the mass balance for diffusional O_2 transport and O_2 consumption:

$$\vec{\nabla}\cdot\vec{J} + M = 0 \qquad (3)$$

where the dot denotes the inner product of both vectors, the gradient

Society for Experimental Biology Seminar Series 51: *Oxygen Transport in Biological Systems*, ed. S. Egginton & H.F. Ross.

operator $\vec{\nabla}$ and the O_2 flux $\vec{J}$. Here, these two equations can be combined to yield:

$$\mathscr{P}\nabla_r^2 p = M;\ \nabla_r^2 p = (1/r)\mathrm{d}(r\,\mathrm{d}p/\mathrm{d}r)/\mathrm{d}r \tag{4}$$

since the gradient operator $\vec{\nabla}$ in this 2-dimensional radial symmetry is only its radial component, ∇_r – and consequently the flux vector $\vec{J}$ is J, equal to J_r, also its radial component. The Krogh–Erlang equation, Eqn 1, is the solution for the boundary conditions $p = p_c$ at $r = r_c$ and $J = 0$ at $r = R$ (no flux through the cylinder border).

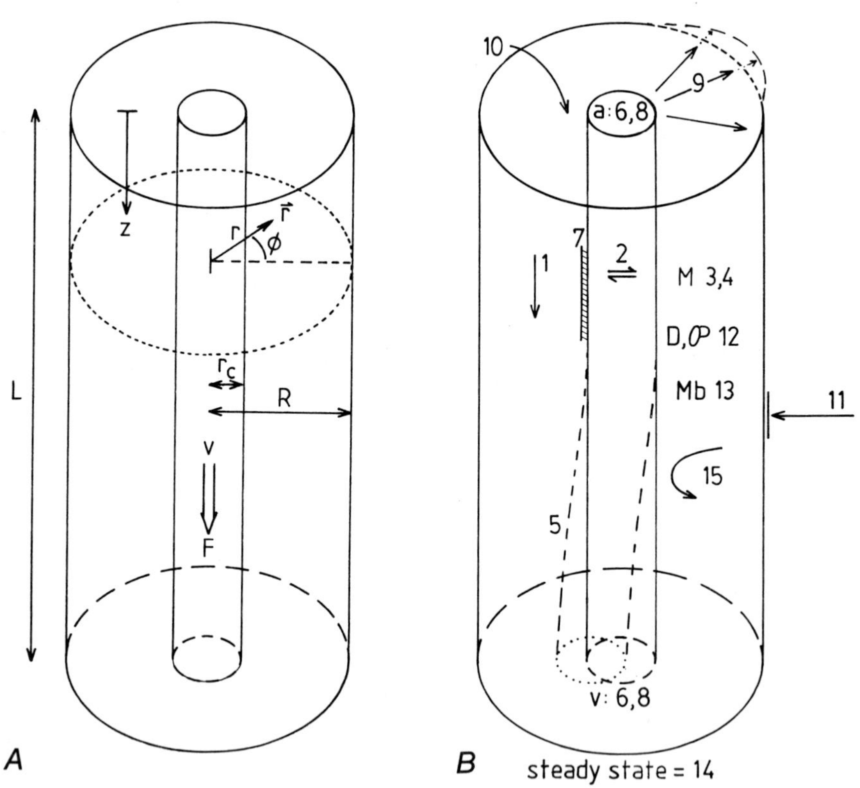

Fig. 1. Outline of the Krogh model (A) and schematic indication of where the assumptions apply (B). A: $\vec{r} = (r, \phi, z)$, cylindrical coordinates; r_c, capillary radius; R, cylinder radius; L, cylinder length; F, v, blood flow and velocity, respectively. B: a, arteriolar, v, venular end; M, oxygen consumption; D, $\mathscr{P}$, oxygen diffusion coefficient and permeability, respectively; Mb, myoglobin. The numbers refer to the enumeration by Kreuzer (1982).

Equation 1 is a rather unusual form of the Krogh–Erlang equation; written because the term $MR^2/4\mathscr{P}$ is of unit pressure. It gives an order of magnitude for the pressure drops involved. For example, at the border of a Krogh cylinder of radius five times the capillary radius, i.e. $r = R = 5r_c$, the term between braces equals 2.26, so the pressure drop from capillary to border is 2.26 $(MR^2/4\mathscr{P})$; half-way, at $r = 3r_c$, it is 1.88.

For the calculation of p at location (r,z), the capillary oxygen pressure p_c must be given for the plane at axial distance z. The model was expanded by Kety (1957) who realised that $p_c(z)$ could be calculated from $p_c(0)$, the pressure at which the blood enters the capillary, and F, the capillary blood flow. In each slab, the capillary loses as much oxygen as is consumed in that plane, leading to:

$$\mathrm{d}c_{O_2}/\mathrm{d}z = -(M/F)\pi(R^2 - r_c^2) \quad (5)$$

where c_{O_2} is the *total* oxygen concentration of the capillary blood, i.e. both physically dissolved and bound to haemoglobin:

$$c_{O_2} = \alpha p + c_{Hm}S \quad (6)$$

where the physically dissolved oxygen is proportional to p according to Henry's law, the O_2 solubility α being the proportionality constant, and c_{Hm} is the concentration of haem groups, being the O_2-binding elements in the haemoglobin. Virtually all the O_2 is present in this bound form. The oxygen **saturation** S is the fraction of Hb loaded with O_2 and depends on p. Integrating Eqn 5 from 0 up to z and applying Eqn 6 for the conversion between c_{O_2} and p yields capillary oxygen pressure at axial distance z, i.e. $p_c(z)$.

Equation 5 is derived from equating the oxygen loss in the flowing capillary blood, $-F\Delta c_{O_2}$, to the amount consumed in the corresponding tissue slab, $M\pi(R^2 - r_c^2)\Delta z$. Again, note the combined parameter, M/F, indicating that the axial drop in p_c is determined by this quotient rather than by the individual parameters of consumption and flow. An even better combined parameter is $M/(Fc_{Hm})$, determining the capillary gradient in oxygen saturation S because of Eqn 6 when the term αp is unimportant. This parameter combination implies that, for instance, an increase in oxygen consumption M or a decrease in blood oxygen-binding capacity c_{Hm} can be met by a corresponding increase in blood flow F.

This combined Krogh model has been the basis for many estimations of the (in)adequacy of tissue oxygen supply. However, one has to realise that it is a very simplified model. Consequently, it has been the object of many studies investigating its limitations (see also Groebe, this volume). A valuable overview was given by Kreuzer (1982) who listed 15 simplifying assumptions as:

1. There is only radial but no longitudinal or axial diffusion.
2. The intracapillary chemical reactions in the blood are neglected, i.e., the oxygen concentration is the same over the capillary cross section, and chemical equilibrium is assumed between oxygen and haemoglobin.
3. The oxygen consumption in the tissue does not depend on the local oxygen pressure (zero-order reaction).
4. The cells may be represented as an homogeneous volume distribution of minute sinks of oxygen independent of time and position.
5. The capillaries are straight, run parallel, have a unidirectional blood flow, and are homogeneously distributed.
6. Capillary radius and length are constant, implying, with constant blood flow, a constant transit time.
7. The capillary wall does not present any resistance to oxygen diffusion.
8. The capillary blood flow is constant. With unchanging oxygen consumption this implies the same venous oxygen concentration in all capillaries (Fick principle).
9. The flow of oxygen from the capillary is cylindrically symmetric.
10. The oxygen exchange occurs only in the capillary, not in arterioles and venules.
11. The oxygen does not diffuse out of the tissue cylinder.
12. The diffusion coefficient is the same throughout the tissue.
13. There is no facilitated diffusion of oxygen, e.g. by myoglobin in muscle.
14. The whole configuration is independent of time (steady state).
15. The transfer is by diffusion only (no stirring).

These points are schematically visualised in Fig. 1*B*. The author adds:

> This is an impressive list of assumptions, which in turn raises the question as to how realistic and meaningful the calculations according to the Krogh model might be.

Consequently, a number of studies have investigated these assumptions by modifying the Krogh layout. Amazingly, most of them show little influence on the calculated oxygen pressure distributions (e.g. Fletcher, 1978; Popel, 1989). As we will see further on, this is *not* equivalent to saying that the Krogh model does hold, for muscle tissue.

In latter years, research has explored how the Krogh treatment of muscle tissue oxygenation could be extended in order to yield a more complete picture. Some of these extensions can be incorporated in the Krogh model itself, others cannot.

Capillary oxygen pressure

This concerns assumption 2 in Kreuzer's list and in Fig. 1*B*. In fact, the Krogh–Erlang equation, Eqn 1, can be formally taken to be independent of intracapillary oxygen since it is an exact solution of 2-dimensional diffusion into and within a homogeneous circular area, matched for a certain p, here p_c, at a certain location, here r_c. If this p_c is taken to be the oxygen pressure just outside the capillary, and r_c this capillary rim, the solution is valid no matter what happens inside the capillary. However, the oxygen source and the basis for oxygen supply is the erythrocyte, loaded with oxygen in the lung and arriving at the capillary bed with a certain value of p, and the modelling objective, of course, is to relate p to that 'delivery' pressure. When unloading begins, gradients arise to drive the transport of O_2 from this source to the capillary rim. Such intracapillary transport involves two basic mechanisms:

(a) O_2 release from the haemoglobin;
(b) O_2 transport, in erythrocyte, plasma and capillary wall.

The whole problem is extremely difficult, mainly because from a modelling standpoint the circumstances are insufficiently known. Following the subdivision above:

(a) The reaction of haemoglobin with oxygen has been extensively studied. The Hb molecule can reversibly bind 4 molecules of O_2 (not atoms!), with binding rates depending on the O_2 concentration, and influenced by many other molecular and ionic species present in the blood. There is massive literature about this and other aspects of haemoglobin (see Bellelli & di Prisco, this volume); however, at present, modelling the mere O_2 release from Hb in the erythrocyte is still a very complicated problem. As an approximation, however, resolving only part of the problems, it can be assumed that the haemoglobin–oxygen reaction is in chemical equilibrium (Bouwer, 1987) and that the relationship between saturation S and partial pressure p is given by its **saturation curve**.
(b) Capillary transport involves both convection and diffusion. The erythrocytes, with their typical biconcave form in human blood in the larger vessels, are squeezed through the capillaries, wherein they will not fit in their undistorted form. The squeezing process must cause mixing of the erythrocyte content but, once in the capillaries, form and orientation of the erythrocytes could be stationary. This leaves the question open whether there is intraerythrocytic convection or not. On the other hand, plasma convective mixing is conceivable and even probable; the erythrocytes can be observed flowing through the microcirculation as a train (Duling *et al.*, 1983), pushing the plasma in between while at the same time it is dragged by the capillary wall.

In addition to convection, there is diffusion of all molecular species, each with its own diffusion coefficient, for each respective phase, while the erythrocyte and plasma concentrations are linked through exchange processes in the erythrocytic membrane. The latter is by no means a trivial barrier; it very effectively can separate some species, like K^+ and Na^+, let pass others, like HCO_3^-, and maintain a pH- and an electric potential gradient between inside and outside. Although there seems to be no impeding influence for the O_2 molecule itself (Kreuzer & Yahr, 1960; Kreuzer & Hoofd, 1987) it will influence O_2 exchange and thus O_2 transport.

Finally, there is diffusional transport in the capillary wall. Although quite thin, of the order of a micrometre, all the O_2 has to pass through this wall and consequently the local oxygen flux is high (some literature emphasises this by use of the term 'high flux density' but this is misleading since flux is correctly defined as molecular transport rate per unit area, and consequently is a density as such. A better expression might be 'high density of flux lines' – see also Fig. 4 below). For instance, according to the Krogh–Erlang formula, Eqn 1, in the former example of $R = 5r_c$, the term between braces when taken at $r = 1.4r_c$ amounts to 0.63, and taken at $r = 1.6r_c$, 0.88. This implies that $0.63/2.26 = 28\%$ of the total pressure drop occurs within the initial 10% of the diffusion distance and as much as $0.88/2.26 = 39\%$ in the first 15%, i.e. just where the capillary membrane is. The problem of this initial oxygen drop might be especially awkward if O_2 diffusion is hampered by the capillary wall, as sometimes suggested in the literature. In view of the oxygen delivery task of the capillary, however, this would seem a paradoxical dysfunctionality.

A simplified model

Most models of capillary O_2 release take the erythrocytes as Hb-containing cylinders fitting in within a straight and regular cylindrical capillary. The erythrocytes are moving in an array through the capillary with plasma filling the gaps in between, and possibly a tiny layer around, between the erythrocyte cylinder and the capillary wall; see Fig. 2*A*. A simplified model is presented here in order to investigate the possible influence of intracapillary oxygen release. Plasma convection is left out of consideration and the erythrocytes are treated as homogeneous oxygen sources. This is depicted in Fig. 2*B*: the plasma does not deliver oxygen, but once an erythrocyte passes by it releases oxygen at a constant rate, μ. Such release is the mathematical opposite of the tissue consumption M, so that, for the capillary, an equation holds similar to Eqn 3:

$$\vec{\nabla}\cdot\vec{J} - \mu(z) = 0;\ \mu(z) = \begin{cases} \mu & \text{for } |z| < \tfrac{1}{2}h_e \\ 0 & \text{for } \tfrac{1}{2}h_e < |z| < \tfrac{1}{2}d_e \end{cases} \tag{7}$$

where h_e is the erythrocyte length and d_e is the distance between the erythrocyte centres; the pattern repeats for each erythrocyte. Because the function $\mu(z)$ is periodic, it can be written as a Fourier series:

$$\mu(z) = \sum_{n=0}^{\infty} \mu_n \cos(2\pi nz/d_e)$$
$$\mu_0 = \mu h_e/d_e; \; \mu_n = 2\mu/(\pi n)\sin(\pi nh_e/d_e) \tag{8}$$

so that inserting Eqns 2 and 8 into Eqn 7 yields the descriptive equation for the capillary oxygen pressure:

$$\mathscr{P}\nabla^2 p = -\sum_{n=0}^{\infty} \mu_n \cos(2\pi nz/d_e) \tag{9}$$

where $\nabla^2 = \vec{\nabla} \cdot \vec{\nabla}$ is the Laplace operator.

There is one further problem: the oxygen permeability $\mathscr{P}$. In both phases, erythrocyte and plasma, its value is different. Permeability of O_2 in the erythrocyte is lower than that in the plasma, because there is a high protein concentration. On the other hand, there is transport of oxygen

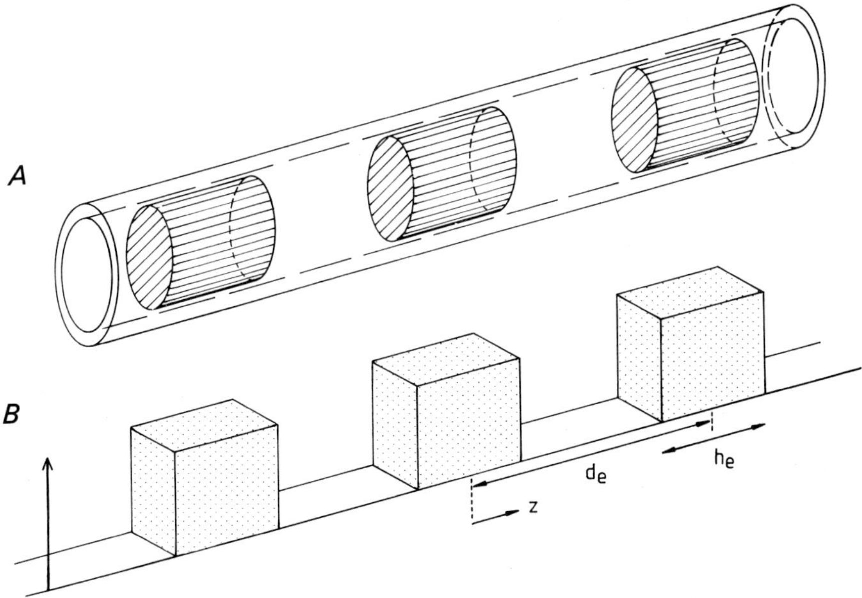

Fig. 2. Erythrocytes represented by cylinders moving through a capillary (A) and corresponding pattern of oxygen release (B; indicated vertically). d_e, distance between erythrocyte centres; h_e, erythrocyte cylinder length; z, axial coordinate.

bound to the haemoglobin (this **facilitated diffusion** is handled in more detail with Mb, below). Together, this renders an *effective* $\mathscr{P}$ that can be either higher or lower than the plasma value. It is possible to choose a range of p where both values are close and thus approximately equal to a mean capillary oxygen permeability $\mathscr{P}_c$. For $\mathscr{P}=\mathscr{P}_c$ overall in the capillary, a solution of Eqn 9 can be constructed also in terms of a Fourier series:

$$p(r,z) = \sum_{n=0}^{\infty} f_{c,n}(r)\cos(2\pi nz/d_e) - f_z z \tag{10}$$

assuming that in the region considered the axial drop in mean p, along the capillary length, is approximately linear – and equal to $f_z z$ in the formulation of the above equation. The Fourier 'coefficients' $f_{c,n}(r)$ now must be functions of the radial distance r. The Laplace operator ∇^2 for this 3-dimensional case is $\nabla_r{}^2 + \partial^2/\partial z^2$ and can be applied to both sides:

$$\begin{aligned} &\mathscr{P}_c \sum_{n=0}^{\infty}\{\vec{\nabla}_r{}^2 f_{c,n}(r) - 4\pi^2 n^2/\mathrm{d_e}{}^2 f_{c,n}(r)\}\cos(2\pi nz/d_e) \\ &= -\sum_{n=0}^{\infty}\mu_n\cos(2\pi nz/d_e) \end{aligned} \tag{11}$$

This identity must hold for each z and consequently for each n in the summation. So, differential equations for each $f_{c,n}(r)$ result, which can be solved as:

$$\begin{aligned} f_{c,0}(r) &= f_{c0} - \mu_0 r^2(4\mathscr{P}_c) \\ f_{c,n}(r) &= f_{cn} I_0(2\pi nr/d_e) + \mu_n d_e{}^2/(4\pi^2 n^2\mathscr{P}_c) \quad n = 1, 2, 3, \ldots \end{aligned} \tag{12}$$

where the f_{cn} are integration constants and $I_0(\ldots)$ is a so-called Modified Bessel function.

A similar procedure can be applied to the tissue, outside the capillary, with Eqn 4 extended for the axial component:

$$\mathscr{P}_t(\vec{\nabla}_r{}^2 + \partial^2/\partial z^2)p = M \tag{13}$$

where now the subscript t refers to the tissue situation. Again, a solution can be found in terms of a Fourier series, similar to Eqn 10:

$$p(r,z) = \sum_{n=0}^{\infty} f_{t,n}(r)\cos(2\pi nz/d_e) - f_z z \tag{14}$$

and solved here as:

$$\begin{aligned} f_{t,0}(r) &= f_{t0} - MR^2/(4\mathscr{P}_t)\{\ln(r^2/r_c{}^2) - (r^2 - r_c{}^2)/R^2\} \\ f_{t,n}(r) &\simeq f_{tn} K_0(2\pi nr/d_e) \quad n = 1, 2, 3, \ldots \end{aligned} \tag{15}$$

where again the f_{tn} are integration constants and $K_0(\ldots)$ is also a Modified Bessel function; the functions $I_0(\xi)$ and $K_0(\xi)$ are different in that $I_0(\xi)$ is increasing and $K_0(\xi)$ decreasing with increasing ξ:

$$\begin{aligned} &I_0(\xi \rightarrow 0) \rightarrow 1;\ I_0(\xi \rightarrow \infty) \rightarrow \infty \\ &K_0(\xi \rightarrow 0) \rightarrow \infty;\ K_0(\xi \rightarrow \infty) \rightarrow 0 \end{aligned} \tag{16}$$

where the increase/decrease is roughly exponential. Note the similarity of the zeroth term of Eqn 15 with the Krogh–Erlang solution Eqn 1. The solution for the higher-order terms, $n = 1, 2, 3, \ldots$, is approximate since, to fulfil the requirements exactly that the radial flux becomes zero at $r = R$, terms $I_0(2\pi nr/d_e)$ also should be added. However, these terms become comparable with $K_0(2\pi nr/d_e)$ only close to the border $r = R$ and, in view of the roughly exponential behaviour of $I_0(\ldots)$ and $K_0(\ldots)$, consequently are negligible elsewhere; particularly near the capillary border $r = r_c$.

The capillary and tissue solutions can be coupled by equating p and radial flux J_r at the boundary $r = r_c$. Following Eqn 2 with $\vec{\nabla} = \partial/\partial r$ for the radial component, inside, for the capillary, it can be expressed as, according to Eqn 10:

$$J_r = -\mathscr{P}_c \partial p/\partial r = -\sum_{n=0}^{\infty} \mathscr{P}_c (\mathrm{d}f_{c,n}/\mathrm{d}r) \cos(2\pi nz/d_e) \tag{17}$$

where, when substituting Eqn 12:

$$\begin{aligned} &\mathscr{P}_c(\mathrm{d}f_{c,0}/\mathrm{d}r) = -\tfrac{1}{2}\mu_0 r \\ &\mathscr{P}_c(\mathrm{d}f_{c,n}/dr) = (2\pi\mathscr{P}_c n/d_e) f_{cn} I_1(2\pi nr/d_e) \quad n = 1, 2, 3, \ldots \end{aligned} \tag{18}$$

And outside, for the tissue (Eqn 14):

$$J_r = -\mathscr{P}_t \partial p/\partial r = -\sum_{n=0}^{\infty} \mathscr{P}_t (\mathrm{d}f_{t,n}/\mathrm{d}r) \cos(2\pi nz/d_e) \tag{19}$$

where (Eqn 15):

$$\begin{aligned} &\mathscr{P}_t(\mathrm{d}f_{t,0}/\mathrm{d}r) = \tfrac{1}{2}M(r - R^2/r) \\ &\mathscr{P}_t(\mathrm{d}f_{t,n}/\mathrm{d}r) \simeq -(2\pi\mathscr{P}_t n/d_e) f_{tn} K_1(2\pi nr/d_e) \quad n = 1, 2, 3, \ldots \end{aligned} \tag{20}$$

So, all the coefficients f_{cn}, f_{tn} and μ can be related to d_e, h_e, p_c, r_c, M, $\mathscr{P}_c$, $\mathscr{P}_t$ and R; e.g.:

$$\begin{aligned} &\mu = (d_e/h_e) M (R^2/r_c^2 - 1) \\ &f_{t0} = p_c;\ f_{c0} = p_c + M(R^2 - r_c^2)/(4\mathscr{P}_c) \end{aligned} \tag{21}$$

Then, the full p profile can be calculated, inside and outside the capillary, for a given situation. An example is shown in Fig. 3, using basic data of Groebe (1990), for skeletal muscle. Oxygen partial pressure p is shown, in the vertical axis, for the front half of the capillary ($|r| < r_c$), from the edge of an erythrocyte ($z = -\frac{1}{2}h_e$) to half-way up the next erythrocyte ($z = +\frac{1}{2}d_e$). The shaded plane is at the average erythrocyte O_2 pressure p_e; also indicated by a thick arrow. The other arrow indicates, on the p axis, the averaged O_2 pressure level at the capillary rim, $r = r_c$. It is considerably lower than p_e: 4.9 kPa vs 9.3 kPa. It has to be emphasised that this is a rather extreme example, for an intererythrocytic gap of three times the erythrocyte length ($d_e/h_e = 4$), but it clearly shows the possible importance of intraerythrocytic O_2 release. The same peaks of p as shown here were found by the author (Groebe, 1990, particularly his Fig. 3), who calculated the O_2 pressure field for an arrangement of four capillaries.

Also, it is important to stress here that intracapillary convection was disregarded. This is often justified in the literature through reference to work of Aroesty & Gross (1970). These authors numerically solved the

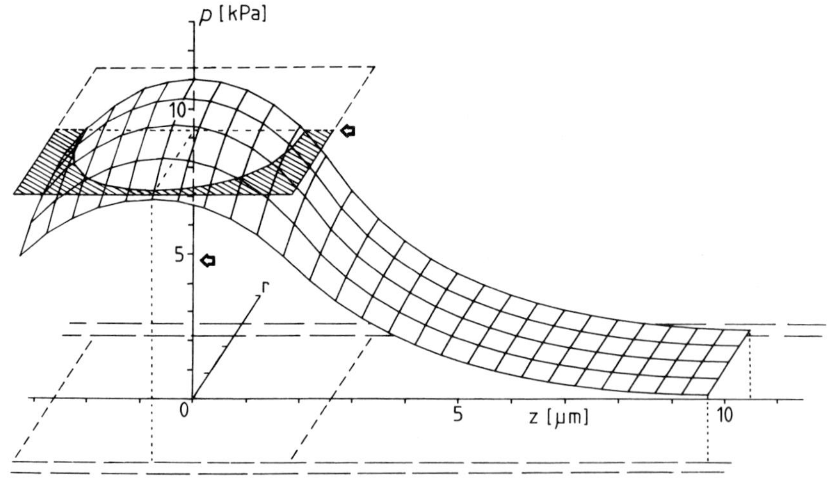

Fig. 3. Oxygen partial pressures (p; vertical axis) calculated for the front half of a capillary; r is radial, z is axial distance. Capillary and erythrocyte outlines are shown projected on the base plane (long and short broken lines, respectively). The shaded area is at the mean erythrocyte pressure. The arrows indicate mean erythrocyte and mean capillary rim pressure (p_e, p_c, respectively). Dotted vertical lines drawn at maximal and minimal capillary rim pressure and at minimal capillary centre pressure.

plasma transport equations in between the (cylindrical) erythrocytes, studied the influence on O_2 transport and found negligible effects. However, their boundary condition assumption was of constant O_2 pressure along the capillary rim; clearly at variance with the profile shown in Fig. 3. Plasma mixing is expected to flatten the p profile outside the shaded area in Fig. 3 and thus lift up the level of the right part. This will yield lower pressure drops than calculated here and higher p_c, for the tissue. How large this effect is cannot be easily estimated.

The additional problem of diffusion in the capillary wall can be handled in two ways. Either the capillary p and $\vec{J}$ can be incorporated in the theoretical treatment, as an extra layer formulated similarly as Eqn 10, etc., or the capillary wall can be treated equal to the tissue.

Extensions which may be incorporated into the Krogh model

Some extra features can be incorporated in the Krogh model without changing its basic concept, but producing a more realistic description of the *in vivo* conditions.

Extraction pressure

The difference between erythrocytic O_2 pressure p_e and capillary pressure p_c in the Krogh model can be formally accounted for by assuming one value for it, in a particular situation, for the whole axial course of the cylinder. Different names were proposed for such a $p_e - p_c$ value: **Capillary Barrier**, CB (Hoofd *et al.*, 1989; Turek *et al.*, 1989) or **Extraction Pressure**, EP (Hoofd *et al.*, 1990*a*). Transport resistance attributable to oxygen release looks like an O_2 barrier; without any *physical* barrier being present (also see Honig *et al.*, 1984). To avoid this latter misunderstanding the term Extraction Pressure was introduced, since the pressure drop is needed solely to extract the oxygen from the erythrocyte source. A fixed value is chosen for EP for a particular situation, e.g. along the full axial course of a Krogh cylinder. While a possibly crude treatment, that obviously is better than assuming no pressure drop at all. The periodic variation in $p(r_c,z)$ along the z axis, as seen in Fig. 3 above, is not of fundamental importance here. This variation is reflected in the terms $f_{tn}K_0(\ldots)\cos(\ldots)$ in the treatment of the former section: see Eqns 14 and 15. The functions $K_0(\ldots)$ decrease exponentially with increasing r so these variations fade out quickly when receding from the capillary. This means that in fact the tissue 'sees' a mean level, being p_c, and consequently a mean EP.

The conceptual advantage of EP (or CB) is realised when exporting it to other situations. The reasoning is that since EP in the above model is a gradient-driven phenomenon, it will be linearly proportional to the delivered O_2 flux, J_r. This flux in turn is directly proportional to the total O_2 consumption in the tissue slab, $M\pi(R^2 - r_c^2)$. So, for other consumption levels M' and/or other Krogh radii R' a new, adjusted EP can be calculated according to:

$$\mathrm{EP}' = \mathrm{EP}\{M'(R'^2 - r_c^2)\}/\{M(R^2 - r_c^2)\} \tag{22}$$

A final remark: EP critically depends on the erythrocyte spacing d_e, or better, on d_e/h_e. The example above was for $d_e/h_e = 4$, leading to EP = 4.4 kPa; for $d_e/h_e = 2$, a much lower value of 1.1 kPa is calculated, and for $d_e/h_e = 3$, 2.6 kPa. Such an effect is also nicely shown in the figures by Groebe & Thews (1989). Also note that for $d_e/h_e = 2$ the mean blood haemoglobin concentration is twice as high as for $d_e/h_e = 4$; in the Krogh–Kety model, with its combined parameter $M/(Fc_{Hm})$, that is fully counteracted by halving the flow, but that makes no difference for EP, and as such breaking the balance between M, F and c_{Hm}.

O_2-dependent oxygen consumption

The Krogh model assumes uniform, zero-order consumption (assumptions 3 and 4 in Kreuzer's list and Fig. 1*B*). The mitochondria, where the bulk of the oxygen is consumed, are capable of maintaining maximal oxygen consumption down to very low p (Starlinger & Lübbers, 1972). So, from these findings it seems justified to represent tissue O_2 consumption with a constant uniform M as long as there is O_2 present. If there is insufficient oxygen available, for a too low capillary oxygen pressure p_c, it will all be consumed in a region around the capillary, and beyond the tissue will be devoid of oxygen. In fact, this is a situation with a decreased Krogh radius $R' < R$. At this distance $r = R'$, both flux and pressure vanish:

$$0 = p_c - MR'^2/(4\mathscr{P})\{\ln(R'^2/r_c^2) - (1 - r_c^2/R'^2)\} \tag{23}$$

as can be seen from Eqn 1, modified accordingly. Equation 23 allows calculation of R'.

There is, however, a question about whether oxygen consumption *in vivo* is of zero order or not. De Koning *et al.* (1981) and Hoofd (1987), from mathematical analysis of O_2 diffusion in slabs of chicken gizzard smooth muscle, found clear evidence that local O_2 consumption did decrease at low to moderate values of p in their preparation. They proposed the symbol q_{50} to indicate the pressure where consumption had fallen to 50% of its maximum value (it could not be named p_{50} because that applies to the 50% saturation of Mb or Hb). A consequence of p-dependent O_2 consumption is that oxygen pressure does not become zero at any location, but it can become vanishingly small in regions distant from the capillary.

Hoofd (1987) analysed three different models of p-dependent consumption and found only small differences in calculated pressures and fluxes. Fletcher (1978) came to a similar conclusion for numerical calculations in Krogh cylinders. So, as an example, we will take a linear model of O_2 consumption:

$$m(p) = Mp/(2q_{50}) \tag{24}$$

where the O_2-dependent consumption is denoted by $m(p)$. Since consumption cannot exceed its maximum value M, the linear part cannot go further than $2q_{50}$. The describing differential equation must be modified from Eqn 4:

$$\mathcal{P}\nabla_{\mathrm{r}}^2 p = m(p) \tag{25}$$

For the linear part of $m(p)$ – so, for $p_c \leqslant 2q_{50}$, an exact solution can be found:

$$p = p_c\{I_0(r/\lambda)/I_1(R/\lambda) + K_0(r/\lambda)/K_1(R/\lambda)\}/\{I_0(r_c/\lambda)/I_1(R/\lambda) + K_0(r_c/\lambda)/K_1(R/\lambda)\}$$

$$\lambda = \sqrt{(2\mathcal{P}q_{50}/M)} \tag{26}$$

again in terms of the Modified Bessel functions $I_n(\ldots)$ and $K_n(\ldots)$.

For other models no analytical solution can be given and a numerical solution must be constructed (Hoofd, 1987; Turek *et al.*, 1989).

Asymmetric diffusion

Radial symmetry is assumption 9 (and also 11) in Kreuzer's list, and in Fig. 1*B*. Removing this constraint, the general solution of the combination of Eqns 2 and 3 now including the angular variable ϕ reads:

$$p = p'_c - MR^2/(4\mathcal{P})$$

$$\{\ln(r^2/r_c^2) - (r^2 - r_c^2)/R^2 + \sum_{n=1}^{\infty}[f'_n r^n \cos(n\phi - n\phi'_n) + \tag{27}$$

$$+ f''_n r^{-n}\cos(n\phi - n\phi''_n)]\}$$

where the coefficients f'_n, f''_n, ϕ'_n, ϕ''_n must follow from the boundary conditions.

Rakušan *et al.* (1984) followed the opposite strategy to investigate the effect of breaking the radial symmetry. They did not solve the $f'_n, f''_n, \phi'_n, \phi''_n$ of Eqn 27 for a particular boundary condition, but constructed the boundaries of the diffusion region for different choices of the coefficients f'_n, f''_n. Once one point of this boundary is found, it can be tracked since there is no flux across, so proceeds along $\vec{J}$. For each n, the coefficients f'_n, f''_n were chosen as zero, except for one pair. As examples here, we will only retain one f'_n; then, the angles ϕ'_n and ϕ''_n become irrelevant and can be taken as zero also, leading to oxygen supply areas as shown in Fig. 4, top. The most important case turned out to be for f'_1 non-zero:

$$p = p'_c - MR^2/(4\mathscr{P})\{\ln(r^2/r_c^2) - (r^2 - r_c^2)/R^2 + f'_1 r\cos(\phi)\} \quad (28)$$

which leads to 'egg-shaped' oxygen supply areas where the capillary no longer is at the centre; see Fig. 4*B*,*C*. Oxygen pressure distribution becomes increasingly different from the Krogh case, Fig. 4*A*, with more elongated areas, when increasing f'_1, where this coefficient also can be expressed in terms of the difference ΔR between longest and shortest radial distance ($\cos(\phi) = 1$ or -1, respectively):

$$f'_1 = 2\Delta R/R^2 \quad (29)$$

This example shows that the assumption of radial symmetry in the Krogh model can be a critical one – as we will see later, it is impossible to fill a plane of a realistic capillary distribution with Krogh circles; when trying to do so, some overlap and there are empty spaces between others. However, the deviations are not large for low values of f'_1. Also, the differences with the Krogh situation are not large for examples with higher terms non-zero, e.g. f'_2 non-zero, especially since such low-p areas are small, e.g. only the corner zones in Fig. 4*F*. In Fig. 4 the oxygen flux is followed coming from the capillary, along **flux lines**, for different choices of f'_1 (corresponding with $\Delta R/R = 0.5$ and $\Delta R = 1$), f'_2 or f'_3. The oxygen supply area encompasses all these tracks. The corresponding p patterns along the $\phi = 0$ (and opposite $\phi = \pi$) axis are shown also, for a p'_c of 6 kPa, which is a value close to the venular one in normal tissue. Note that the profiles shown, travelling to the shortest and the longest axis, respectively, are the extremes raised and lowered as compared with the Krogh situation. Though it is possible to extend the Krogh model in this way, it is not of much practical help because it is unknown how the extensions should be made for a given tissue situation.

Myoglobin

This concerns assumption 13 in Kreuzer's list, and Fig. 1*B*. The actual tissue for which the Krogh model was devised, skeletal muscle, usually

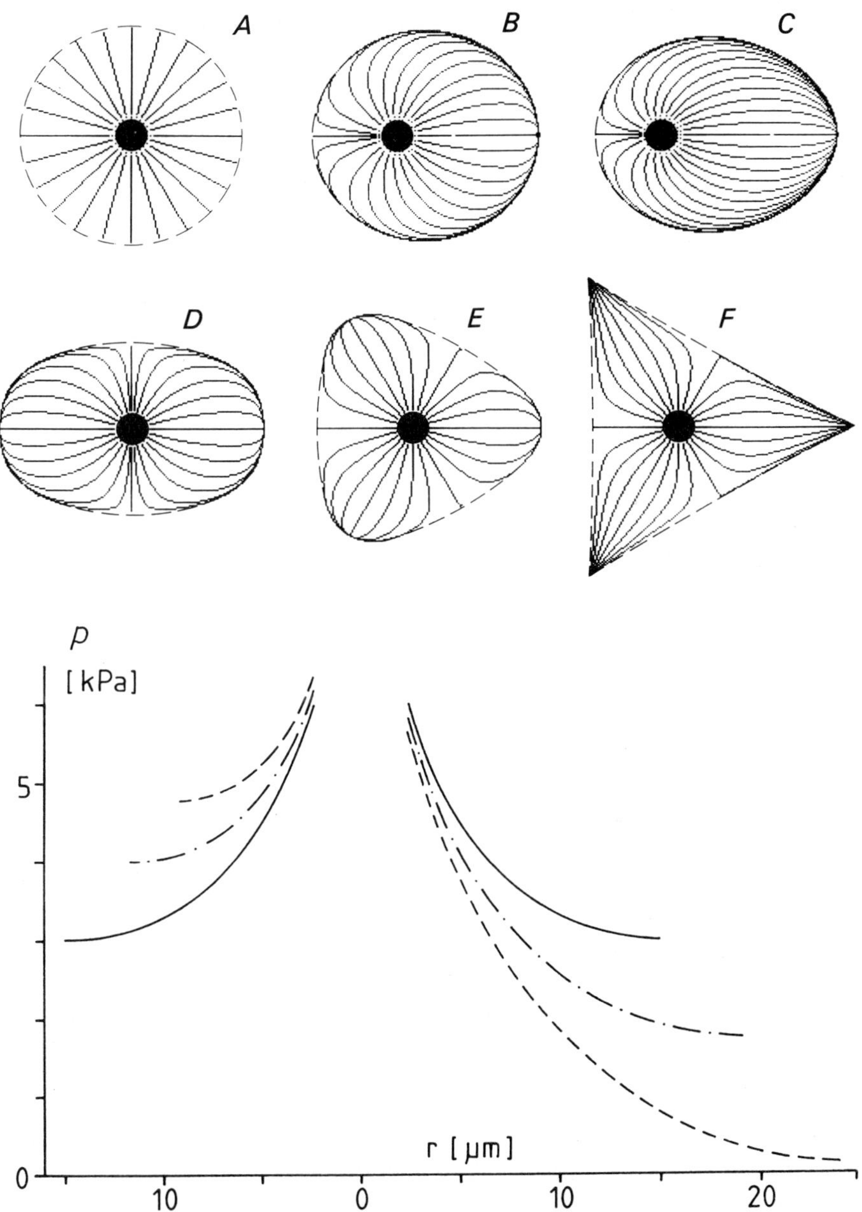

Fig. 4. Angular-dependent oxygen supply. Top: computer-drawn flux lines, and encompassing O_2 supply areas (broken lines), for: *A*: Krogh circular area, *B*, *C*: non-zero f'_1, *D*: non-zero f'_2, *E*, *F*: non-zero f'_3. Lower panel shows p against radial distance r for $\phi = 0$ (right side) and $\phi = \pi$ (left side) non-zero f'_1.

contains an oxygen-carrying molecule, myoglobin (Mb). This molecule, as Hb in the erythrocyte, can take up oxygen at high O_2 pressure and release it at low O_2 pressure. When the loaded molecule has travelled a distance in between we speak of **facilitated transport**. Contrary to Hb, the Mb contains only one, O_2-binding, haem group; thus the Mb–O_2 reaction simply reads: $Mb + O_2 \rightleftarrows MbO_2$ and the chemical equilibrium between Mb and O_2 obeys:

$$[MbO_2]/([Mb] + [MbO_2]) = k'(O_2]/(k + k'[O_2]) \tag{30}$$

with reaction rates k and k', backward and forward, respectively. The left-hand side of this equation is the definition of the *myoglobin* oxygen saturation; the right-hand side is rewritten such that, using $[O_2] = \alpha p$:

$$S = p/(p + p_{50}) \tag{31}$$

where $p_{50} = k/(k'\alpha)$ is the O_2 pressure at which myoglobin is 50% saturated, at chemical equilibrium.

The total flux of oxygen now has to include the diffusion of MbO_2 and Eqn 2 is modified accordingly:

$$\vec{J} = -\mathscr{P}\vec{\nabla}p - D_{Mb}\vec{\nabla}[MbO_2] \tag{32}$$

assuming MbO_2 diffusing with diffusion coefficient D_{Mb}. The equation can be written as:

$$\begin{aligned} \vec{J} &= -\mathscr{P}\vec{\nabla}(p + p_F S) \\ p_F &= D_{Mb}c_{Mb}/\mathscr{P} \end{aligned} \tag{33}$$

where p_F is called the **facilitation pressure**, and c_{Mb} is the total myoglobin concentration ($[Mb] + [MbO_2]$). Incorporating Mb-facilitated diffusion into models of oxygen transfer would be as simple as that, were it not for the fact that also the myoglobin oxygen saturation S must be known. As with Hb in the erythrocyte, S can be taken to be in equilibrium with p, and the equilibrium relation Eqn 31 applied. The combined parameter p_F is the maximum pressure difference that can be added to the driving force of p in diffusional transport (Hoofd & Kreuzer, 1980; de Koning *et al.*, 1981). It is this driving force that sets up a gradient to deliver O_2 at a certain location. For example, for $p = 10$ kPa and $p_F = 2$ kPa, and $S \simeq 1$, the total driving force will be 12 kPa. This implies that the oxygen reaches into the tissue more easily than without Mb.

An example is provided in Fig. 5, along with a scheme of the process. It is for a very low capillary O_2 pressure of 3 kPa to emphasise the influence of p_F, taken to be either 0 or 2 kPa. Myoglobin $p_{50} = 0.5$ kPa, $M/(4\mathscr{P}) = 0.005$ kPa μm^{-2}, $r_c = 2.4$ μm; approximate data for rat heart

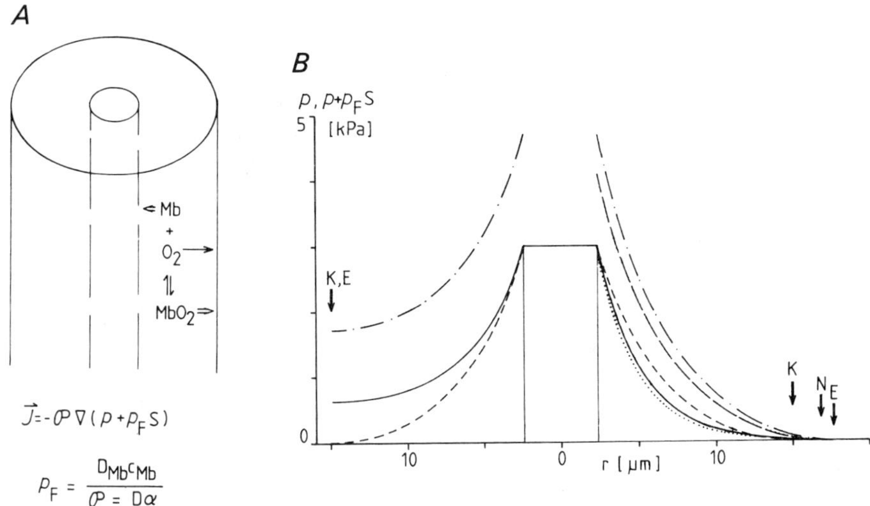

Fig. 5. Principles of facilitated diffusion (*A*) and examples of profiles of O_2 pressure and O_2 driving force against radial distance r (*B*), for $p_c = 3$ kPa (level in the capillary, indicated by vertical lines). Short-dashed line: p, no facilitation; dash-dotted line: $p + p_F S$, equilibrium; solid line: resulting p; long-dashed line: $p + p_F S$, non-equilibrium; dotted: resulting non-equilibrium p. The vertical arrows indicate the outer border of the O_2 diffusion range, for: *K*: no facilitation, *E*: equilibrium, *N*: non-equilibrium.

(Hoofd *et al.*, 1990*a*). Without facilitation, $p_F = 0$, the O_2 can supply a Krogh region up to $R = 15\,\mu m$ (short-dashed lines), as indicated by the arrows labelled K. With facilitation, $p_F S$ must be added to the driving force (Eqn 33) where S_c (S at the capillary rim) is equal to $3/(3 + 0.5) = 0.86$ (Eqn 31). So, the total driving force (not the pressure!) is raised from $p_c = 3$ up to $p_c + p_F S_c = 3 + 2 \times 0.86 = 4.72$ kPa, and the dash-dotted lines pertain for $p + p_F S$ and, according to Eqn 31, the solid lines for p. When taking the same Krogh area as before, $R = 15\,\mu m$ (left side in Fig. 5*B*), driving force and pressure reach lowest values of 1.72 kPa and 0.62 kPa, respectively, at $r = R$. Alternatively, as shown at the right side in Fig. 5*B*, the oxygen now could supply a Krogh region up to $R = 17.7\,\mu m$, as indicated by the rightmost arrow labelled E. This is a distance of 18% larger, or a supply area of even 40% larger!

But there is no role for myoglobin facilitation at high p. For a flux enhancement there must be a gradient in S, according to Eqn 33, and there will be no gradient if p is high everywhere (since then, S is close to 1 everywhere). So, Mb only can help tissue O_2 transport at low oxygen pressures, and only when loaded to sufficiently high saturations. Consequently, the role of Mb in the tissue as a whole is only a moderate one (de Koning *et al.*, 1981; Hoofd, 1987; Hoofd *et al.*, 1987; Turek *et al.*, 1991).

In Fig. 5 also a case is shown for *non*-equilibrium in the chemical reaction between Mb and O_2; $p + p_F S$ is shown as a long-dashed line, starting at a lower value as compared with equilibrium, and the corresponding p as a dotted line. The problem of non-equilibrium facilitated diffusion, a both mathematically and computationally interesting case, is satisfactorily solved in the literature (see Schultz *et al.*, 1974; Hoofd & Kreuzer, 1979; Hoofd, 1987; Kreuzer & Hoofd, 1987; Popel, 1989). S is lower at $r = r_c$ and consequently the O_2 reaches less far; up to 16.8 µm (arrow labelled N).

Layered solutions

It should be mentioned here that, from a mathematical viewpoint, it is no problem to solve Eqn 4 also for circular concentric regions. For each such region n, with oxygen consumption M_n and O_2 permeability $\mathscr{P}_n$, a solution for p can be written as:

$$p = p_n - \Xi_n \ln(r^2/r_c^2) + \{M_n/(4\mathscr{P}_n)\}r^2 \tag{34}$$

where the coefficients p_n and Ξ_n must be chosen so that the boundary conditions at $r = r_c$ and $r = R$ are met and that the circular regions fit together. Fitting together implies both p and radial flux J_r, which for the nth region, according to Eqn 2, is:

$$J_r = 2\mathscr{P}_n \Xi_n / r - \tfrac{1}{2} M_n r \tag{35}$$

As an example, for a 2-zone Krogh area let us assume that O_2 is consumed only further away from the capillary. So there is a first region around the capillary, up to $r = \varrho$, where (almost) no oxygen is consumed, i.e. $M_1 \simeq 0$, and a second, where $M_2 \neq 0$; also, $\mathscr{P}_1$ may differ from $\mathscr{P}_2$. The solution is:

$$p = \begin{cases} p_c - M_2(R^2 - \varrho^2)/(4\mathscr{P}_1)\ln(r^2/r_c^2) & r_c < r < \varrho \\ p_c - M_2(R^2 - \varrho^2)/(4\mathscr{P}_1)\ln(\varrho^2/r_c^2) - & \\ \quad - M_2 R^2/(4\mathscr{P}_2)\{\ln(r^2/\varrho^2) - (r^2 - \varrho^2)/R^2\} & \varrho < r < R \end{cases} \tag{36}$$

Similarly, regions where O_2 just is consumed close to the capillary can be modelled (e.g. Mainwood & Rakušan, 1982); and linear consumption where the boundary $p = 2q_{50}$ is somewhere in between r_c and R.

Extensions beyond the Krogh model

The Multicylindrical model

From one glance at a muscle tissue cross-section, it is obvious that the tissue cannot be represented by one single Krogh cylinder. Some capillaries lie close together, others are farther separated (see Fig. 6*B* and Fig. 8*A*, below). The idea behind the Multicylindrical model (Turek & Rakušan, 1981; Turek *et al.*, 1989) is that tissue can be represented by a *set of* Kroghian tissue cylinders with various radii so as to represent the heterogeneity in capillary spacing. A portraying picture of such a subdivision is shown in Fig. 6*A*. Around each capillary a Krogh cylinder is constructed, and in each of these cylinders the entire p field is calculated using the mathematical treatments as described above.

The input data of the Multicylindrical model are the same as for the Krogh case with its extensions, but now different values of all variables could be chosen for each of the Krogh cylinders involved. Obviously, this is not practical and the literature has been restricted to R, the Krogh radius, and $LM(R^2 - r_c^2)/(Fc_{Hm})$, the O_2 consumption-to-supply ratio (see above). The distribution of Krogh radii R, or Krogh areas πR^2, can be obtained from analysis of tissue capillarity. A straightforward approach is the use of capillary domains (Hoofd *et al.*, 1985). Such domains are polygons geometrically constructed around capillary centres; an example is provided in Fig. 6*B*. Their area distribution is taken equal to the Krogh area distribution (Turek *et al.*, 1987). For the consumption-to-supply ratio, various schemes have been used, relating O_2 consumption to, for example, muscle mass or to venous oxygen pressure.

One might argue that a set of circular Kroghian cylinders does not represent the tissue either, bearing in mind the non-circular domain areas in Fig. 6*B*. However, from Fig. 4 above it can be seen that the calculated p values only become very different from the Krogh ones for quite distorted supply areas. Some parts of the supply area have a lowered p, others with raised p, the extremes in the examples of Fig. 4 (an already large area!) not extending beyond 3 kPa. Similarly, Turek *et al.* (1986) found 1.4 kPa with a maximum of 2.2 kPa for a very extreme case. When the goal is, as here, to calculate O_2 distribution in a whole tissue, with a full scale of possible p values, it might not cause such of a problem taking the subdivision as Krogh-like cylinders.

Results of tissue calculations like here using the Multicylindrical model are mostly presented in terms of an **oxygen pressure histogram**. At a number of locations (chosen beforehand: e.g. a regular pattern, which is expected to be random with respect to the capillaries), p is calculated and

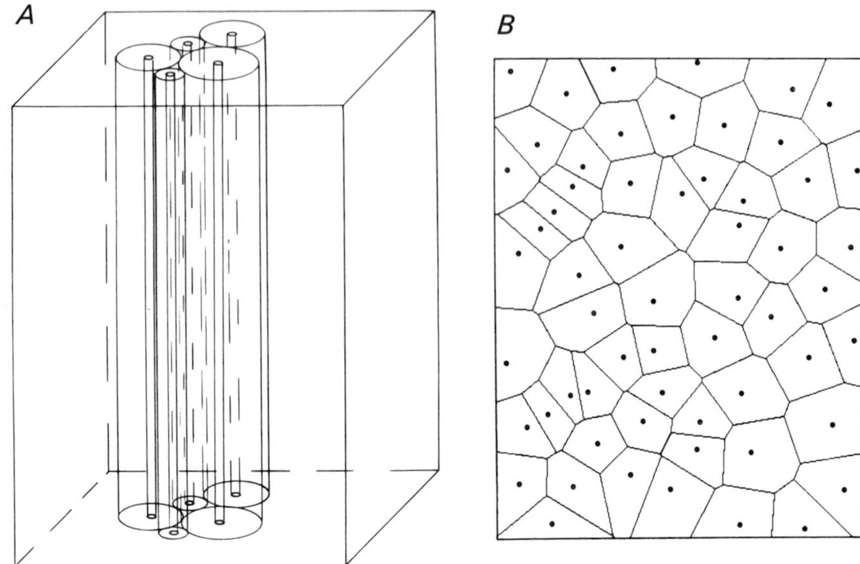

Fig. 6. Impression of the Multicylindrical model (*A*) and computer printout example of a domains subdivision of tissue cross-section (polygons in *B* around capillary centres indicated with filled circles). The circular areas of the cylinders in the model are taken such that they match the distribution of the domain areas.

sampled in classes, e.g. a value of 3.618 kPa comes in the class 3.0–4.0 kPa. The total number found in each class is represented or, changing only the axis, the fractional number or the percentage. Alternatively, a **cumulative histogram** can be represented, adding all lower classes, increasing up to 100%. Such histograms are also a common way of presenting measurements, sampled at a number of locations in a tissue. Likewise, the tissue volume where p falls within a certain class may be calculated, which is equivalent to sampling at an infinite number of locations. Quite a few situations and possibilities have been calculated in the literature, too much to be covered here, and the reader is referred to that and to Egginton & Ross (this volume) where an example is shown taken from the most recent reference (Turek *et al.*, 1991); also, capillary domains and their relation with tissue features are handled in that chapter.

The Multicapillary model

The Multicapillary model also starts from capillary distribution as it is measured, but does not use Krogh cylinders any more. Instead, the Krogh–Erlang formula itself is extended. How this is done can be seen by examining further Eqn 1; particularly the terms between the braces. Already the last term alone, containing r^2, is a solution of the describing differential equation, Eqn 4, whereas the first, logarithmic, term is a **homogeneous** solution, of the equation $\mathscr{P}\nabla_r{}^2 p = 0$. So, the last term could be called an **oxygen consumption field** term, whereas the first term is an **oxygen supply** term; the radial flux at distance r equals $-\frac{1}{2}MR^2/r$, in all directions, so the O_2 delivery is $2\pi r$ times $\frac{1}{2}MR^2/r = \pi MR^2$. This is the amount that goes into an area πR^2, the Krogh circular area. Concluding, the Krogh–Erlang equation might be represented as:

$$\text{Krogh } p = \text{Constant} + \text{Source} + \text{Field} \tag{37}$$

and generalisation is natural (Hoofd *et al.*, 1988, 1989; Clark *et al.*, 1989):

$$\text{Generalised } p = \text{Constant} + \text{Sources} + \text{Field} \tag{38}$$

The term $\ln(r^2/r_c{}^2)$ in Eqn 1 is a source term located at $r=0$. The logarithm as a source term in a 2-dimensional geometry is well known in physics, notably electromagnetism. For a generalised model, these source terms now are at various different locations, where the capillaries are, and Eqn 38 must be elaborated in terms of the location vector r:

$$p(r) = C_p - M/(4\mathscr{P}) \left\{ \sum_{n=1}^{N} [(A_n/\pi) \ln(|r - r_n|^2/r_{cn}{}^2)] - \Phi(r) \right\} \tag{39}$$

where r_n is the location of the nth source and $\Phi(r)$ is what was called above the consumption field; there are N sources. C_p and A_n are integration constants; r_{cn} is added to make the logarithm term dimensionless and could be chosen freely: here, equal to the capillary radius of the nth capillary. A_n is the area where oxygen is supplied to; as can be derived from analysing the flux from that n^{th} source, similarly as done above for the single source in the Krogh case.

The term $\Phi(r)$ is a more difficult case. This consumption field term already *alone* must obey the combination of Eqns 2 and 3, so it describes the effect of consumption on p. Consequently, the boundary conditions for solving $\Phi(r)$ must be imposed not by capillary conditions but only by shape and dimensions of the consuming area (for constant M). So, for a circular area, we expect the same cylindrically symmetric solution as for the Krogh circle:

$$\Phi(r) = |r|^2 = r^2 \quad \text{circular symmetry, } r \leqslant R \tag{40}$$

Hoofd *et al.* (1990*a*) also supplied a solution for rectangular areas; this solution, now in terms of Cartesian coordinates $r=(x,y)$, where (0,0) is at the centre, can be (re)written as:

$$\Phi(r)=|r|^2+(1/\pi)\sum_{n=1}^{4}\{\Delta x_n \Delta y_n \ln(\Delta x_n{}^2+\Delta y_n{}^2)+ \\ +(\Delta x_n{}^2-\Delta y_n{}^2)\,\mathrm{arctg}(\Delta y_n/\Delta x_n)\} \tag{41}$$

where the sum is over the four corners $r_{s1},\ldots,r_{s4}$ and $(\Delta x_n,\Delta y_n)=r-r_{sn}=(x-x_{sn},y-y_{sn})$.

Also, the boundary conditions for Eqn 39 are not trivial. The boundary conditions for the Krogh–Erlang equation were:

(a) $p=p_c$ at $r=r_c$
(b) $J_r=0$ at $r=R$

Condition (a) cannot be taken unaltered for the situation here because a constant p over the capillary rim will be rare; see Eqn 39. So, for each capillary, n, some chosen or averaged p_{cn} must be used instead (Hoofd *et al.*, 1989), for example:

$$p(r_n)=p_{cn}=C_p-M/(4\pi\mathscr{P}r_{cn}{}^2)\oint \mathrm{d}r \\ \left\{\sum_{n=1}^{N}[(A_n/\pi)\ln(|r-r_n|^2 r_{cn}{}^2)]-\Phi(r)\right\} \tag{42}$$

where the circular integral is over the capillary rim, radius r_{cn}.

Neither can condition (b) be adopted here unaltered. But it can be restated as:

(b) All oxygen from the source is consumed in an area πR^2

and then the counterpart here is easy: all oxygen, from all sources, is consumed in an area equal to the total area considered, A (Hoofd *et al.*, 1989):

$$\sum_{n=1}^{N}A_n=A \tag{43}$$

Moreover, this is a realistic condition. Eqn 43 will allow O_2 to flow out of the area considered, e.g. when a capillary is near the border, but elsewhere O_2 will flow in, just as in real tissue.

Using these $N+1$ boundary conditions, the constants C_p, A_1, A_2, . . ., A_N can be solved and in turn p calculated at any location in the flat plane, using Eqn 39. But the mere calculation of the A_n suffices to apply the Kety approach, Eqn 5, for each individual capillary:

$$dc_{O_2n}/dz = -(M/F_n)(A_n - \pi r_{cn}^{\ 2}) \tag{44}$$

where the index n is for the nth capillary. In this way again, layers can be piled, now for a tissue cylinder (Eqn 40) or block (Eqn 41) of parallel capillaries; see Fig. 7, left side. In principle, one could choose a different pattern of capillaries in any plane, allowing distortions, shorter capillary courses and even branching as depicted in the right side of the block in Fig. 7, but it is unclear how severely this would violate the assumption of no (appreciable) axial diffusion.

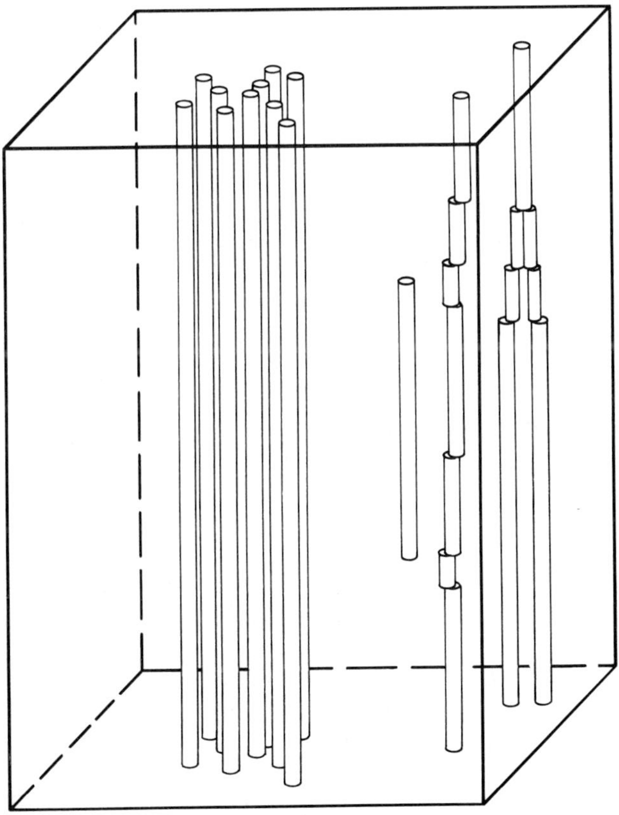

Fig. 7. Parallel capillaries in a tissue block, left side, with distribution of capillary distances as measured in a cross-section, used as basis for the Multicapillary model. No cylindrical or other form of O_2 supply region is pre-assumed in this model. In the right side of the block, some alternative patterns of capillary course are shown.

Two examples, for two different types of capillary distribution, are presented here. One is for normal rat heart tissue where a typical photomicrograph was taken and capillary locations read into a computer; this case is represented in Fig. 8*A*. There were 71 capillaries. In order to appreciate the differences in the calculated results with the Krogh model predictions a second case was also considered where these 71 capillaries were regularly spaced (with the same mean supply area, of course) as depicted in Fig. 8*B*. The supply areas will be hexagons in this example, leading to negligibly small differences with the Krogh case (angular-dependent model $f'_6 \neq 0$, $\Delta R/R = 0.144$; see also Groebe, this volume). The data were taken for rat heart (Hoofd *et al.*, 1990*a*), including Mb-facilitated diffusion, for fixed values of L, F and c_{Hm} and for EP = 2 kPa where the individual EP_n were calculated in similar fashion to Eqn 22:

$$\mathrm{EP}_n = \mathrm{EP}(A_n - \pi r_{cn}^2) / \left\{ \left(A - \sum_{m=1}^{N} \pi r_{cm}^2 \right) / N \right\} \tag{45}$$

Both cases were calculated for $z = 0$–$250\,\mu\mathrm{m}$, only half the capillary length, where some capillaries started at the arterial $p_c = 13.3$ kPa and the others 'half-way', such that $p_c = 6.4 \pm 0.7$ kPa, i.e. a distribution that the other capillaries reached at $z = 250\,\mu\mathrm{m}$. Furthermore, the top and side $10\,\mu\mathrm{m}$ regions were omitted from the calculations to avoid unrealistic p values because of being too close to the borders. The resulting histograms are shown in Fig. 9: *A*, realistic capillary distribution, *B*, regular spacing. The differences are evident. Taking only one Krogh cylinder to represent the tissue leads to a much more peaked O_2 histogram (steeper cumulative histogram) and severely underestimates the percentage of regions with low oxygen pressures.

Discussion

This chapter presented a survey of the various different possibilities of Krogh model extensions. Within the Kroghian scope, extensions like facilitated diffusion and Extraction Pressure are easily incorporated.

Fig. 8. Capillary distributions in a cross-section as used in the calculations of the Multicapillary model. *A*: 71 capillaries, locations read in from a typical photomicrograph of normal rat myocardium; *B*: 71 capillaries, regularly distributed. The areas were subdivided, as indicated by the long-dashed lines, in six zones, three of them (top left and right; bottom middle) starting with arteriolar p_c, the others already 'half-way' to the venular p_c. The side portions as indicated by short-dashed lines were not included in the O_2 histograms.

A

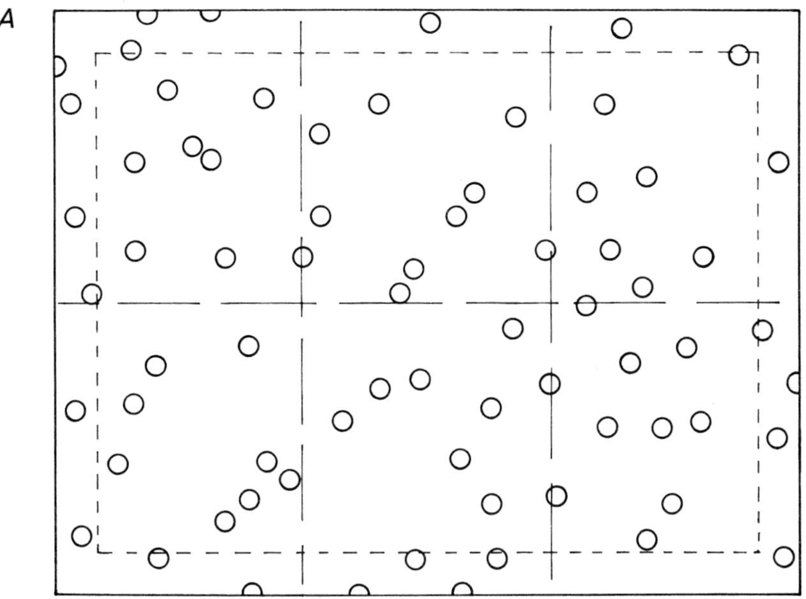

B

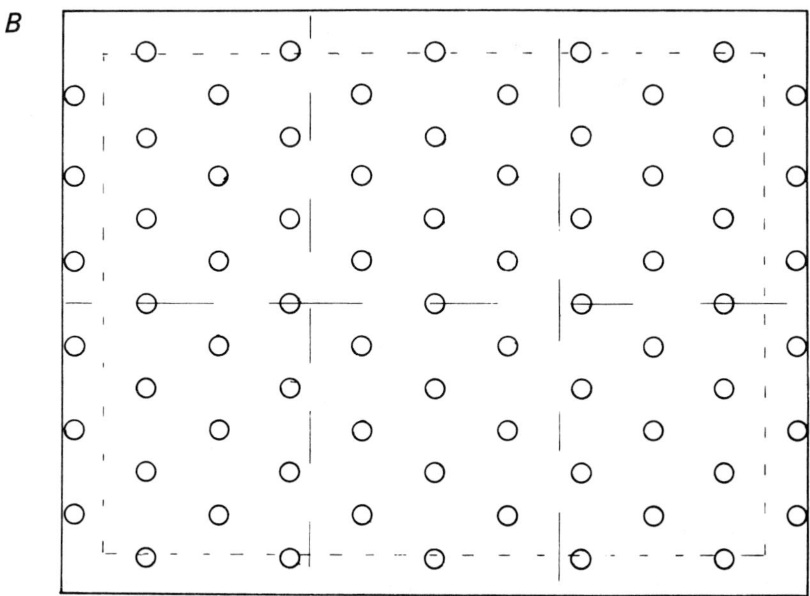

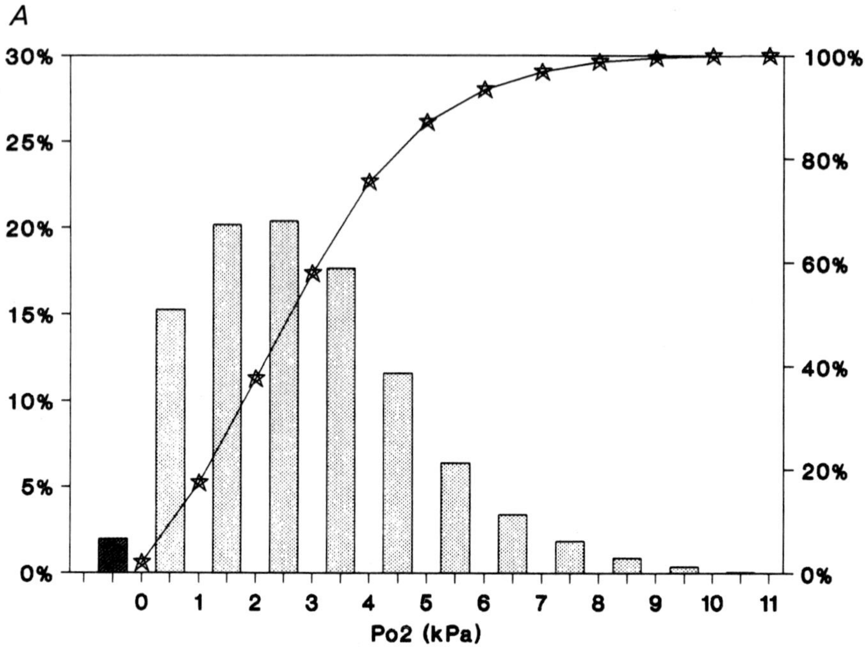
A
30%
25%
20%
15%
10%
5%
0%
100%
80%
60%
40%
20%
0%
0 1 2 3 4 5 6 7 8 9 10 11
Po2 (kPa)

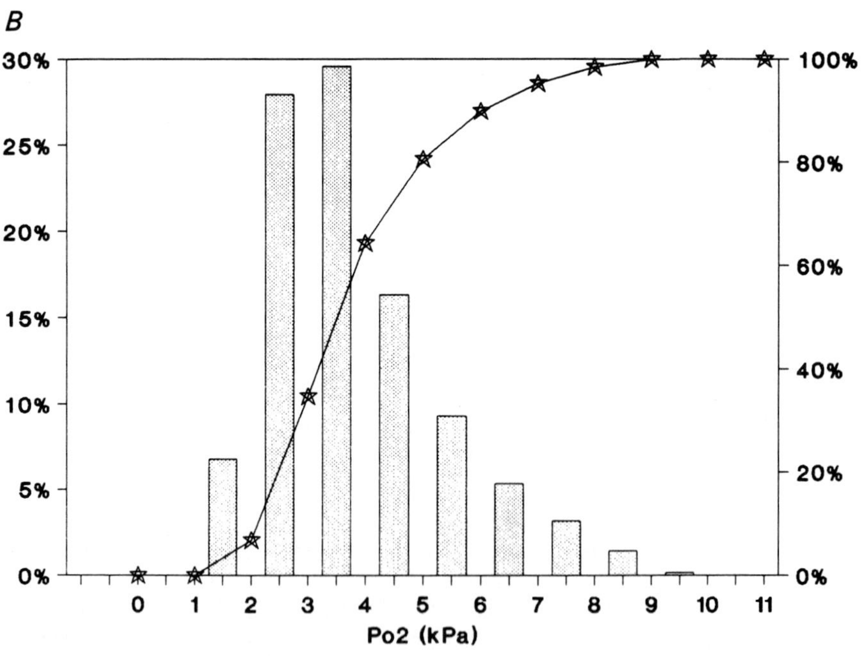
B
30%
25%
20%
15%
10%
5%
0%
100%
80%
60%
40%
20%
0%
0 1 2 3 4 5 6 7 8 9 10 11
Po2 (kPa)

Extending the scope, heterogeneous models like the Multicylindrical model and the Multicapillary model were developed. Those models offer the possibility not only of incorporating realistic tissue capillarity, but also blood heterogeneities of flow and haematocrit, and consequently are capable of accounting for new experimental data on these determinants of tissue oxygenation. This might refine the predictions of the models when data of heterogeneities of capillary length, flow and haematocrit become available as coupled to the available data.

The Multicylindrical model is the more informative one. For example, for a quick tissue calculation one could take a few sample Krogh cylinders and calculate the resulting p values, even with a pocket calculator. The model is capable of incorporating any extensions as described above for the Krogh cylinder, notably different consumption models, like p-dependent O_2 consumption.

The Multicapillary model is the more sophisticated one. It no longer depends on some unrealistic assumptions of the Krogh model. However, a computer is needed to obtain results, so that no 'quick calculation' can be done – at least not for many capillaries. Also, there is, as yet, no way to incorporate other than zero-order consumption. Features like distorted or branching capillaries cannot be used until investigated thoroughly. The model must use a realistic tissue situation, contrary to the Multicylindrical model which only needs data of the distribution of capillary distances (or domain areas), e.g. in terms of the two parameters median distance and logarithmic standard deviation (Turek & Rakušan, 1981; Rakušan *et al.*, 1984).

All these models rely on the absence of any appreciable axial diffusion, i.e. in the z direction. Investigations mostly on the Krogh model show that, under certain conditions, this is a valid assumption (see Kreuzer, 1982; Hoofd *et al.*, 1990*a*; Groebe, this volume). One might argue that numerical models of tissue oxygenation will be superior. This is a dubious and maybe meaningless statement, even apart from the considerable difficulties of proper program design. For common deterministic models, in view of the locally high gradients and small dimensions, one would need a very small step, say 1 μm as an upper limit, whereas the tissue block to be handled must be quite large (Hoofd *et al.*, 1989), say 100 μm × 100 μm × 100 μm, resulting in 10^6 (!) calculation points where a

---

Fig. 9. O_2 histograms calculated in a tissue block of 177 μm × 140 μm × 250 μm, for *A*, heterogeneous and *B*, regular capillary distributions of Fig. 8. Bars, class frequencies (left axis); ★ and connecting lines, cumulative histogram (right axis). The dark bar is for values of p calculated below zero.

3-dimensional second-order differential equation must be handled. In addition there are the unresolved problems of blood flow distribution, the O_2 release in erythrocytes and the fact that a trial-and-error method must be used since boundary conditions are known only very locally (namely, at the arteriolar ends). As to non-deterministic models, for stochastic models Groebe & Thews (1989) came to the conclusion that these were practically not applicable here. With this in mind, and an extensive choice of different tissue situations that might be covered, one might opt for an easier model.

The Krogh model alone has its value for judging particular situations, but clearly is unable to predict O_2 distribution for a whole tissue (see Fig. 9 above; and Egginton & Ross, this volume). Consequently, the same must be presumed for all models using regular plans, including 3-dimensional ones, even when numerically solved (e.g. Grunewald, 1973; Metzger, 1973). It is sometimes argued that other models might do better than the Krogh model, or might be more applicable; in particular, the so-called concentric models, where O_2 diffuses into a solid tissue cylinder from outwardly located capillaries (Groom *et al.*, 1984; Piiper & Scheid, 1986). In view of a whole tissue situation, however, such a statement becomes redundant; whether tissue is modelled differently should not influence its O_2 distribution, at least not appreciably. Correct comparison of both models confirms this view (Rakušan *et al.*, 1984; Hoofd *et al.*, 1987). But the concentric models are able to incorporate other features, like some forms of tissue heterogeneities (Hoofd *et al.*, 1990*b*). The models presented here have been applied to investigate various tissue situations, although not as yet so much for the newer Multicapillary model (Hoofd & Turek, 1991). The reader is referred to the relevant literature.

Acknowledgement

The author wishes to thank Dr Z. Turek for valuable advice and discussions.

Appendix: Symbols and abbreviations

A	Area
C_p	Constant in p field
CB	Capillary Barrier
c	(Total) Concentration
D	Diffusion coefficient
d	distance

EP	Extraction Pressure
F	Flow
f	(Fourier) Coefficients
$f(\ldots)$	Fourier functional coefficients
Hb	Haemoglobin
h	length
$I_n(\ldots)$	Modified Bessel function of order n
J	1-dimensional flux
$\vec{J}$	flux vector
$K_n(\ldots)$	Modified Bessel function of order n
k	backward reaction rate
k'	forward reaction rate
L	Cylinder length
M	(Maximal) Oxygen consumption
Mb	Myoglobin
$m(p)$	O_2-dependent O_2 consumption
N	Total number
n	Index for counting
$\mathscr{P}$	Oxygen permeability
p	Oxygen partial pressure
p_{50}	Oxygen pressure for 50% saturation
p_F	Facilitation pressure
q_{50}	Oxygen pressure for 50% consumption
R	Krogh cylinder radius
r	Radial coordinate
$\vec{r}$	Location vector
S	Saturation with oxygen
v	Velocity
x, y	Cartesian coordinates
z	Axial coordinate

Greek symbols

α	Oxygen solubility
ΔR	Diffusion length difference
Δr	Radial step
$\Delta x, \Delta y$	Corner-to-location distances
λ	Characteristic length
μ	Oxygen release
$\mu(z)$	z-dependent oxygen release
Ξ	Zone coefficient
ξ	Dummy variable

ϱ	Zone boundary
$\Phi(\vec{r})$	Consumption field term
ϕ	Angle in cylindrical coordinate system

Mathematical symbols

∂	Partial differential operator
$\vec{\nabla}$	Gradient operator (vector)
∇	Gradient operator component
$\cdot$	Inner product operator
$\lvert\ldots\rvert$	Absolute value of ...
$[\ldots]$	Concentration of species ...

Special subscripts

c	Capillary
e	Erythrocyte
Hm	Haem
$0, 1, 2, \ldots, m, n$	Indices
r	Radial component
s	Corner
t	Tissue
z	Axial component

Special superscripts

$'$	Different situation
$''$	Different situation

References

Aroesty, J. & Gross, J.F. (1970). Convection and diffusion in the microcirculation. *Microvascular Research* **2**, 247–67.

Bouwer, S.Th. (1987). Facilitated oxygen diffusion through hemoglobin solutions. Measurement of diffusion and reaction parameters. Dissertation Thesis, University of Nijmegen, The Netherlands.

Clark, P.A., Kennedy, S.P. & Clark, A., Jr (1989). Buffering of muscle tissue P_{O_2} levels by the superposition of the oxygen field from many capillaries. *Advances in Experimental Medicine and Biology* **248**, 165–74.

de Koning, J., Hoofd, L.J.C. & Kreuzer, F. (1981). Oxygen transport and the function of myoglobin. Theoretical model and experiments in chicken gizzard smooth muscle. *Pflügers Archiv* **389**, 211–17.

Duling, B.R., Damon, D.N., Donaldson, S.R. & Pittman, R.N. (1983). A computerized system for densitometric analysis of the microcirculation. *Journal of Applied Physiology* **55**, 642–51.

Fletcher, J.E. (1978). Mathematical modeling of the microcirculation. *Mathematical Biosciences* **38**, 159–202.

Groebe, K. (1990). A versatile model of steady state O_2 supply to tissue. Application to skeletal muscle. *Biophysical Journal* **57**, 485–98.

Groebe, K. & Thews, G. (1989). Effects of red cell spacing and red cell movement upon oxygen release under conditions of maximally working skeletal muscle. *Advances in Experimental Medicine and Biology* **248**, 175–85.

Groom, A.C., Ellis, C.G. & Potter, R.F. (1984). Microvascular geometry in relation to modeling oxygen transport in contracted skeletal muscle. *American Review of Respiratory Disease* **129** (Suppl.), S6–9.

Grunewald, W. (1973). Computer calculation for tissue oxygenation and the meaningful presentation of the results. *Advances in Experimental Medicine and Biology* **37B**, 783–92.

Honig, C.R., Gayeski, T.E.J., Federspiel, W., Clark, A., Jr & Clark, P. (1984). Muscle O_2 gradients from hemoglobin to cytochrome: new concepts, new complexities. *Advances in Experimental Medicine and Biology* **169**, 23–38.

Hoofd, L. (1987). Facilitated diffusion of oxygen tissue and model systems. Dissertation Thesis, University of Nijmegen, The Netherlands.

Hoofd, L. & Kreuzer, F. (1979). A new mathematical approach for solving carrier-facilitated steady state diffusion problems. *Journal of Mathematical Biology* **8**, 1–13.

Hoofd, L. & Kreuzer, F. (1980). Facilitation of oxygen diffusion in hemoglobin solutions: new theoretical aspects. In *Proceedings of the 21st Dutch Federation Meeting*, Abstract 183. Nijmegen: Federation of Medical Scientific Societies.

Hoofd, L., Olders, J. & Turek, Z. (1988). A simple method to calculate oxygen fields in a tissue cross-section with arbitrary capillary distribution. In *Proceedings of the 29th Dutch Federation Meeting*, Abstract 175. Nijmegen: Federation of Medical Scientific Societies.

Hoofd, L., Olders, J. & Turek, Z. (1990*a*). Oxygen pressures calculated in a tissue volume with parallel capillaries. *Advances in Experimental Medicine and Biology* **277**, 21–9.

Hoofd, L. & Turek, Z. (1992). pO_2 Histograms calculated in a block of rat heart tissue. In *Abstract book ISOTT Conference 1991 – Willemstad, Curaçao*. Rotterdam: Erasmus University Rotterdam, Department of Anaesthesiology (in press).

Hoofd, L., Turek, Z. & Egginton, S. (1990*b*). Concentric oxygen diffusion in tissue with heterogeneous permeability and consumption. *Advances in Experimental Medicine and Biology* **277**, 13–20.

Hoofd, L., Turek, Z., Kubat, K., Ringnalda, B.E.M. & Kazda, S. (1985). Variability of intercapillary distance estimated on histological sections of rat heart. *Advances in Experimental Medicine and Biology* **191**, 239–47.

Hoofd, L., Turek, Z. & Olders, J. (1989). Calculation of oxygen pressures and fluxes in a flat plane perpendicular to any capillary distribution. *Advances in Experimental Medicine and Biology* **248**, 187–96.

Hoofd, L., Turek, Z. & Rakušan, K. (1987). Diffusion pathways in oxygen supply of cardiac muscle. *Advances in Experimental Medicine and Biology* **215**, 171–7.

Kety, S.S. (1957). Determinants of tissue oxygen tension. *Federation Proceedings* **16**, 666–70.

Kreuzer, F. (1982). Oxygen supply to tissues: the Krogh model and its assumptions. *Experientia* **38**, 1415–26.

Kreuzer, F. & Hoofd, L. (1987). Facilitated diffusion of oxygen and carbon dioxide. In *Handbook of Physiology*, Section 3, Vol. IV, *The Respiratory System: Gas Exchange*, ed. L.E. Fahri & S.M. Tenney, pp. 89–111. Bethesda, MD: American Physiological Society.

Kreuzer, F. & Yahr, W.Z. (1960). Influence of red cell membrane on diffusion of oxygen. *Journal of Applied Physiology* **15**, 1117–22.

Krogh, A. (1919). The number and distribution of capillaries in muscle with calculations of the oxygen pressure head necessary for supplying the tissue. *Journal of Physiology (London)* **52**, 409–15.

Mainwood, G.W. & Rakušan, K. (1982). A model for intracellular energy transport. *Canadian Journal of Physiology and Pharmacology* **60**, 98–102.

Metzger, H. (1973). Geometric considerations in modeling oxygen transport processes in tissue. *Advances in Experimental Medicine and Biology* **37B**, 761–72.

Piiper, J. & Scheid, P. (1986). Cross-sectional P_{O_2} distributions in Krogh cylinder and solid cylinder models. *Respiration Physiology* **64**, 241–51.

Popel, A.S. (1989). Theory of oxygen transport to tissue. *Critical Reviews in Biomedical Engineering* **17**, 257–321.

Rakušan, K., Hoofd, L. & Turek, Z. (1984). The effect of cell size and capillary spacing on myocardial oxygen supply. *Advances in Experimental Medicine and Biology* **180**, 463–75.

Schultz, J.S., Goddard, J.D. & Suchdeo, S.R. (1974). Facilitated transport via carrier-mediated diffusion in membranes I. Mechanistic aspects, experimental systems and characteristic regimes. *AIChE Journal* **20**, 417–45.

Starlinger, H. & Lübbers, D.W. (1972). Methodical studies on the polarographic measurement of respiration and 'critical oxygen pressure' in mitochondria and isolated cells with membrane-covered platinum electrodes. *Pflügers Archiv* **337**, 19–28.

Turek, Z., Hoofd, L. & Rakušan, K. (1986). Myocardial capillaries and tissue oxygenation. *Canadian Journal of Cardiology* **2**, 98–103.

Turek, Z., Hoofd, L. & Rakušan, K. (1987). A comparison of the methods for assessment of the heterogeneity of myocardial capillary spacing. *Advances in Experimental Medicine and Biology* **215**, 13–19.

Turek, Z., Olders, J., Hoofd, L., Egginton, S., Kreuzer, F. & Rakušan, K. (1989). P_{O_2} histograms in various models of tissue oxygenation in skeletal muscle. *Advances in Experimental Medicine and Biology* **248**, 227–37.

Turek, Z. & Rakušan, K. (1981). Lognormal distribution of intercapillary distance in normal and hypertrophic rat heart as estimated by the method of concentric circles: its effect on tissue oxygenation. *Pflügers Archiv* **391**, 17–21.

Turek, Z., Rakušan, K., Olders, J., Hoofd, L. & Kreuzer, F. (1991). Computed myocardial P_{O_2} histograms: effects of various geometrical and functional conditions. *Journal of Applied Physiology* **70**, 1845–53.

K. GROEBE

Factors important in modelling oxygen supply to red muscle

Introduction

Oxygen supply to oxidative muscle is a highly complex process in which convectional and diffusional transport processes are involved, and which depends on a variety of physiological and morphological factors. However, in modelling O_2 supply simplifying assumptions on the transport system are inevitable. Therefore, it is of interest to identify those factors which are particularly important, the impacts of which must not be neglected. While the significance of heterogeneities in capillary geometry and flows has been dealt with elsewhere (Egginton & Ross, this volume), this chapter analyses the importance of various factors in individual capillaries and their pertinent capillary domains. Figure 1 displays a schematic drawing of the physiological situation and points out some of the factors that influence O_2 transport: a capillary domain is supplied from a central capillary which is perfused by a given blood flow. Actual capillary domain shapes are variable with location in the muscle and with time, and depend upon capillary network geometry as well as upon dynamic quantities such as the distribution of blood flow rates and capillary P_{O_2} values. Adjacent red blood cells (RBC) within the capillary are separated by plasma-filled gaps. Most of the capillary domain is occupied by the muscle fibres in which myoglobin (Mb) serves as an oxygen carrier. Erythrocytes and muscle fibres are separated by a carrier-free region (CFR) which is made up of a peri-erythrocytic plasma sleeve, capillary endothelium, and interstitial space and which has a thickness of about 1.5 μm (cf. Groebe, 1990 and unpublished data).

Before proceeding to the actual analysis there are two decisions to make:

1. *Choice of factors studied.* Parameters characterising any one of the mentioned factors (such as size of RBC, capillary, etc., O_2 diffusivity, O_2 solubility) exert influence upon O_2 transport amounting to a vast number of parameter influences that

Society for Experimental Biology Seminar Series 51: *Oxygen Transport in Biological Systems*, ed. S. Egginton & H.F. Ross.

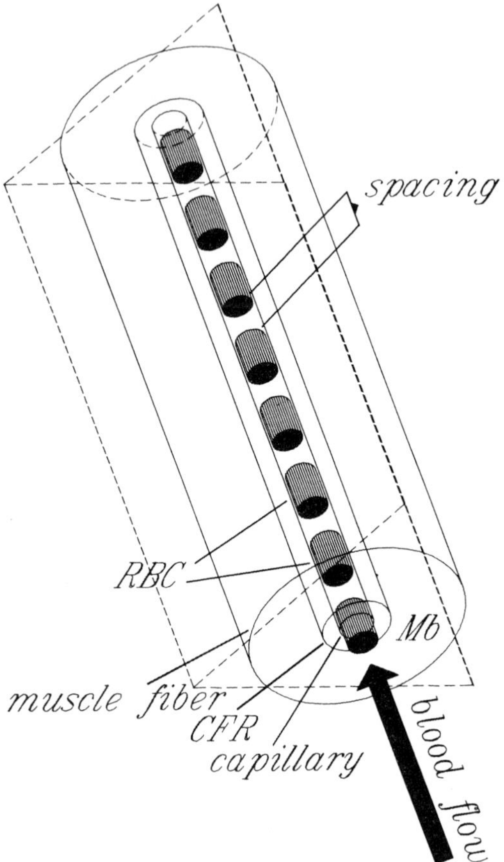

Fig. 1. Schematic drawing of capillary domain geometry in muscle. The capillary domain is represented by a large circular or triangular cylinder. RBC, red blood cells; 'spacing', plasma-filled gaps; Mb, myoglobin; CFR, carrier-free region.

could possibly be studied. In view of former work that has been done on this topic, this study will only deal with two groups of parameters appertaining to

geometric factors and
role of myoglobin.

2. *Choice of basic data sets.* The importance of a given factor may change depending on the data set which is taken as the 'control case', and from which deviations of the respective

Table 1. *List of performance-independent data employed in the calculations. All data refer to 37°C*

Symbol		Description		Value
α_{Blood}	=	O_2 solubility in blood	=	1.4×10^{-6} mol l^{-1} mm Hg^{-1}
α_{CFR}	=	O_2 solubility in the carrier-free layer	=	9.4×10^{-7} mol l^{-1} mm Hg^{-1}
α_M	=	O_2 solubility in the muscle fibre	=	9.4×10^{-7} mol l^{-1} mm Hg^{-1}
α_{RBC}	=	O_2 solubility in the RBC	=	1.5×10^{-6} mol l^{-1} mm Hg^{-1}
C_{Hb}	=	total Hb concentration in the RBC	=	0.0203 mol l^{-1}
C_{Mb}	=	total Mb concentration in the fibre	=	5.4×10^{-4} mol l^{-1}
d_{CFR}	=	thickness of carrier-free region	=	1.5 μm
D_{CFR}	=	O_2 diffusion coefficient in the CFR	=	1.65×10^{-5} cm^2 s^{-1}
D_M	=	O_2 diffusion coefficient in the fibre	=	1.16×10^{-5} cm^2 s^{-1}
D_{Mb}	=	Mb diffusion coefficient in the fibre	=	5.47×10^{-7} cm^2 s^{-1}
D_{RBC}	=	O_2 diffusion coefficient in the RBC	=	9.5×10^{-6} cm^2 s^{-1}
Hct	=	blood hematocrit	=	0.5
k	=	rate constant of O_2 dissociation from HbO_2	=	44 s^{-1}
L	=	capillary length	=	1150 μm
L_{RBC}	=	red cell length	=	5.2 μm
P_{50}	=	half-saturation P_{O_2} of Mb	=	5.3 mm Hg
P_{crit}	=	critical mitochondrial P_{O_2}	=	0.5 mm Hg
R_{RBC}	=	red cell radius	=	2.0 μm
S_A	=	arterial Hb-O_2 saturation	=	0.95
σ_{RBC}	=	RBC surface area	=	65.3 μm^2
V_{RBC}	=	red cell volume	=	66 fl

factors are studied. For the results to be realistic, it is therefore essential that all values in any one parameter set used pertain to one and the same physiological situation. On the other hand, the collection of control cases employed should cover a physiological spectrum. In the present study, all investigations have been performed using five control cases, the parameters of which have been chosen to describe working dog gracilis muscle electrically stimulated at rates between 1 and 8 Hz. Corresponding O_2 consumption rates ($\dot{V}_{O_2}$) range from 3 to 15 ml O_2 100 g^{-1} min^{-1}. All data are compiled in Tables 1 and 2.[1] Their choice has been discussed by Groebe (1990, and unpublished data) where the relevant references are also cited.

[1] The density of perfused muscle capillaries varies with muscle performance and is specified as 'functional capillary density' in Table 2.

Table 2. *List of data specific for the respective performances studied*

		Stimulation frequency (Hz)				
		1	2	4	6	8
$\dot{V}_{O_2}$	= O_2 consumption rate ($ml\ O_2\ 100\ g^{-1}\ min^{-1}$)	3	6	11	13	15
$\dot{Q}$	= blood flow rate (ml blood 100 g^{-1} min $^{-1}$)	26.1	40.3	68.6	84.1	119.1
S_v	= venous Hb saturation	0.45	0.30	0.25	0.275	0.40
fcd	= functional capillary density (mm^{-2})	800	950	1000	950	650
R_K	= radius of capillary domain (μm)	19.9	18.3	17.8	18.3	22.1
f_{RBC}	= fraction of capillary length occupied by RBCs	0.444	0.5	0.571	0.625	0.667
	organ venous blood pH	7.37	7.31	7.26	7.24	7.20
n	= Hill coefficient of Hb	3.02	3.19	3.34	3.41	3.95
P_{50}	= half saturation P_{O_2} of Hb (mm Hg)	29.0	30.2	31.2	31.6	34.0

Methods

The model

The model applied in this study will be described in detail elsewhere (Groebe, unpublished data), and is based on the analytical description of the P_{O_2} distribution in the domain of an individual capillary as shown in Fig. 1. In this process, the diffusion equation is solved separately in capillary, CFR and muscle fibre, and the three solutions are linked together by appropriate interface conditions. In considering RBC spacing, a simplified procedure is used (see 'Results and Discussion', p. 243) which has been shown to yield good approximations to the real situation (Groebe, 1990). Capillary domain size is initially calculated from the given capillary density, and the actual size of well-oxygenated tissue is set to accordingly smaller values if capillary P_{O_2} becomes too low to supply the entire domain. This procedure allows for an estimate of the anoxic tissue portion. The model extends the classical Kroghian approach in several respects:

1. RBC O_2 unloading along the capillary;
2. Effects of the particulate nature of blood;

3. Free and haemoglobin-facilitated O_2 diffusion and reaction kinetics inside RBCs;
4. Free and myoglobin-facilitated O_2 diffusion inside the tissue;
5. Free O_2 diffusion in carrier-free region separating RBC and tissue;
6. Capillary-to-fibre ratio of 1 or, optionally, of 2;
7. Optional diffusional O_2 transport in parallel with the capillary direction.

Presentation of results

In the first instance, O_2 supply models such as the one used in the present study result in spatial distributions of P_{O_2} or Mb–O_2 saturation within the modelled tissue region. Figure 2 shows such a P_{O_2} distribution in a longitudinal section through the capillary domain which is typical for heavily working dog gracilis muscle. If more detailed information is needed this may be obtained from *x–y* plots of P_{O_2} profiles along specified lines through the capillary domain. Figure 3, for example, displays radial P_{O_2} profiles from the 3-dimensional P_{O_2} distribution at a given coordinate along the capillary.

The above forms of graphical presentation, though fairly evident, are not very well suited for comparing different tissue P_{O_2} distributions in entire capillary domains. Therefore, from these spatial P_{O_2} distributions 'probability distributions' of the tissue P_{O_2} (or of Mb–O_2 saturation) have been calculated which specify the likelihood of measuring a given P_{O_2} (or Mb–O_2 saturation) in a randomly chosen tissue location. For presenting several of these distributions in a single diagram, their information content has been further condensed in that only mean and standard deviation of a given P_{O_2} or Mb saturation distribution are displayed as functions of $\dot{V}_{O_2}$ (e.g. Fig. 14*A*). Using the same style, the distribution of P_{O_2} drops (ΔP_{O_2}) between RBCs and their pertinent tissue cross-sections (and the fraction of anoxic tissue) are also presented (e.g. Fig. 11*B*).

Results and Discussion

For interpreting the changes in tissue P_{O_2} associated with variations in individual factors, some principles of muscle O_2 supply need to be discussed first. In working red muscle, P_{O_2} changes with distance from capillary in a fairly typical way. Figure 3 displays radial P_{O_2} profiles halfway down the capillary for low, intermediate and high transcapillary fluxes (i.e. the O_2 fluxes out of the capillary per unit of capillary length).

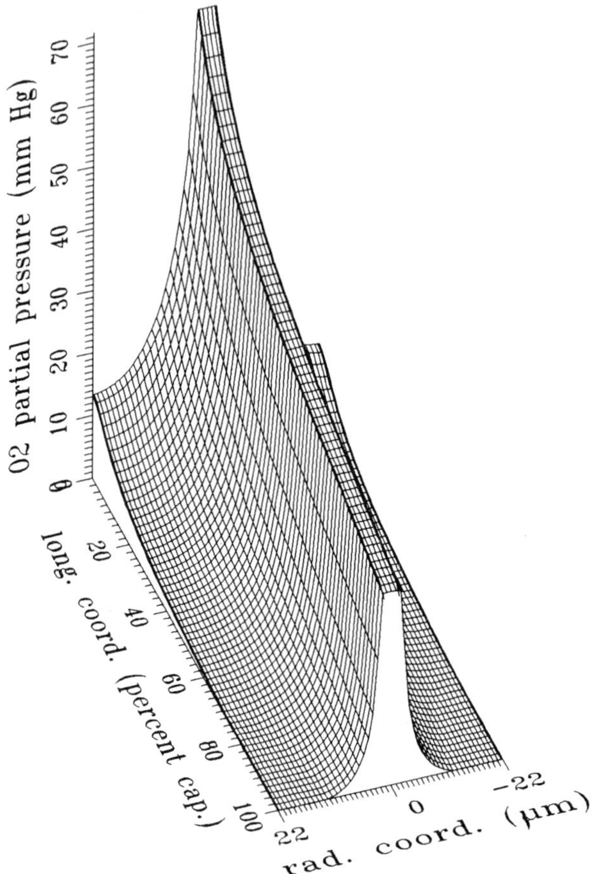

Fig. 2. P_{O_2} profiles in a longitudinal section through a capillary domain of a muscle working at $\dot{V}_{O_2} = 15\,\text{ml O}_2\ 100\,\text{g}^{-1}\,\text{min}^{-1}$. Within the central capillary, mean red cell P_{O_2} is displayed. Note the steep P_{O_2} drops in a vicinity of the capillary and the shallow gradients in the rest of the fibre.

While at $\dot{V}_{O_2} = 3\,\text{ml O}_2\,100\,\text{g}^{-1}\,\text{min}^{-1}$ P_{O_2} closely resembles the value predicted by classical Krogh models, major P_{O_2} gradients at $\dot{V}_{O_2} = 15\,\text{ml O}_2\,100\,\text{g}^{-1}\,\text{min}^{-1}$ are located in the vicinity of the capillary only, and the profile is rather shallow at low P_{O_2} levels in the rest of the fibre.

In order to explain this observation, recall that a P_{O_2} gradient is equal to the O_2 flux density divided by the O_2 conductivity, K_{O_2}. (O_2 flux density is the magnitude of the O_2 flux per cross-sectional area occupied by the flux.) Both factors, K_{O_2} and flux density, contribute towards the

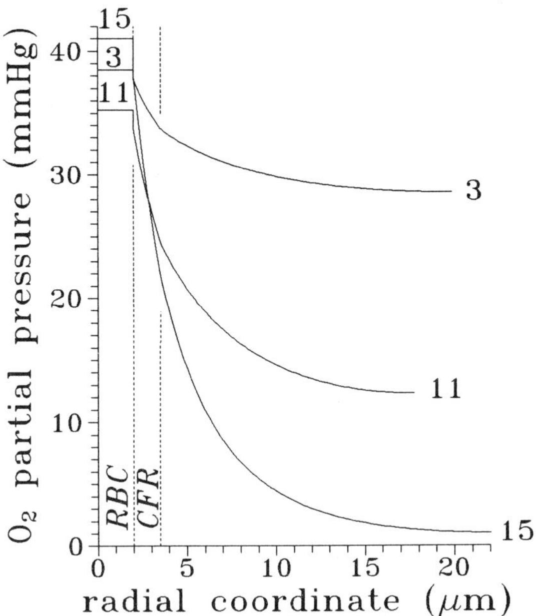

Fig. 3. Radial P_{O_2} profiles half-way down the capillary. 'RBC' and 'CFR' indicate positions of red blood cell and carrier-free region, respectively. Curve labels give O_2 consumption rates of 3, 11 and 15 ml O_2 100 g^{-1} min^{-1}.

characteristic shapes of P_{O_2} profiles. In carrier-free tissue, $K_{O_2} = D\alpha$ where D is the O_2 diffusion coefficient and α is the O_2 solubility. In the case of Mb-facilitated O_2 diffusion, O_2 conductivity has to be replaced by the effective O_2 conductivity which at O_2 partial pressure P is given by

$$K_{O_2} = D\alpha + D_{Mb}C_{Mb}\frac{P_{50}}{(P + P_{50})^2} \tag{1}$$

where D_{Mb}, C_{Mb} and P_{50} are diffusivity, concentration and half-saturation P_{O_2} of myoglobin. Obviously, effective conductivity depends on actual P_{O_2} and on carrier characteristics. The dependence of K_{O_2} on P_{O_2} is displayed in Fig. 4 (solid curve). At high P_{O_2}, effective conductivity is almost equal to that for free O_2 diffusion (dotted horizontal line) and it becomes maximal at very small P_{O_2} values. Accordingly, in dog gracilis muscle, the portion of the conductivity which is mediated by the carrier Mb (dashed line) drops from five times free O_2 conductivity at $P_{O_2} = 0$ mm Hg to zero at high P_{O_2}.

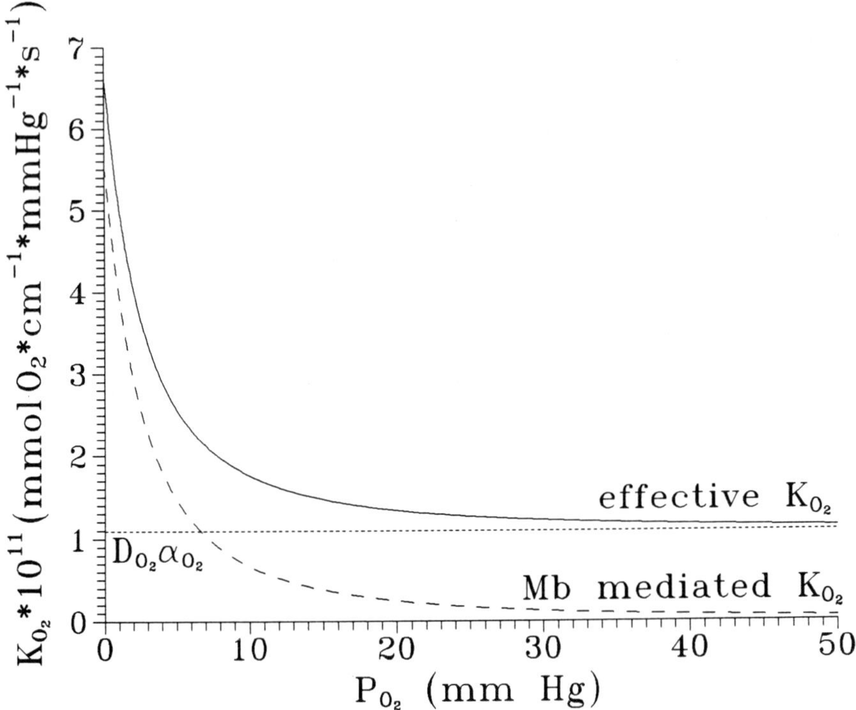

Fig. 4. P_{O_2} dependence of effective O_2 conductivity in the presence of myoglobin ('effective K_{O_2}', solid curve). Effective conductivity at high P_{O_2} is represented by the dotted horizontal line ($D_{O_2}\alpha_{O_2}$) and Mb-mediated conductivity by the broken line.

As stated above, P_{O_2} gradients are proportional to O_2 flux density. Because at any instant all of the oxygen consumed within the fibre emerges from the RBCs present within the capillary domain and because the aggregate volume of these RBCs is about 200 times smaller than the fibre volume, O_2 flux density is largest near the RBC surface. In addition, this is the location where there is no facilitated diffusion and where effective conductivity therefore is smallest. This explains why P_{O_2} falls rapidly in the peri-capillary region. Further remote from the capillary, flux densities become smaller, P_{O_2} is lower and facilitation of O_2 diffusion becomes effective. Therefore, gradients are essentially shallow there.

In the analyses that follow, at first various geometric factors will be discussed, starting with those relating to the entire capillary domain and continuing with the geometry in the capillary and in the peri-capillary region.

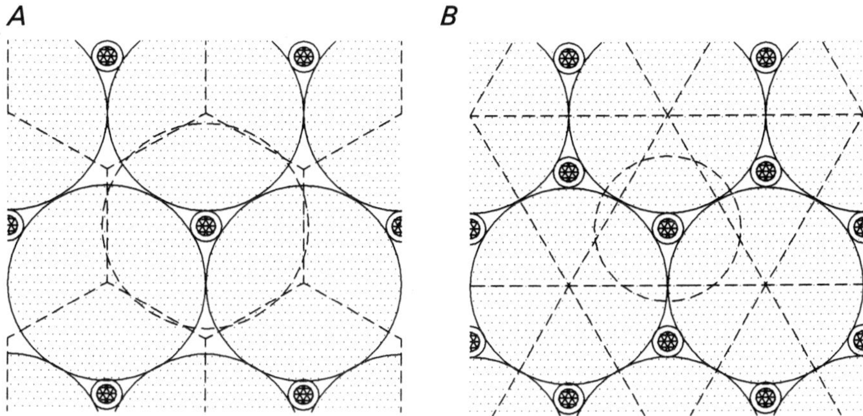

Fig. 5. Cross-sections through idealised muscles exhibiting c/f ratios of 1 (*A*) and 2 (*B*). Large dot-filled circles, muscle fibres; small circles, capillaries containing RBCs (cross-hatched), surrounded by CFRs; dashed hexagons and triangles, capillary domains; dashed circles, Krogh cylinders of same size as capillary domains.

Shape of capillary domain cross-sections

Classical Krogh models assume that capillary domains are of circular shape, which is not only the most convenient geometry for modelling but also the most efficient one for O_2 supply of a given tissue volume by a given number of capillaries. In muscle, capillaries can be positioned only in the spaces between muscle fibres which may enforce other than circular shapes of capillary domain cross-sections. A descriptor holding information on these shapes is the capillary-to-fibre ratio (c/f ratio). This is illustrated in Fig. 5 which displays cross-sections through an idealised muscle. In Fig. 5*A*, the c/f ratio is 1 and capillary domains assume hexagonal shapes. Obviously, these hexagons are well approximated by circles, suggesting that the cylinder geometry should furnish an appropriate description of the physiological situation. In Fig. 5*B*, the muscle fibres are unchanged, the capillary density, however, is doubled. Thus, c/f ratio is 2 and capillary domains are of triangular shape. Apparently, a cylinder of matching size (dashed circle) represents but a poor approximation to the capillary domain. As typical c/f ratios range between 1 and 2 (see e.g. Martin *et al.*, 1932) this discrepancy needs further evaluation.

Figure 6*A* gives the means and standard deviations of the distributions of P_{O_2} drops between RBCs and adjacent tissue for the two capillary domain shapes at the five different muscle performances considered.

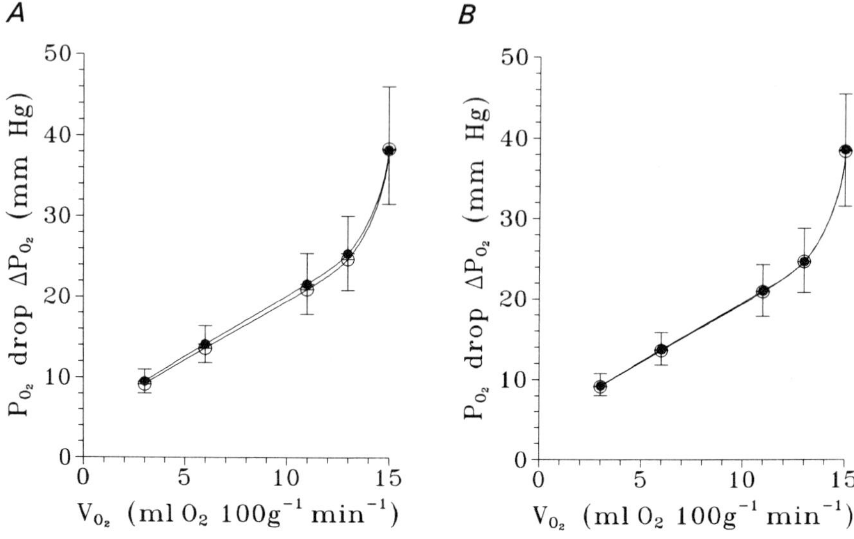

Fig. 6. Means and standard deviations of the distributions of P_{O_2} drops between RBC and adjacent tissue for *A*, different capillary domain shapes (●, triangular domains; ○, axially symmetric domains) and *B*, different modes of diffusion within domains (●, combined radial and longitudinal diffusion; ○, exclusively radial diffusion) as functions of muscle O_2 consumption rate.

Capillary domain cross-sectional area is the same in both cases. Deviations of P_{O_2} distributions in the axially symmetric case from the ones in the triangular case are much less pronounced than could have been expected from the differences in the shapes of capillary domains (Fig. 5*B*). Mean values and standard deviations of corresponding P_{O_2} probability distributions for the two cases differ by 5% or less. As could have been expected from the shapes of the capillary domains, mean values are generally larger and variations are smaller in the axially symmetric case.

In consequence, a c/f ratio of up to 2 generally does not impose any restrictions on the applicability of axisymmetric capillary cylinder models to muscle. Even at a performance of 15 ml O_2 100 g^{-1} min^{-1}, at which point anoxia starts to appear, histograms from both types of models, as well as mean values and standard deviations, are still very similar. It should be noted, though, that typical cell diameters in dog gracilis are about 50 μm (Gayeski & Honig, 1986). This corresponds to functional capillary densities of 509 mm^{-2} for a c/f ratio of 1 and of 1018 mm^{-2} for a c/f ratio of 2. While functional capillary densities of around 650 mm^{-2} are observed at maximum muscle performance in electrically stimulated muscles, values of about 1000 mm^{-2} are typical for stimulation at 4 Hz. Therefore, in the most critical supply situations, c/f ratios are actually closer to 1 than to 2.

Role of O_2 diffusion parallel to capillary direction

A particularly important assumption for keeping O_2 supply models easily tractable is the one of negligible O_2 diffusional flux parallel to the capillary direction. This assumption has been shown to be feasible in the absence of O_2 carriers in the tissue (Thews, 1953, 1960). In the presence of Mb, however, O_2 diffusion is greatly enhanced, particularly at the low values of fibre P_{O_2} present in heavily working muscle. Furthermore, it has been found that O_2 diffusivity parallel to the fibre direction is 2.5 times larger than the one perpendicular to fibres (Homer *et al.*, 1984), and the same is likely to be true for Mb diffusivity also. As a consequence, the importance of longitudinal diffusion has to be reassessed.

Figure 6*B* displays means and standard deviations of the P_{O_2} drops between RBC and tissue for the five levels of performance considered, and for the cases of combined radial and longitudinal diffusion or of exclusively radial diffusion. In good agreement with experimental findings (Gayeski & Honig, 1988), longitudinal P_{O_2} gradients are small and the two cases are virtually identical. Consequently, the assumption of negligible diffusional O_2 flux parallel to the capillary represents an excellent approximation to the real situation.

Role of heterogeneous O_2 consumption

It has been observed that in at least some muscles some of the mitochondria tend to be arranged in clusters located in the vicinity of capillaries (e.g. Hoppeler *et al.*, 1991). The function of these mitochondrial clusters is not yet understood. Under the assumption that at a given energy demand, and hence $\dot{V}_{O_2}$, local O_2 consumption rate is distributed in the same way as mitochondrial density, peri-capillary clustering should have an effect on O_2 transport because average O_2 diffusion paths become shorter. In the present investigation, a distribution of O_2 consumption rate has been studied which in an annulus of 20 μm diameter surrounding the capillary is twice as high as in the rest of the fibre. Radial P_{O_2} profiles half-way down the capillary for three average O_2 consumption rates are displayed in Fig. 7*A*. As expected, P_{O_2} gradients in the fibre are shallower for heterogeneous consumption. Average P_{O_2} drops between RBC and muscle fibre, as functions of (average) O_2 consumption rate, are dis-

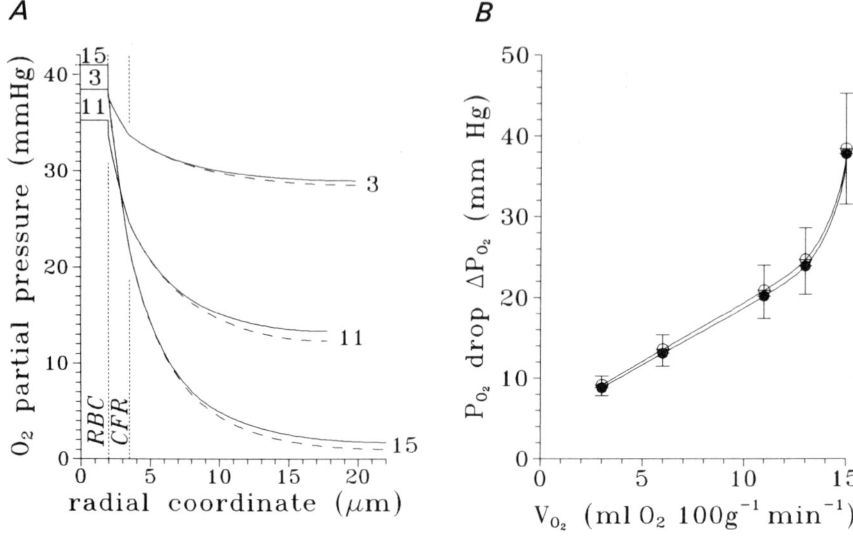

Fig. 7. Effects of heterogeneously distributed O_2 consumption rate. *A*, Radial P_{O_2} profiles half-way down the capillary for average O_2 consumption rates of 3, 11 and 15 ml O_2 100 g^{-1} min^{-1} (curve labels). Solid lines, heterogeneous $\dot{V}_{O_2}$ (i.e. $\dot{V}_{O_2}$ in an annulus of 20 μm diameter surrounding the CFR twice as high as in the rest of the capillary domain); broken lines, homogeneous $\dot{V}_{O_2}$. *B*, Average P_{O_2} drops between RBC and muscle fibre along with their standard deviations as functions of (average) muscle O_2 consumption rate. ●, heterogeneous $\dot{V}_{O_2}$; ○, homogeneous $\dot{V}_{O_2}$.

played in Fig. 7*B*. P_{O_2} drops and their variability become smaller when part of the O_2 consumption is concentrated next to the capillary. However, this effect is very small in spite of the quite massive heterogeneity assumed. While mean P_{O_2} drops decrease by only about 3% their standard deviations fall by 10%. Altogether, heterogeneous O_2 consumption is not a matter to be particularly concerned about in modelling.

Importance of RBC spacing

Hellums (1977) was the first to point out that only those loci of the capillary surface adjacent to RBCs are maximally effective in O_2 exchange, and that because of this observation a higher P_{O_2} drop is required as driving force for a given transcapillary O_2 flux. An upper bound to this increased driving force may easily be obtained by assuming that all along the O_2 diffusion path outward diffusion of oxygen takes

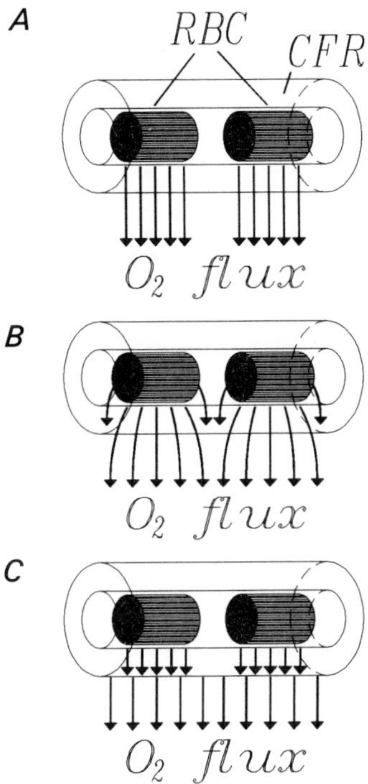

Fig. 8. Different approaches to considering effects of intracapillary RBC spacing. *A*, Upper bound. *B*, Real situation. *C*, Approximation in the present model. For more details see text.

place only adjacent to capillary sites at which RBCs are located (Fig. 8*A*). In this case, the P_{O_2} drop for free O_2 diffusion would be equal to the P_{O_2} drop without spacings divided by the fraction of capillary length containing RBCs. The real situation is different, however, in that the cross-sectional area occupied by the O_2 flux is small next to the RBC surface and grows larger with distance from the capillary lumen (Fig. 8*B*). A precise mathematical treatment of this situation is fairly complex and would involve considering diffusion parallel to the capillary direction in the peri-capillary region (see e.g. Groebe & Thews, 1989). In the present model, a simplified procedure is applied which has been shown to approximate the real situation in working skeletal muscle quite well (Groebe & Thews, 1990): O_2 flux is assumed to be exclusively radial throughout. While within the CFR this flux is modelled to occur immediately subjacent to RBC membranes only, a constant O_2 outward flux all along the capillary is assumed within the muscle fibre (Fig. 8*C*).

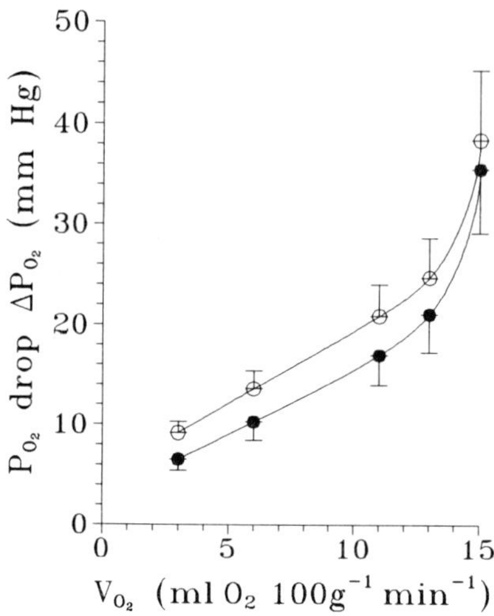

Fig. 9. Effects of neglecting RBC spacing on average P_{O_2} drops between RBC and muscle fibre and their standard deviations (as functions of muscle O_2 consumption rate). ●, without spacing; ○, with RBC spacing.

The consequences of neglecting RBC spacing upon P_{O_2} drops between RBC and fibre are shown in Fig. 9. At all performances, ΔP_{O_2} decreases by 3–4 mm Hg which corresponds to 10–30% of ΔP_{O_2}. Independence of the absolute rather than the relative changes in ΔP_{O_2} upon performance is caused by reduction in the control case spacing with increasing O_2 consumption rate (cf. Table 2). In summary, effects of neglecting RBC spacing are not huge, but are certainly worth consideration.

Importance of carrier-free region

In calculations in this section the carrier-free region (CFR) has been omitted. Figure 10*B* shows mean P_{O_2} drops between RBC and tissue as functions of O_2 consumption rate. As expected, without CFR ΔP_{O_2} is smaller by 2–3 mm Hg. In view of the large fraction of the total P_{O_2} drop that was shown before to occur within the CFR, the relatively small differences in mean ΔP_{O_2} appear to be somewhat surprising. This may be explained by Fig. 10*A*, which displays radial P_{O_2} profiles half-way down the capillary for three O_2 consumption rates. While profiles run essentially parallel at radial coordinates greater than 3.5 μm, there are appreciable differences in P_{O_2} gradients next to the RBC. Yet, profiles without

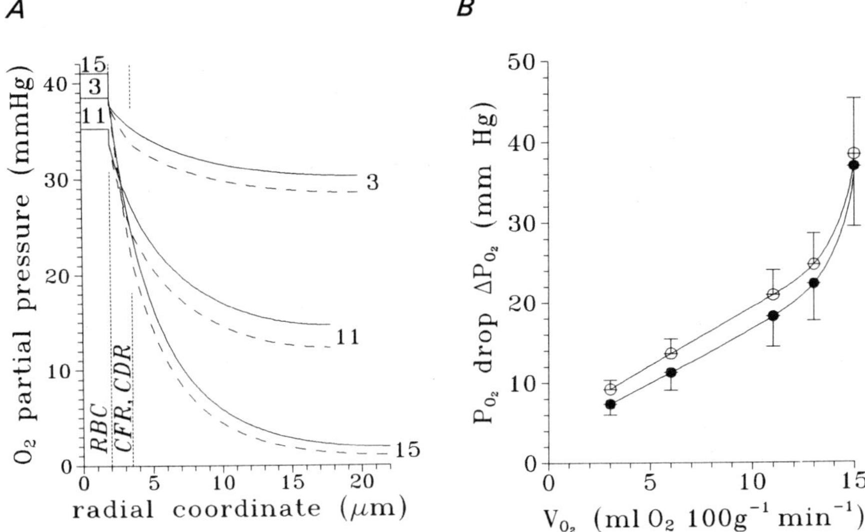

Fig. 10. Effects of neglecting the carrier-free region between erythrocyte and muscle fibre. *A*, Radial P_{O_2} profiles half-way down the capillary for O_2 consumption rates of 3, 11 and 15 ml O_2 100 g^{-1} min^{-1} (curve labels). Solid lines, without CFR; broken lines, CFR = 1.5 μm. *B*, Average P_{O_2} drops between RBC and muscle fibre along with their standard deviations as functions of muscle O_2 consumption rate. ●, without CFR; ○, CFR = 1.5 μm. In the absence of a CFR part of its role is taken over by a P_{O_2}-dependent region deficient of functional carrier surrounding the RBC (CDR).

CFR exhibit the same typical appearance that was discussed above: steep peri-capillary P_{O_2} drops and only minor P_{O_2} variations in the rest of the fibre. This observation is made because peri-capillary P_{O_2} is high and therefore Mb facilitation of O_2 diffusion has little effect. Consequently, a layer of muscle fibre surrounding the capillary acts as if it were a carrier-free region because Mb is fully saturated and non-functional as an O_2 carrier. This region therefore could be termed 'region deficient of functional carrier' or 'carrier-deficient region' (CDR).

Effect of halving RBC diameter

The shapes of red blood cells change with actual haemodynamic conditions in blood vessels (cf. Skalak & Branemark, 1969). Therefore, intracapillary RBC diameter is a dynamic quantity which cannot be assessed from standard histological preparations. As a consequence, there are relatively few data on intracapillary RBC diameter, and it is of interest to study the impacts of this uncertainty for modelling. Figure 11*B* illustrates P_{O_2} drops between RBC and tissue as functions of O_2 consumption rate for RBC radii of 1 and 2 μm. All other data are the same for both cases. P_{O_2} drops increase by 3–7 mm Hg or 30–40% if RBC radius is halved. Effects are much the largest demonstrated so far in this chapter, even though O_2 diffusion distances for any tissue location are not changed by any more than 1 μm! In order to understand the magnitude of this effect, one has to recall the role of O_2 flux densities discussed above: in the pericapillary region P_{O_2} gradients are proportional to O_2 flux densities, the latter being roughly proportional to the inverse distance from the capillary centre. As a consequence, P_{O_2} gradients near RBC surfaces are not only the largest compared with the rest of the tissue, but are also – because of their proximity to the capillary centre – very sensitive to RBC radius. (The smaller the RBC radius, the more sensitive they are.) Because peri-

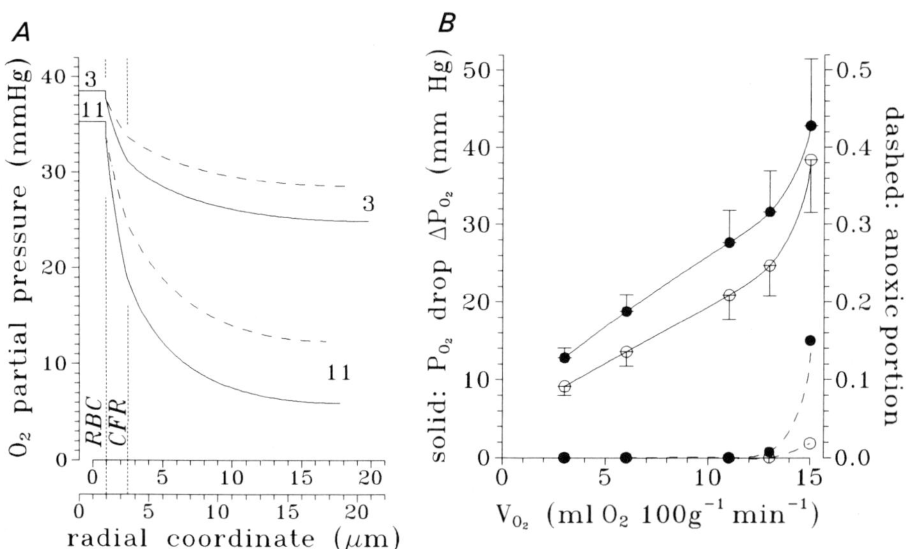

Fig. 11. Effects of variations in RBC radius, R_{RBC}. *A*, Radial P_{O_2} profiles half-way down the capillary for O_2 consumption rates of 3 and 11 ml O_2 100 g^{-1} min^{-1} (curve labels). Solid lines, $R_{RBC} = 1$ μm; broken lines, $R_{RBC} = 2$ μm. *B*, Average P_{O_2} drops between RBC and muscle fibre along with their standard deviations (solid curves) and corresponding anoxic tissue portions (broken curves) as functions of muscle O_2 consumption rate. ●, $R_{RBC} = 1$ μm; ○, $R_{RBC} = 2$ μm.

capillary P_{O_2} drops largely dominate the tissue P_{O_2} profile, the entire profile itself is also extremely sensitive to changes in RBC diameter. This is illustrated in Fig. 11*A*, which displays radial P_{O_2} profiles half-way down the capillary for two O_2 consumption rates (for the sake of clarity, profiles for 15 ml O_2 100 g^{-1} min^{-1} have been omitted). In order to keep the edges of RBC and CFR aligned for both RBC radii, the (upper) abscissa applicable to the case of an RBC radius of 1 μm has been shifted by 1 μm to the right. At corresponding performances, there are pronounced differences in P_{O_2} gradients next to the RBC surface for different RBC radii, whereas further into the fibre shapes of profiles are very similar, although at different P_{O_2} levels.

Expressed as percentage of ΔP_{O_2}, the effect of halving RBC radius is much the same for all submaximal performances. The apparently smaller effect at a consumption rate of 15 ml O_2 100 g^{-1} min^{-1} is explained by a fall in transcapillary O_2 flux which is brought about by a sudden rise of the anoxic tissue portion to 15%. This is because at near maximum performance tissue P_{O_2} is very low even in the control case, and therefore the ability to maintain transcapillary O_2 flux by increasing ΔP_{O_2} becomes exhausted.

In the following subsections, two aspects of myoglobin function will be considered.

Effect of omitting Mb facilitation of O_2 diffusion

Changes in Mb concentration have the same effects as proportional changes in Mb diffusivity. Setting Mb concentration or diffusivity to zero is equivalent to neglecting facilitation of O_2 diffusion. Figure 12*B* displays ΔP_{O_2} as a function of O_2 consumption rate for the control cases and for the cases without Mb facilitation. ΔP_{O_2} values without Mb are larger by only 0.5–3 mm Hg, which amounts to about 10% of ΔP_{O_2} at the most. Standard deviations rise by 5–30%. As opposed to these moderate effects, neglecting Mb facilitation ensures a pronounced increase in anoxic tissue portion which starts to arise at $\dot{V}_{O_2}$ = 11 ml O_2 100 g^{-1} min^{-1} and becomes as large as 20% at maximum performance.

These effects may be understood by considering the radial P_{O_2} profiles half-way down the capillary, shown in Fig. 12*A*. At $\dot{V}_{O_2}$ = 3 ml O_2 100 g^{-1} min^{-1} the presence of Mb shows hardly any effect, corresponding to poor facilitation of O_2 diffusion at high P_{O_2} values, as discussed above. At 15 ml O_2 100 g^{-1} min^{-1} and without Mb, half-way down the capillary no more than about 80% of the capillary domain is well supplied with oxygen. Furthermore, profile shape has changed significantly: peri-capillary gradients are less pronounced (because of the

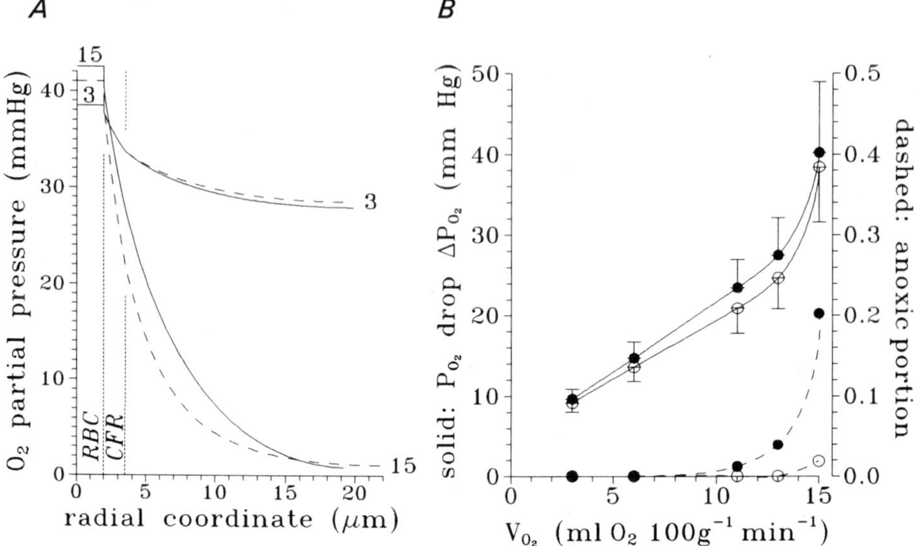

Fig. 12. Effects of neglecting facilitation of O_2 diffusion by Mb. *A*, Radial P_{O_2} profiles half-way down the capillary for O_2 consumption rates of 3 and 15 ml O_2 100 g^{-1} min^{-1} (curve labels). Solid: no Mb, dashed: $C_{Mb} = 0.54$ mM. *B*, Average P_{O_2} drops between RBC and muscle fibre along with their standard deviations (solid curves) and corresponding anoxic tissue portions (broken curves) as functions of muscle O_2 consumption rate. ●, No Mb; ○, $C_{Mb} = 0.54$ mM.

reduction in transcapillary O_2 flux) and considerable P_{O_2} gradients persist almost up to the outer edge of the capillary domain.

In Fig. 12*A*, the profiles for 15 ml O_2 100 g^{-1} min^{-1} nicely illustrate the principle of myoglobin function. At high P_{O_2} (i.e. at low performance, or close to the capillary), Mb is hardly functional at all. As P_{O_2} falls to low values facilitation becomes effective, and once the oxygen has passed the peri-capillary region with its high P_{O_2} and steep P_{O_2} drops it is transported further into the fibre almost free of diffusional resistance. Along this latter part of the O_2 diffusion path P_{O_2} gradients consequently become very shallow – a change in profile shape which does not occur in the absence of Mb. Hence, Mb may be viewed as a reserve in diffusional O_2 transport capacity which is activated whenever P_{O_2} falls to critically low values. In consequence, it is not too surprising that presence of Mb does not change tissue P_{O_2} to any great extent. Rather, it represents a means of maintaining oxygen flux at P_{O_2} levels at which anoxia would inevitably arise without Mb.

Effect of halving Mb half-saturation P_{O_2}

The dependency of effective O_2 conductivity, K_{O_2}, upon P_{O_2} has been discussed above. Mb half-saturation P_{O_2}, P_{50}, is a parameter of this dependence (Eqn 1). Particularly in the older literature, measured P_{50}s are substantially smaller than the one of 5.3 mm Hg given by Gayeski & Honig (1986). Figure 13 displays effective K_{O_2} for Mb P_{50} of 5.3 and 2.65 mm Hg as functions of P_{O_2}. At low P_{O_2}, K_{O_2} for $P_{50} = 2.65$ mm Hg is almost twice the value for $P_{50} = 5.3$ mm Hg. With increasing P_{O_2}, K_{O_2} for the latter (control) case becomes slightly larger than K_{O_2} for the case of halved P_{50}.

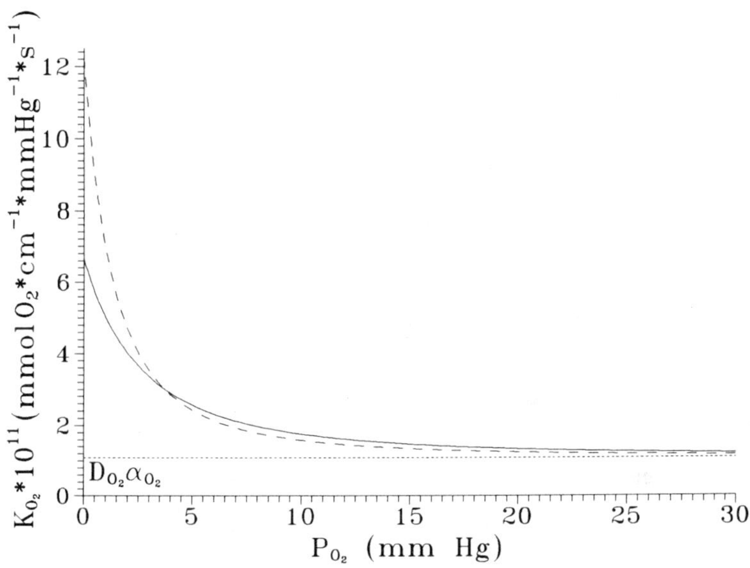

Fig. 13. Effective O_2 conductivities, K_{O_2}, for Mb $P_{50} = 5.3$ mm Hg (solid curve) and for $P_{50} = 2.65$ mm Hg (broken curve) as functions of P_{O_2}. The dotted line ('$D_{O_2}\alpha_{O_2}$') indicates free O_2 conductivity.

The effects of halving myoglobin P_{50} upon muscle Mb–O_2 saturation, upon radial P_{O_2} profiles half-way down the capillary and upon the distribution of P_{O_2} drops between RBC and tissue are illustrated by Fig. 14: saturation (*A*) as well as ΔP_{O_2} (*C*) are shown as functions of O_2 consumption rate. At all consumption rates, differences in fibre Mb–O_2 saturation between the two cases are quite pronounced and amount to about 10% saturation. By comparison, P_{O_2} drops between RBC and fibre as well as P_{O_2} profiles hardly change at all when halving P_{50}.

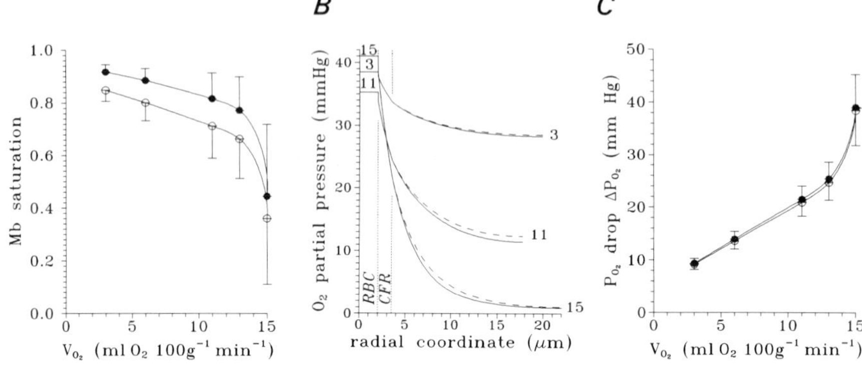

Fig. 14. Effects of variations in myoglobin half-saturation P_{O_2}, P_{50}. *A*, Fibre Mb–O_2 saturations and standard deviations as functions of muscle O_2 consumption rate. *B*, Radial P_{O_2} profiles half-way down the capillary for O_2 consumption rates of 3, 11 and 15 ml O_2 100 g^{-1} min^{-1} (curve labels). *C*, Average P_{O_2} drops between RBC and muscle fibre and standard deviations as functions of muscle O_2 consumption rate. Solid or ●, $P_{50} = 2.65$ mm Hg; dashed or ○, $P_{50} = 5.3$ mm Hg.

The corresponding radial P_{O_2} profiles are given in Fig. 14*B* for three O_2 consumption rates. As mentioned in the above discussion of Figs 4 and 13, facilitation is small at high P_{O_2} values and becomes even smaller for halved P_{50}. Peri-capillary P_{O_2} drops are correspondingly somewhat steeper in the latter case. On the other hand, the high K_{O_2} for $P_{50} = 2.65$ mm Hg at low P_{O_2} renders profiles more shallow in the periphery of capillary domains, which is particularly obvious for $\dot{V}_{O_2} = 15$ ml O_2 100 g^{-1} min^{-1}. As a consequence, at $\dot{V}_{O_2} = 15$ ml O_2 100 g^{-1} min^{-1}, P_{O_2} at the outer edge of the capillary domain is the same for both P_{50} values. The lack of impact of lowering P_{50} on ΔP_{O_2} values, as well as on their standard deviations, suggests that the opposing effects on P_{O_2} gradients at high and low P_{O_2} roughly cancel. This could be interpreted in such a way that the overall capacity of myoglobin for facilitating O_2 diffusion in a given physiological situation depends on Mb concentration and diffusivity and not on Mb P_{50}.

Conclusion

Some principles governing O_2 transport in myoglobin containing muscle tissue have been discussed. It has been shown that diffusion parallel to the capillary direction, variations in the shapes of capillary domain cross-sections of given area, and clustering of mitochondria around capillaries have very little effect on P_{O_2} distributions in red muscle and may safely be neglected. Intracapillary RBC spacing and carrier-free region between RBC and muscle cell are important to consider in modelling, even though in the absence of a CFR part of its role is taken over by a P_{O_2}-dependent region deficient of functional carrier surrounding the RBC. The most pronounced effects on P_{O_2} drops between RBC and tissue, and/or on anoxic tissue portion, were seen when RBC radius was halved or Mb facilitation of O_2 diffusion was abolished. The former result is particularly important because currently there is considerable uncertainty about intracapillary RBC diameters. Having myoglobin half-saturation P_{O_2}, on the other hand, had large impacts on tissue Mb–O_2 saturation, but not on tissue P_{O_2} distribution.

References

Gayeski, T.E.J. & Honig, C.R. (1986). O_2 gradients from sarcolemma to cell interior in red muscle at maximal $\dot{V}_{O_2}$. *American Journal of Physiology* **251**, H789–99.

Gayeski, T.E.J. & Honig, C.R. (1988). Intracellular P_{O_2} in long axis of individual fibers in working dog gracilis muscle. *American Journal of Physiology* **254**, H1179–86.

Groebe, K. (1990). A versatile model of steady state O_2 supply to tissue. Application to skeletal muscle. *Biophysical Journal* **57**, 485–98.

Groebe, K. & Thews, G. (1989). Effects of red cell spacing and red cell movement upon oxygen release under conditions of maximally working skeletal muscle. *Advances in Experimental Medicine and Biology* **248**, 175–85.

Groebe, K. & Thews, G. (1990). Role of geometry and anisotropic diffusion for modelling P_{O_2} profiles in working red muscle. *Respiration Physiology* **79**, 255–78.

Hellums, J.D. (1977). The resistance to oxygen transport in the capillaries relative to that in the surrounding tissue. *Microvascular Research* **13**, 131–6.

Homer, L.D., Shelton, J.B., Dorsey, C.H. & Williams, T.J. (1984). Anisotropic diffusion of oxygen in slices of rat muscle. *American Journal of Physiology* **246**, R107–13.

Hoppeler, H., Matthieu-Costello, O. & Kayar, S.R. (1991). Mitochondria and microvascular design. In *The Lung: Scientific Foundations*, ed. R.G. Crystal, J.B. West *et al.*, pp. 1467–77. New York: Raven Press.

Martin, E.G., Wooley, E.C. & Miller, M. (1932). Capillary counts in resting and active muscles. *American Journal of Physiology* **100**, 407–16.

Skalak, R. & Branemark, P.-I. (1969). Deformation of red blood cells in capillaries. *Science* **164**, 717–19.

Thews, G. (1953). Über die mathematische Behandlung physiologischer Diffusionsprozesse in zylinderförmigen Objekten. *Acta Biotheoretica* **10**, 105–37.

Thews, G. (1960). Die Sauerstoffdiffusion im Gehirn. Ein Beitrag zur Frage der Sauerstoffversorgung der Organe. *Pflügers Archiv* **271**, 197–226.

P.M. BECKETT and W. ARMSTRONG

The modelling of convection- and diffusion-driven aeration in plants

Introduction

Although the rhizomes and seeds of some wetland species may survive quite lengthy periods of anoxia, and may even effect a degree of anoxic growth (Brandle & Crawford, 1987; Brandle, 1990), most plants at most times require oxygen both for maintenance respiration and growth. Despite their bulk, higher plants have not evolved any of the complex metabolically pump-driven distribution systems found in animals; carrier molecules such as haemoglobin, although perhaps widespread, are usually present in trace amounts only, and may function as oxygen sensors rather than carriers or significant reservoirs (Appleby *et al.*, 1988). Plants are usually permeated by a labyrinth of gas space, and it is gas- and liquid-phase diffusion that are thought to account for the bulk of the gas movements both within the plant and with the environment. Recently, however, several types of pressurised gas flow have been discovered (for review see Armstrong *et al.*, 1990*b*). Although none of these involves an active expenditure of energy on the part of the plant (the pressurisations are diffusion-driven), they are undoubtedly of considerable benefit for long-distance gas transport in plants such as water-lilies and the common reed.

The study of aeration in plants and animals has been helped enormously by the development of both physical and mathematical models. This chapter concerns the development and application of mathematical models as an aid to the study of both the diffusion and pressurised gas-flow aeration in roots and rhizomes. It is hoped that some of the principles dealt with may be helpful to those involved in the modelling of oxygen supply in other systems such as blood capillaries in animals. Oxygen diffusion in root nodules is not dealt with; information on that topic may be obtained elsewhere (Sheehy *et al.*, 1985; Hunt *et al.*, 1988; Dakora & Atkins, 1989).

Aeration in plants is somewhat complicated by the fact that they simultaneously occupy two environments.

Above ground, the atmosphere provides the aerial parts with an

Society for Experimental Biology Seminar Series 51: *Oxygen Transport in Biological Systems*, ed. S. Egginton & H.F. Ross.

abundant source of O_2 and, at night, a very effective sink for CO_2. Gas exchange between shoot and atmosphere is facilitated by plentiful gaseous paths (stomata and lenticels) across the leaf and stem surfaces, by the thinness of leaves and their abundant internal gas space, and in aerial stems, by the peripheral location of the most active tissues which are either permeated by, or are relatively close to, gas spaces. Consequently, little attention has been given to the likelihood of oxygen stresses in aerial shoot systems unless they become wholly or partially submerged. However, low leaf oxygen pressures play an important part in the physiology of at least one group of plants: the oxygenase activity of the Rubisco enzyme in C4 plants is limited by low oxygen pressures deep within the specialised vascular bundle sheaths. The carboxylating activity of the enzyme is thus enhanced, and photorespiration reduced, resulting in a greater degree of carbon fixation. Aeration in partially or wholly submerged aquatic macrophytes will always be influenced by photosynthesis; Bowes (1987) and Robe & Griffiths (1990) provide useful reference lists relating to aquatic plant photosynthesis. It will exert an influence both through diffusion and by the generation of convections, although the pressurisations may at times have little influence on the aeration process (Sorrel, 1991). Since modelling is at an early stage, shoot aeration is considered here only in so far as it influences root and rhizome aeration by its involvement in driving convective gas flows unconnected with photosynthesis.

Below ground, the degree of aeration afforded to roots and rhizomes by the soil environment can vary enormously and with far-reaching consequences. In *waterlogged soils*, for example, plants can obtain little if any oxygen from the soil; indeed, such soils are usually anoxic in all but the surface layers. Anaerobic microbial metabolism in these soils helps create nutrient deficiencies, and high concentrations of phytotoxins including materials such as Fe^{2+}, Mn^{2+}, sulphides and the lower fatty acids (Ponnamperuma, 1984). The concentration of the gaseous plant hormone ethene which is constantly generated by roots may also rise because (i) its escape to the soil is hindered by the water, (ii) more of the precursor may be generated in the roots by anaerobic conditions in central non-porous tissue and (iii) ethene may be produced in the soil itself during the anaerobic decomposition of organic matter. Although plants often succumb to waterlogging, at least two successful strategies for long-term growth and survival can be identified (Armstrong *et al.*, 1991). Both involve anoxia avoidance, although some degree of hypoxia/anoxia may occur accommodated by anaerobic metabolism (Brandle, 1990; Drew, 1991). They are (i) superficial rooting, in which fine roots develop by utilising the dissolved oxygen in the surface layers of the soil and (ii) reliance upon gas transport from the aerial shoot system. This is achieved to varying degrees by cortical gas space which extends from and within the shoot to within a few micrometres of the root/root cap junction. The success of many wetland species can be attributed to enhanced gas-space (aerenchyma) development in response to soil flooding. The process is stimulated by the accumulation of the phytohormone ethene (Jackson *et al.*, 1985; Atwell *et al.*, 1988; Justin & Armstrong, 1991*a*).

The effectiveness of internal transport from the aerial parts depends chiefly upon the degree and distribution of two factors: the physical resistance to transport, and the oxygen demands along the diffusion path. The physical resistance is a function of path length, porosity (fractional volume of gas space) and path tortuosity (Armstrong, 1979, 1982). The oxygen demand comprises (i) the respiratory uptake by the root tissues and (ii) oxygen leakage from the roots into the soil. This oxygen leakage (radial oxygen loss, ROL) to the rhizosphere is a most important spinoff from the enhanced gas-space provision, since it helps in the oxidative removal of phytotoxins (Armstrong, 1982; Laan *et al.*, 1991) and may support aerobic microbial populations beneficial to the plant (Hansen & Andersen, 1981; Hofmann, 1990).

In *well-drained soils*, the soil air connects directly with the aerial atmosphere and conditions are thus more favourable for root aeration. However, since the soil air is relatively stationary, and because there is particulate matter and water to impede transport, oxygen concentrations in the soil air decline with distance from the surface in response to the oxygen demands of the roots, the soil fauna and microbes. In general the oxygen deficits developed in the soil air should not be great ($\leqslant 5$ kPa partial pressure: Currie, 1962), but in aggregated soils the microporous structure may be water-filled and the aggregates low in oxygen (or even anoxic) (Currie, 1961; Greenwood & Goodman, 1967; Smith, 1980). Even in a freely draining soil, the water content and structure of the soil may be such that some roots might lie wholly or in part at a critical distance from gas-filled pore space. Moreover, since non-woody root surfaces generally lack surface pores, oxygen transfer from the soil is impeded by a liquid-phase path through the epidermal and, often, the hypodermal layers as well as surface water films. Oxygen supply to the roots is further restricted by the competing demands of root surface and rhizosphere microbial populations. In aerated soil, in solution culture or in the bathing fluid of respirometers, oxygen transport can be largely or exclusively radial and directed towards the centre of the root. Within the non-porous tissues – stele, outer cortex (sometimes), the hypodermis and epidermis (wall layers) – and within boundary layers external to the root, gas movement is essentially liquid phase and slow (e.g. D_{oxygen} in water-

$= 2.1 \times 10^{-5}\,cm^2\,s^{-1}$ at 20 °C). In porous regions of the cortex, gas-phase transport usually predominates and will be much faster since the resistance to flow should normally be substantially less in the cortex than elsewhere (e.g. D_{oxygen} in air $= 0.201\,cm^2\,s^{-1}$ at 20 °C). The consequences for aeration/anoxia of these radial differences in resistance to gas exchange, and the variation in tissue dimensions, permeabilities and respiratory demand can be effectively analysed by mathematical modelling.

The various tissue regions (cylinders) of graminacean (grass) roots are typified by sections of maize root (Fig. 1*A–C*). The sections illustrate how some tissues are porous and others non-porous: only the cortex (often only inner cortex) and the inner stele (medulla or pith) have any significant gas space. In sections *B–D* the gas-space volume of the inner cortex has become enlarged by cell collapse (lysigeny); the resulting tissue is termed aerenchyma. The porosity of the various tissues (i.e. the percentage volume or cross-section occupied by gas space) may be determined in various ways from photographs; the porosity of whole root segments is often measured by pycnometry methods (Justin & Armstrong, 1987).

Early, and very useful, attempts to analyse the effectiveness of this radial aeration in roots by modelling treated the root as a homogeneous cylinder, in terms of both structure and respiratory demand (Kristensen & Lemon, 1961; Lemon, 1962*a*,*b*; Armstrong, 1979; De Willegan & Van Noordwijk, 1984). More complex models, one mathematical and one electrical, incorporating longitudinal transport from the shoot as well as radial transport from the soil, but again based on structural and metabolic homogeneity, have also proved useful (Armstrong & Wright in Armstrong, 1979; De Willegan & Van Noordwijk, 1989), as did earlier models designed primarily to study the aeration potential of wetland root systems (Luxmoore *et al.*, 1970; Armstrong, 1979). Those models which combined consideration of internal transport from the shoot with internal transport to or from the soil have demonstrated that, even in 'well-drained' soils, oxygen transport from the shoot may still play an important role in root aeration in some circumstances. Latterly, however, models designed to mimic the multicylindrical nature of roots have been adopted (Armstrong & Beckett, 1985, 1987; Armstrong *et al.*, 1990*c*), and it is chiefly on these models that we shall concentrate here. Such models have recently been extended to encompass the effects of waterlogging, accommodating both the longitudinal gas-phase diffusion from the shoot system, oxygen consumption by the various tissue cylinders and radial oxygen loss to the soil. They also take account of longitudinal and radial variability of respiratory and diffusion characteristics in roots. For the future, however, a multicylindrical model which incorporates axial

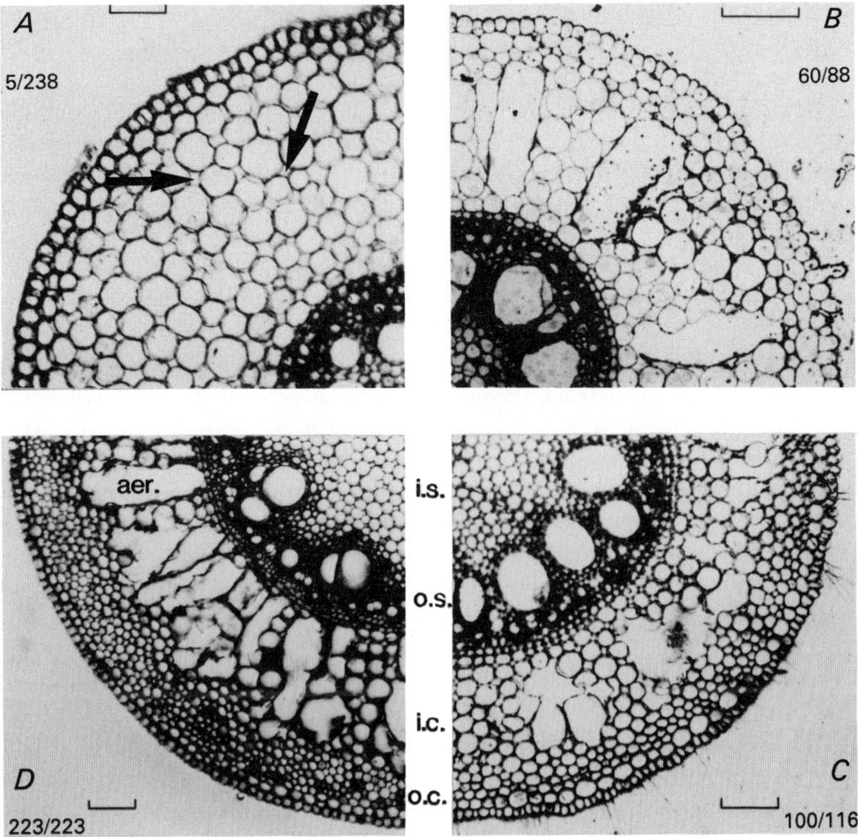

Fig. 1. Transverse sections of adventitious maize roots grown in stagnant ¼-strength Hoagland's solution, showing different relative sizes of the concentric tissue cylinders and the development of an aerenchymatous cortex: o.c., outer non-porous cortex and epidermis (wall layers); i.c., inner porous cortex; o.s., outer non-porous stele (pericycle, phloem and xylem); i.s., inner porous stele (medulla/pith); aer., aerenchymatous space. Note that section *A* has very little outer non-porous cortex; the bulk of the cortex here is permeated by small non-aerenchymatous intercellular gas spaces (examples marked by arrows), which may later become enlarged by cell collapse to form aerenchyma. These spaces run longitudinally through the root and may or may not be interconnected radially and circumferentially. Bar = 0.1 mm. Fraction distances: distance from the root apex/total root length (mm) are shown for each section (after Justin & Armstrong, 1991*b*).

transport from the shoot together with radial movement of oxygen from, as well as to, the soil is desirable. Other models relevant to plant aeration may be found in Leuning (1983) and Glinski & Stepniewski (1985).

Regarding convection, it has been generally assumed that the gas movements from aerial shoot to root or rhizome will be chiefly diffusive, but in theory both diffusive and convective (bulk) flows are possible. If, in a submerged plant, a significant proportion of the CO_2 released in respiration were to escape laterally to the submerging waters in the case of leaves (or radially to the rhizosphere), or to the xylem stream, rather than volatilising to reciprocate the deficit of oxygen partial pressure in the cortex, the resulting suction pressure could induce convection. The velocity of this convection would be greatest where the organ connected with the atmosphere and least (zero) at the furthest point, or at whatever distance the organ became anoxic. It should be noted, however, that anaerobic production of CO_2 or photosynthetic production of oxygen could reverse the convection. A mathematical model, described later, has provided a useful means of assessing the likely significance of this *non-throughflow* convection. The additional convections discovered in water-lilies (Dacey, 1981; Große *et al.*, 1991) and reeds (Armstrong & Armstrong, 1989, 1991; Armstrong *et al.*, 1991) may be classified as *throughflow*. Air enters one part of the aerial shoot system, is pressure-driven to submerged parts and the spent gases, enriched in carbon dioxide (and products such as methane from the sediments), are vented back to the atmosphere from other areas of the shoot system. The latter should not be confused with the mass flow which is inevitably generated within any equimolar diffusion process. The detailed mathematics underlying the physics of the pressurisation processes are beyond the scope of this chapter but a model is described which may be used to assess the consequences for aeration of these convections. Fundamental to our modelling is the assumption that respiration at any locus will be independent of oxygen concentration unless it becomes anoxic, when respiration at that point will cease. This is a very useful assumption for simplifying the mathematical analysis and has a tolerable approach to reality because of the very high oxygen affinity of cytochrome oxidase.

The modelling process

In cases in which a significant degree of convection operates in addition to diffusion, the theoretical analysis for calculating the O_2 and CO_2 concentrations is based on equations which describe the simultaneous convection and diffusion of these gases. When dealing with *throughflow* convection we will assume that the velocity at each and every point in the plant system can be found; for example, using the Poiseuille formula for flow of a viscous fluid through circular pipes, using the D'Arcy law if we have flow through a porous medium or using some empirical formula. In such cases we are trying to find the movement of gases, mainly oxygen and carbon dioxide, as they are carried with the bulk flow and simultaneously diffuse relative to the flow. On the other hand, *non-throughflow* convection rates must be calculated in order to be mutually consistent with the diffusion process.

Since the bulk velocity at any point within the plant has both magnitude and direction it must be described by a vector, denoted here by **q**, which can be resolved into the mutually perpendicular components which are directed parallel to the local coordinate directions. When using Cartesian coordinates we write $\mathbf{q} = (u,v,w)$, with u, v and w as the velocity components parallel to the x-, y- and z-axes, respectively, the equation which governs both the convection and diffusion of an element takes the form

$$\frac{\partial C}{\partial t} + u\frac{\partial C}{\partial x} + v\frac{\partial C}{\partial y} + w\frac{\partial C}{\partial z}$$

$$= \frac{\partial}{\partial x}\left(D_x\frac{\partial C}{\partial x}\right) + \frac{\partial}{\partial y}\left(D_y\frac{\partial C}{\partial y}\right) + \frac{\partial}{\partial z}\left(D_z\frac{\partial C}{\partial z}\right) \pm Q \qquad (1)$$

where $C(x,y,z,t)$ is the concentration of the element at the point (x,y,z) at time t: D_x, D_y and D_z are the diffusivities in the coordinate directions and Q is a source term describing the rate at which the element is added (for example, O_2 as a result of photosynthetic action) or removed (for example, O_2 in order to meet the respiratory demand). The appearance of the diffusivities inside the first differential is necessary if the diffusivity varies in the domain of interest, as is frequently the case in plants where variations in the cell structure or cell configuration within tissues have a strong influence on the diffusion characteristics. For cases in which each diffusivity is the same, and constant, throughout the domain then the diffusivity can be taken outside the differential operator, and the diffusion terms appear as $D\vec{\nabla}^2 C$ where $\vec{\nabla}^2$ is the Laplacian operator. In Eqn 1 the first term describes the rate of change in concentration at a fixed point as time changes, the next three terms account for convection in the directions of the coordinate axes, the first three terms on the right-hand side model the diffusion and Q is the source or sink effect. Given u, v and w, together with the appropriate diffusion coefficients and the strength of the source, we have an equation from which to calculate concentration. In general Eqn 1 is difficult to solve in all but a few simple geometries and some simplifying assumptions must be made. The technique of mathematical modelling involves retaining only those terms which relate to what are perceived to be the dominant physical effects in a particular problem, and deriving a simpler equation to which the solution is a good approximation to the physical state. It is relevant to note that several familiar formulae used in diffusion theory are merely solutions of simplified versions of Eqn 1; for example, the expression $L = (2DC_0/Q)^{1/2}$ which gives the distance which can be aerated by 1-dimensional diffusion in a homogeneous system in which Q is largely independent of concentration and one end maintained at concentration C_0.

When dealing with problems involving convective flow and diffusion along tubes it is much simpler, from a mathematical point of view, to use cylindrical polar variables r, θ and z. In cases of axial symmetry, in which there is no dependence on the azimuthal coordinate θ, the corresponding equation which governs the simultaneous convection and diffusion is

$$\frac{\partial C}{\partial t} + q_r \frac{\partial C}{\partial r} + q_z \frac{\partial C}{\partial z} = \frac{1}{r}\frac{\partial}{\partial r}\left(D_r r \frac{\partial C}{\partial r}\right) + \frac{\partial}{\partial z}\left(D_z \frac{\partial C}{\partial z}\right) + Q \qquad (2)$$

where q_r and q_z are the components of velocity in the radial and axial directions. If it is not possible to assume symmetry around the axis it is necessary to include additional convective and diffusion terms involving partial differentiation with respect to θ; this is the case, for example, when roots are in contact with wet soil over part of their circumference (sectoral blocking) (Van Noordwijk & De Willegan, 1984; Armstrong & Beckett, 1985). In addition to Eqn 2 being the basis for much theoretical analysis of convection and diffusion in plants, it is also highly relevant to problems associated with O_2 transport by blood as it flows through the capillaries in animal tissue (Kreuzer, 1982; see also Hoofd, this volume).

A central theme of this chapter is that Eqns 1 or 2 can be used as the starting point for numerous theoretical analyses. In later sections (pp. 261–72) we ignore any convection and concentrate on diffusion; moreover, despite the fact that in many cases, particularly in wetland plants but also in some well-drained soils, it is necessary to accommodate axial diffusion down the cortex. Initially we consider only radial diffusion throughout the root. In cases where it is not directly applicable, the radial model will still be relevant as it will be the basis from which to accommodate diffusion parallel to the axis.

We begin (p. 261) by considering diffusion within, and around, a single homogeneous root. The following section (p. 267) describes the method by which non-homogeneity is incorporated, and since the idea of introducing multicylindrical annuli has numerous other applications, considerable mathematical detail is included in order to make clear how the equations can be used. Extensions to time-dependent problems are then briefly outlined (p. 272).

The effects of convection are introduced on p. 272, comparing the contribution of the two modes of convection with reference to the governing mathematical equation. The final sections (pp. 275–88) are devoted to the role of convection in two specific plants, namely rice and *Phragmites australis*; the latter also has relevance to the aeration of water-lilies.

Diffusion in roots

Radial diffusion in homogeneous roots

Although we have already emphasised the importance of the multicylindrical nature of roots we are including this section both by way of introduction, but also because the method which will be used to identify anoxia within the multi-shelled roots can, we believe, be used to considerable advantage in other models: for example, for the identification of anoxia within individual cylinders, for the identification of the limits of oxygenated rhizospheres around roots in flooded soils and for determining the extent of the oxygenation of tissue around a blood capillary.

First, consider a homogeneous root embedded in an aerobic environment, e.g. a well-drained soil, which is able to sustain an O_2 concentration C_0 at the root surface, and seek the radial variation of concentration within the root. If Q is the averaged respiratory activity and D the radial diffusivity within the root, the radius of which is R, then Eqn 2 reduces to

$$0 = D\frac{1}{r}\frac{\partial}{\partial r}\left(r\frac{\partial C}{\partial r}\right) - Q \qquad (3)$$

The general solution of this is

$$C(r) = \frac{Q}{4D}r^2 + B\log_e r + A \qquad (4)$$

where A and B are arbitrary constants of integration the values of which are deduced by imposing the boundary conditions. For a completely aerobic root the concentration must remain finite when $r = 0$, which requires that $B = 0$ to avoid a singularity, and imposing $C(R) = C_0$ yields $A = C_0 - QR^2/4D$. Thus, we have the particular solution

$$C(r) = C_0 - \frac{Q}{4D}(R^2 - r^2) \qquad (5)$$

Note that it is this solution which gives the well-used formula $QR^2/4D$ for the concentration drop between the surface of the root and its axis.

Now if $QR^2/4D > C_0$ the root cannot be totally aerobic and in that case it becomes necessary to calculate the radius of the anoxic core r_a. The first point which must be emphasised in this respect is that it is *not* possible to deduce r_a immediately from Eqn 5 by setting $C(r) = 0$ and deducing the value of r, namely $r = R(1 - 4DC_0/QR^2)^{1/2}$; this is because the associated O_2 distribution would lead to diffusion into the anoxic core. Instead, it is necessary to find a value for r_a which corresponds to having mutually consistent conditions of zero concentration and zero flux at the radius r_a. When the root is not completely aerobic the general solution is still given by Eqn 4 but now the domain is $[r_a, R]$, and with r_a as yet unspecified there is a total of three unknowns which requires three conditions. We still have $C(R) = C_0$ added to which we now have $C = 0$ and $(\partial C/\partial r) = 0$ when $r = r_a$. Imposing these on the general solution yields three equations from which to find values for the three unknowns A, B and r_a, namely

$$\frac{Q}{4D}R^2 + B\log_e R + A = C_0 \tag{6.1}$$

$$\frac{Q}{4D}{r_a}^2 + B\log_e r_a + A = 0 \tag{6.2}$$

and

$$\frac{Q}{2D}r_a + \frac{B}{r_a} = 0 \tag{6.3}$$

Eliminating A and B leads to the following transcendental equation for r_a:

$$\frac{Q}{4D}(R^2 - {r_a}^2) + \frac{Q}{2D}{r_a}^2\log_e(r_a/R) = C_0 \tag{7}$$

This equation is presented in an alternative form elsewhere (Armstrong, 1979). The solution can be found using one of several straightforward iterative mathematical techniques, or by using Fig. 2 below.

As the second example we consider diffusion away from a single root into an anoxic, but oxygen-demanding, environment. Here the problem is to find the extent of the oxygenated rhizosphere; that is, the radius r_∞ beyond which the environment remains anoxic when there is a constant concentration C_0 maintained at the root surface. In this case we are concerned with radial diffusion into the soil from the root surface; the concentration is again governed by Eqn 3 where $C(r)$ is now the O_2 concentration in the soil, D the radial diffusivity in the soil and Q the respiration rate; the general equation is again given by Eqn 4 and is relevant to the domain $[R,r_\infty]$. In this case we impose $C(R) = C_0$ at the root surface together with $C = 0$ and $(\partial C/\partial r) = 0$ when $r = r_\infty$ which yield the following equation for r_∞:

$$\frac{Q}{4D}(R^2 - r_\infty{}^2) + \frac{Q}{2D} r_\infty{}^2 \log_e(r_\infty/R) = C_0 \tag{8}$$

In connection with roots, this equation was first used (Armstrong 1970) to refute claims (Bouldin, 1967) that wetland plants could not maintain an oxygenated rhizosphere in anoxic sediments. Note that apart from the change in the variables this is precisely the same as Eqn 7. Consequently, it is possible to deduce the values of r_a/R and r_∞/R by the same technique; if we omit the subscripts a and ∞ each equation can be expressed in the form

$$1 - \left(\frac{r}{R}\right)^2 + 2\left(\frac{r}{R}\right)^2 \log_e\left(\frac{r}{R}\right) = \frac{4DC_0}{QR^2} \tag{9}$$

and hence for any prescribed combination of the variables D, C_0 and R we can associate solutions in which $r/R < 1$ with an interior anoxic region, while if $r/R > 1$ then we are finding the extent of an aerobic region exterior to a cylinder. Since the possible existence of internal anoxia in a root is one main focus of this type of research it is important to have an accurate method of delineating such regions; the conditions of zero concentration and zero flux characterise all such boundaries and in consequence we will keep returning to equations which have a very similar form to Eqn 9.

Before outlining a method for solving Eqn 9 it is also relevant to comment briefly on the so-called Krogh solution (see Hoofd, this volume) which is commonly used to find the O_2 concentration in tissue external to a cylindrical capillary of radius R, and similar problems. The solution for the concentration $C(r)$, based on maintaining a constant concentration C_0 in the capillary and the assumption of radial diffusion in the tissue annulus $R < r < R^*$ *with zero diffusive flux at the outer surface* $(r = R^*)$ can be deduced from Eqn 4 and is

$$C(r) = C_0 + \frac{Q}{4D}(r^2 - R^2) - \frac{Q}{2D} R^{*2} \log_e(r/R) \tag{10}$$

An immediate deduction is that $C(R^*) \neq 0$ for an arbitrarily chosen value of R^*. Only if by putting $r = R^*$ in Eqn 10 to yield $C(r) = 0$ is there no potential for diffusion beyond the confines of the tissue cylinder; this yields precisely the same equation from which we found r_∞. Moreover, if the tissue cylinder is sufficiently large, the concentration at $r = R^*$ predicted by Krogh's solution could be negative. These are not new observations because these and other limitations of Krogh's solution are well documented by Kreuzer (1982); however, despite the well-known nature of the limitations there seems little understanding of the remedy. What is needed is a method of choosing R^* in order to have a tissue cylinder in which all the O_2 is consumed and without diffusion beyond the outer boundary; that is, a method of solving Eqn 9.

Only with the use of an iterative technique can we find a solution to any prescribed accuracy, but for most practical purposes the following graphical method will suffice – the only limiting factor is the scale of the diagram. Plotting y on the vertical axis and the ratio r/R on the horizontal axis we first draw the graph of $y=\phi(r/R)$, where $\phi(r/R)$ is the expression on the left-hand side of Eqn 9. Figures 2*A* and *B* show the graph drawn on two different scales; *B* shows the global form with the function monotonically increasing as r/R increases if $r>R$, while *A* emphasises the form for $r/R \leqslant 2$ which emphasises the form relevant for the calculation of internal anoxia. In the case of roots we can deduce the size of the rhizosphere and any internal anoxia as follows. Given values for R, D, Q and C_0 we calculate the value of the expression $4DC_0/QR^2$. The intersection(s) of a horizontal line $y=4DC_0/QR^2$ with $y=\phi(r/R)$ gives values of r/R which satisfy Eqn 9. Clearly, there will always be a point of intersection with that part of the curve where $r/R>1$; this corresponds to the extent of the rhizosphere because $r_\infty > R$. However, there will only be an intersection with the curve for $r/R<1$ provided $4DC_0/QR^2<1$; since values of r which are less than R are associated with stelar anoxia there is only a solution for r_a if the concentration $C_0 < QR^2/4D$, precisely as stated above. For values of C_0 greater than this value the root will be totally aerobic.

By way of final comment in this section attention is drawn to the fact that the graph of $y=\phi(r/R)$ can be used to assess the inaccuracy induced by finding the value of r which satisfies $C(r)=0$ in Eqn 5, that is, $r=R(1-4DC_0/QR^2)^{1/2}$, and taking this as the radius of the anoxic core as compared with the value of r_a which comes from Eqn 7. Figure 3*A* shows the graphs of $y=\phi(r/R)$ and $y=[1-(r/R)^2]$ on the domain [0,1], so the value of r_a satisfying Eqn 9 and r given by $r=\mathrm{R}(1-4DC_0/QR^2)^{1/2}$ can be found by locating the intersections of $y=4DC_0/QR^2$ with the two curves. For example, a value of C_0 which yields the value for $4DC_0/QR^2$ which is indicated by α ($\alpha=0.62029$) on the vertical axis, the correct anoxic core radius is shown as r_a ($r_a=0.35$) on the horizontal axis, while r^* ($r^*=0.61620$) is the incorrect value resulting from the formula $r=R(1-4DC_0/QR^2)^{1/2}$. The error in the radius is approximately 76% which corresponds to an error in the anoxic cross-sectional area of 210%. Figure 3*B* illustrates the concentration profile which satisfies both the

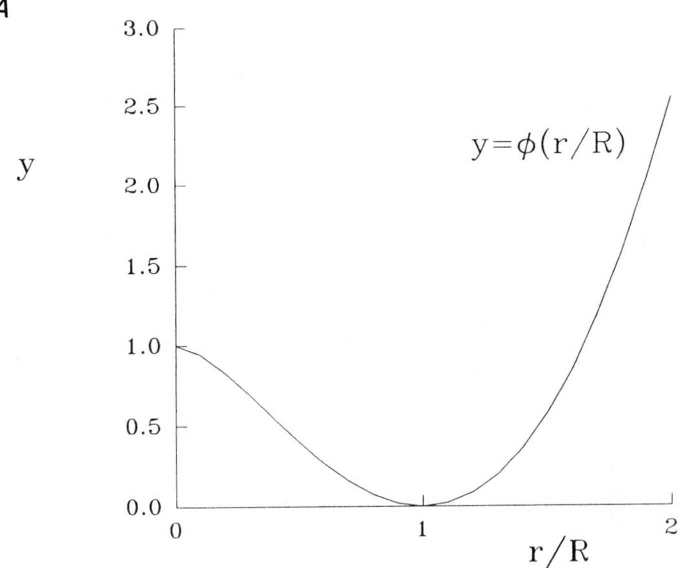

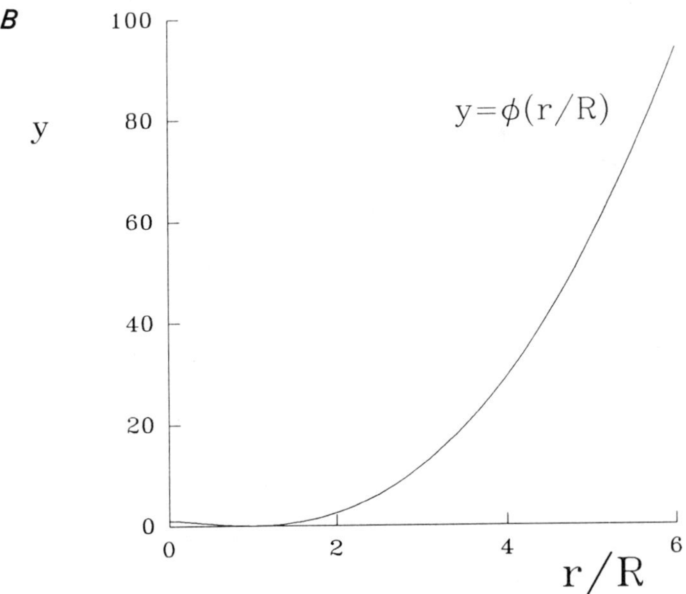

Fig. 2. The graph of $y = \phi(r/R) = 1 - (r/R)^2 + 2(r/R)^2 \log_e(r/R)$, *A* for the domain $r/R \in [0,2]$, *B*, $r/R \varepsilon [0,6]$.

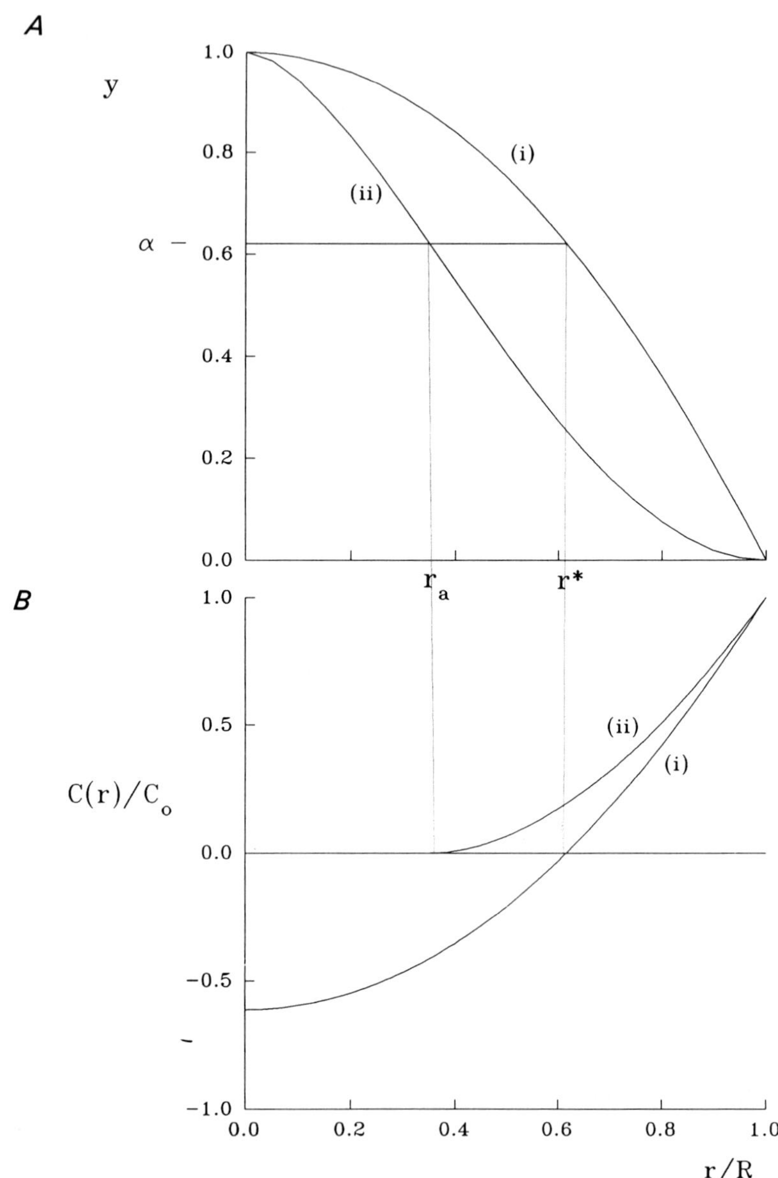

Fig. 3. *A*, Graphs of (i) $y = 1 - (r/R)^2$, (ii) $y = \phi(r/R)$ for $r/r \in [0,1]$. *B*, O_2 concentration profiles in a partially anoxic root predicted by (i) the 'aerobic' solution, (ii) the correct anoxic solution.

conditions of zero concentration and zero flux (the correct one) and that obtained from the fully aerobic solution by truncation at the point where $C(r) = 0$; the latter clearly shows a concentration gradient which would generate diffusion into what is taken to be the anoxic region.

Diffusion in multicylindrical roots

Although there is considerable appeal, from a modelling standpoint, to consider a root to be a single homogeneous cylinder, the structures described above show that any homogeneity is at best restricted to within well-defined annular regions (Fig. 1). We can therefore model the root as a multicylindrical structure comprising annuli within which the diffusion and respiration rates can be considered to be constant. Typically, portions with dense packing of cells (low porosity) may exhibit high respiratory demand, and diffusive impedance will also be high. At the other extreme, aerenchymatous regions (high porosity) will permit rapid diffusion whilst respiratory demand is necessarily low (Fig. 4 in Armstrong *et al.*, 1990*c*). The potential for respiration is modelled as an appropriate O_2 sink distributed uniformly in each annulus. This approach is relevant for the analysis of roots embedded in *aerated* soil, in which case the O_2 diffusion is entirely inward, and for *anoxic* environments when the O_2 transport is from the cortex outward to the soil and inward to the stele. In the latter case steady-state radial diffusion is only possible if the concentration of O_2 in the cortex is maintained at a constant level by longitudinal diffusion down the cortex from the aerial parts; radial diffusion in the stele and wall layers is therefore coupled to axial diffusion down the cortex in a manner which is outlined below (p. 270).

The diffusion within each cylindrical annulus is again governed by Eqn 2, but because of the potential for different diffusivities, etc., in each annulus the equation for the ith annulus is written in the form

$$0 = D_i \frac{1}{r} \frac{\partial}{\partial r}\left(r \frac{\partial [C_i]}{\partial r} \right) - Q_i \tag{11}$$

where $[C_i](r)$ is the O_2 distribution, D_i the diffusivity and Q_i the respiration in the ith annulus. For example, $i = 1$ gives the inner stele, $i = 2$ may give the outer stele and $i = 3$ the cortex, while the modelling of the wall layer(s) (e.g. epidermis) would determine the index relating to these annuli and to the soil (or water) region. The thickness of each annulus is defined by its outer radius; e.g. r_1 being the radius of the inner stele, r_2 the outer radius of the outer stele, etc., hence the thickness of the outer stele would be given by $r_2 - r_1$, etc. The general solution of Eqn 11 has the form

$$[C_i](r) = \frac{Q_i}{4D_i} r^2 + B_i \log_e r + A_i \tag{12}$$

where A_i and B_i are arbitrary constants which must be determined from the simultaneous equations which express the conditions of continuity in O_2 concentration and O_2 flux at the boundaries between the annuli. Once we have solved these equations it is possible to deduce everything else.

It may be noted, however, that it is not necessary to solve for all the arbitrary constants in order to deduce a great deal about the O_2 distribution. In particular, if we know only the values of the constants B_i it is possible to deduce the concentration drop across each annulus. For a typical annulus Eqn 12 yields

$$[C_i](r_i) - [C_i](r_{i-1}) = \frac{Q_i}{4D_i}(r_i^2 - r_{i-1}{}^2) + B_i \log_e(r_i/r_{i-1}) \tag{13}$$

For the case of a totally aerobic root the values of B_i can easily be determined by equating the total respiratory demand of all the annuli within the cylinder with radius r_i and the diffusive flux into that region. For example, equating the flux (per unit height of cylindrical annulus) at $r = r_2$ with the respiratory demand in the two parts of the stele yields

$$2\pi r_2 D_2 \left(\frac{\partial [C_2]}{\partial r}\right)_{r=r_2} = Q_1 \pi r_1{}^2 + Q_2 \pi (r_2{}^2 - r_1{}^2) \tag{14}$$

and hence, after differentiating Eqn 12 for $i = 2$, we can deduce that B_2 is given by

$$B_2 = \frac{1}{2D_2}(Q_1 r_1{}^2 - Q_2 r_1{}^2) \tag{15}$$

Therefore the fall in O_2 concentration across the outer stele, from Eqn 13, is

$$\frac{Q_2}{4D_2}(r_2{}^2 - r_1{}^2) + \frac{1}{2D_2}(Q_1 r_1{}^2 - Q_2 r_1{}^2) \log_e(r_2/r_1) \tag{16}$$

Once we know all the constants B_i it is possible to deduce the concentration drop across each annulus in a similar manner. An alternative for calculating the deficits is given by Armstrong *et al.* (1990*c*).

Roots embedded in aerobic soils

For the case of inward diffusion from an aerated soil the above procedure can be continued outward through the cortex and wall layers and finally to the soil layer, which is itself simply another annulus. Denoting the annulus for the soil layer by the index s we have

$$[C_2](r) = \frac{Q_s}{4D_s}r^2 + B_s \log_e r + A_s \tag{17}$$

with B_s known. At this stage it is possible to confirm that the root is indeed fully aerobic by checking that the sum of the concentration drops is equal to or less than the O_2 concentration in the soil. If the total decrease exceeds this value there must be some anoxia present and we can proceed as follows.

First, assume a value for r_a, the radius of the anoxic core, which is less than r_1 which presumes the anoxia is restricted to within the inner stele. We then follow the method for calculating the constants B_i with the modification that the respiratory demand (per unit length) within the stele is now $Q_1\pi(r_1^2 - r_a^2)$. This guarantees that there will not be any diffusion into the anoxic region but does *not* necessarily yield a solution in which the concentration falls to zero when $r = r_a$; only if the sum of the decreases across the annuli (including the portion of the inner stele for which $r > r_a$) is precisely equal to the concentration in the soil do we have a consistent situation with zero concentration and zero flux at $r = r_a$. If, by means of trial and error, it is possible to find a value of r_a which is less than r_1 then we have solved the problem. If, however, the condition requires that $r_a > r_1$ then our assumption that anoxia is restricted to the inner stele is incorrect and we next assume it extends to the outer stele and repeat the argument.

As an alternative, but effectively the same, method all the constants A_i and B_i are found from the boundary conditions to yield a single equation for r_a in much the same manner as A and B were eliminated from Eqns 6.1–6.3 to yield Eqn 7 in the case of the homogeneous root. Indeed, the equation for r_a will still have the same inherent structure as Eqn 7, namely r_a satisfies.

$$r_a^2 \log_e(r_a) + \alpha r_a^2 + \beta = 0 \tag{18}$$

where α and β are constants that can be expressed (in a somewhat complicated combination) in terms of the various diffusivities, respiratory demands, annular radii and external O_2 concentration. An example of the technique has been presented, as an appendix, by Armstrong & Beckett (1985) for the case of a single cylinder representing the stele.

Although it is often possible to maintain steady-state respiration by drawing on the available O_2 in the soil it is also possible, even in aerobic soils, to have simultaneous diffusion down the cortex. Space prohibits a detailed analysis of the manner in which this is modelled mathematically, but in essence it is then necessary to include the differential with respect to z which appears on the right-hand side of Eqn 2, and regard the radial diffusion as contributing to a sink effect within the cortex at each section. A very similar approach is, however, always necessary in the case of anoxic environments and a fuller account of that procedure is included in the following section.

The multicylindrical analysis for inward radial diffusion has been used recently to demonstrate the likely form of the radial oxygen profiles in soil-aerated roots, has highlighted the potential for some stelar anoxia, and has shown that critical oxygen pressure (COPR) for respiration (as viewed from outside the root) can be expected to vary enormously, not only from root to root but even along the length of an individual root (Armstrong *et al.*, 1990*c*, 1991): predictions for the COPR of maize roots are that it might vary from *c.* 10 kPa to >21 kPa, and could even be >100 kPa. Also highlighted is the critical role played by the non-porous outer annuli (epidermal and hypodermal cells), and the external soil solution (and soil particle) boundary layers (Armstrong & Beckett, 1985).

Roots in anoxic environments

When a root is surrounded by an anoxic medium the maintenance of any steady-state respiration must depend on an O_2 supply from somewhere other than the soil. The anatomy of many roots (Justin & Armstrong, 1987) is such that the cortex provides the potential for axial diffusion (in the z direction) from a known O_2 value at the base of the root and this sustains the respiratory demand within the cortex, and that of the other sections of the root and its environment according to the previous assumption of purely radial diffusion in these sections. If the O_2 concentration at any section z is presumed known, and assuming purely radial diffusion within the cortex, it is possible to use the preceding analysis for the two annuli comprising the stele; again, we begin by assuming a totally aerobic stele but now introduce an anoxic core if the concentration decrease exceeds the cortical concentration at the section in question. The detail is presented by Armstrong & Beckett (1987).

The outward flow of O_2 from the cortex, through the wall layer(s) and into the environment, will always require the calculation of the distance, r_∞, at which the concentration falls to zero. The method of solution is therefore very similar to that used to find r_a; that is, we assume a value for r_∞, calculate the concentration decreases across the wall layers and the aerobic layer within the soil, and perturb the value of r_∞ until we have the consistent conditions of zero concentration and zero O_2 flux when $r = r_\infty$. In the model presented by Armstrong & Beckett (1987) the multicylindrical structure within the wall layer itself was suppressed into a single layer with averaged properties; the analysis of that situation is also available in the appendix of that paper.

In this model there is no variation of O_2 across the cortex, the only variation being parallel to the axis due to longitudinal diffusion. The relevant equation for $[C_3](z)$ is then

$$0 = D_3 \frac{\partial^2 [C_3]}{\partial z^2} - Q_3 - \text{Flux}_{\text{St}} - \text{Flux}_{\text{W}} \tag{19}$$

where Q_3 is the respiration per unit volume within the cortex itself, Flux_{St} is the total flux into the outer stele at $r = r_2$ averaged across the cross-sectional area of the cortex and Flux_{W} the flux away from the cortex averaged in the same manner. These averaged fluxes Flux_{St} and Flux_{W} which are modelled as additional sink activity within the cortex depend on the O_2 concentration at the vertical location z, hence Eqn 19 is highly non-linear, and its solution is necessarily found by an iterative technique. This involves the calculation of the radial diffusion for some presumed axial O_2 distribution, the calculation of the associated O_2 fluxes from the cortex into the stele and away through the wall layers, after which Eqn 12 is solved with the new values of Flux_{St} and Flux_{W} at each z section. The cycle is repeated until there is consistency between the axial variation of $[C_3](z)$ and the associated radial distribution inward and outward from the stele. Again, this procedure is described in the appendix to the paper by Armstrong & Beckett (1987).

The results obtained by applying this model to submerged graminean (grass) or cyperacean (sedge) root systems highlighted the potential for both stelar anoxia and some epidermal and hyperdermal anoxia. The results suggested the possibility that often where roots are described in the literature as hypoxic because of ethanol production, raised alcohol dehydrogenase activity and reduced energy charge, these symptoms may often be indicative only of some sub-apical stelar anoxia caused by basipetally increased diffusive impedance in the endodermis. Subsequent experimental investigations stimulated by these modelling results (Thomson & Greenway, 1991) has confirmed the existence of sub-apical stelar anoxia in maize roots in which the only oxygen supply was from the shoot via the root cortex gas space. The results of the modelling data obtained so far may be further summarised as follows. They show that (i) stelar anoxia should raise oxygen levels elsewhere in the root and enhance root extension, (ii) with constant diffusivity in the stele, anoxia should first occur close to the root apex and spread basally and laterally with further root extension; if stelar diffusivity should decline basipetally, stelar anoxia will arise first in some sub-apical region, (iii) there will be competition between the oxygen demands of the root and rhizoshpere, (iv) root extension growth may be reduced by radial oxygen losses into the rhizosphere, but the sub-apical decline in root wall permeability in wetland species safeguards against this effect, (v) differences in critical oxygen pressure for extension growth (COPE) and respiration (COPR) may arise because of sub-apical stelar anoxia development and (vi) survival of roots may be helped by the manner in which anoxia spreads in roots with aerobic conditions persisting in the phloem and pericycle until only traces of oxygen remain in the cortex. It is also evident that the aerobic conditions may provide the means for metabolising ethanol produced in an anoxic stele or meristem, and provide materials for transfer to the stele to sustain its viability.

Information provided by this type of model has also been of value in the applied field, by predicting the amounts and major sites of oxygen release by the common reed *Phragmites* in its role as a promoter of domestic and agricultural effluent purification in wetlands constructed especially for this purpose. The model identifies narrow lateral roots as the major site of oxygen release to the sediment, and predicts that sediment oxygenation rates of up to $20\,g\,m^{-2}\,day^{-1}$ are possible (Armstrong *et al.*, 1990*a*).

Non-steady problems

The preceding analysis has concerned steady-state diffusion relevant to situations in which the source O_2 concentration, that at the edge of the boundary layer in aerated soils or at the root base for submergence in an anoxic environment, is maintained at a constant level. In practice either of these may vary, and it becomes necessary to solve the time-dependent diffusion equation in each annulus; such solutions are necessarily found numerically using finite difference representations of the equations. For roots in anoxic environments it has been observed experimentally that a sudden reduction in basal O_2 concentration leads to damped periodic variations in apical O_2 concentration (T. Webb & W. Armstrong, unpublished data); this cannot, however, be explained by a simple pure diffusion model because the parabolic nature of the system generates asymptotic behaviour without any oscillatory nature. Oscillatory variation can be induced by modifying the assumption of a constant respiration to accommodate a 'time-lagged capacitance'. A detailed assessment of this effect is not presented here because as yet the mechanism is not fully understood. What is clear, however, is that comparison of the experimental results with those for a range of theoretical assumptions for the respiratory activity show the importance of mathematical modelling in the detective work associated with understanding such phenomena.

Convection allied to diffusion

The calculations of O_2 transport down the root cortex presumed a known O_2 concentration at the base of the root. This section is devoted to ways in which wetland plants maintain such O_2 levels when the shoot is partially submerged, or there is a buried rhizome system. Clearly, there is some scope for diffusion through that submerged portion of the plant which links the roots to the aerial parts, but it may be insufficient to meet the demands of the root system and plants have evolved to take advantage of a range of fluid mechanisms which enhance this diffusive contribution. Later (pp. 275, 285) we describe the ways in which two wetland plants, namely rice (*Oryza* sp.) and the reed *Phragmites australis*, enhance the O_2 supply to the roots; in each case there is potential for convection. In this section we highlight the differences between the convective flows which are involved by reference to a simple 1-dimensional mathematical model; the motivation for considering this particular model is described more comprehensively by Beckett *et al.* (1988).

First, it is necessary to note that Eqn 1 must be modified (in order to satisfy mass conservation) for a 1-dimensional problem if the velocity component varies with position. Taking flow in the z-direction so the velocity is $U(z)$ the modified equation for convection and diffusion in the presence of a uniform sink term is

$$\frac{\mathrm{d}}{\mathrm{d}z}(UC) = D\frac{\mathrm{d}^2C}{\mathrm{d}z^2} - Q \tag{20}$$

From this we can consider *throughflow* convection by taking $U(z)$ to be constant, and *non-throughflow* convection by taking $U(z) = U_0(1 - z/\ell)$ where ℓ is the length which can be aerated. In each case ℓ is found by solving the equation for $C(x)$ subject to the boundary conditions

$$C = C_0 \text{ at } z = 0 \quad \text{plus } C = 0 \text{ and } \frac{\mathrm{d}C}{\mathrm{d}z} = 0 \text{ at } x = \ell \tag{21}$$

Here the three boundary conditions are necessary because although the equation is only of second order we do not know *a priori* the value of ℓ.

When $U(z)$ is constant, i.e. $U(z) = U$, the general solution of Eqn 20 is

$$C(z) = \alpha + \beta \exp\left(\frac{U}{D}z\right) - \frac{Q}{U}z \tag{22}$$

and subject to Eqn 21 this yields the particular solution

$$C(z) = C_0 - \frac{QD}{U^2}\exp\left(-\frac{U}{D}\ell\right) + \\ + \frac{QD}{}\exp\left(-\frac{U}{}(\ell - z)\right) - \frac{Q}{}z \tag{23}$$

in terms of ℓ, where ℓ is given by the transcendental equation

$$\exp\left(-\frac{U}{D}\ell\right)+\frac{U}{D}\ell=1+\frac{C_0U^2}{QD} \tag{24}$$

It may be noted that this equation also yields formulae for the lengths of channel which can be aerated either by pure diffusion or by pure convection; these are $\ell=(2DC_0/Q)^{1/2}$, which is obtained by seeking the limit as $U\rightarrow 0$, and $\ell=C_0U/Q$ by setting $D=0$.

In our *non-throughflow* model the velocity decreases linearly along the channel under the assumption that respiration removes O_2 from the channel in a uniform manner. Considerations of global conservation also mean that the mass inflow at $z=0$ must equate to the total respiration in the channel; hence U_0 is given by $\varrho U_0=\ell Q$, where $\ell=$ density.

In this case the solution for $C(x)$, in terms of the aerobic length ℓ, is

$$C(z)=\varrho\left\{1-\exp\left(-\frac{Q[\ell-x]^2}{2D\varrho}\right)\right\} \tag{25}$$

while ℓ is given by

$$\ell=\left\{-\frac{2D\varrho}{Q}\log_e\left(1-\frac{C_0}{\varrho}\right)\right\}^{\frac{1}{2}} \tag{26}$$

We can again deduce the formula for the length ℓ which is aerated by pure diffusion on taking the limit as $(C_0/\varrho)\rightarrow 0$. We can also deduce that it is impossible to aerate any length of the channel if there is no diffusion; setting $D=0$ in Eqn 26 yields $\ell=0$. The consequence of this last observation is central to the argument of Beckett *et al.* (1988), when asserting that in *non-throughflow* convection the role of convection is secondary to that of diffusion.

In both cases the effects of diffusion and convection are not simply additive: Tables 1 and 2 show the lengths which can be aerated by combined convection and diffusion and compare these with the lengths which would be aerated by diffusion or convection alone. The contribution by convection in the case of *non-throughflow* is seen to be only a small fraction of that which can be aerated by diffusion alone. It should be emphasised that Table 2 displays results for a range of respiration rates; it is not meaningful to stipulate a range of velocities, as is the case for *throughflow* convection, because given any specific value of Q there is a *unique* convection rate which must be consistent with the 'suction' caused by the respiratory activity. Some typical respiration rates were chosen for

Table 1. *The length of a channel which can be aerated by diffusion and/or convection in* throughflow *convection; Q = 100 ng cm^{-3} s^{-1}*

U (cm min^{-1})	Length by diffusion alone (mm)	Length by convection alone (mm)	Length by convection and diffusion (mm)
0.01	332.1	4.5	333.6
0.10	332.1	44.8	347.7
1.00	332.1	448.3	570.1
10.00	332.1	4483.3	4495.6
100.00	332.1	44833.3	44834.6

Table 2. *The length of a channel which can be aerated by diffusion or diffusion/convection in* non-throughflow *convection for a range of respiration rates*

Respiration Q (ng cm^{-3} sec^{-1})	Length by convection and diffusion (mm)	Length by diffusion alone (mm)	Velocity at $z = 0$ U_0 (cm min^{-1})	Length by convection with U_0 (mm)
10	1109.3	1050.2	0.049	221.9
50	496.1	469.7	0.111	99.2
100	350.8	332.1	0.156	70.2
200	248.0	234.8	0.221	49.6
250	221.9	210.0	0.247	44.4
300	202.5	191.7	0.271	40.5
500	156.9	148.5	0.350	31.4

Table 2 while the velocities cover a range which includes the value arising in Table 1.

Aspects of dark-aeration in partially or wholly submerged rice

Interest in *non-throughflow* convection was generated recently by claims that it was the major mechanism of aeration in deep-water rice, a group of far-eastern varieties that are adapted to cope with flash floods which may bring water depths of *c.* 2 m or more (Raskin & Kende, 1983, 1985). In all rice varieties and in other graminean species, a convective pathway from the atmosphere is provided by gas films maintained along the sub-

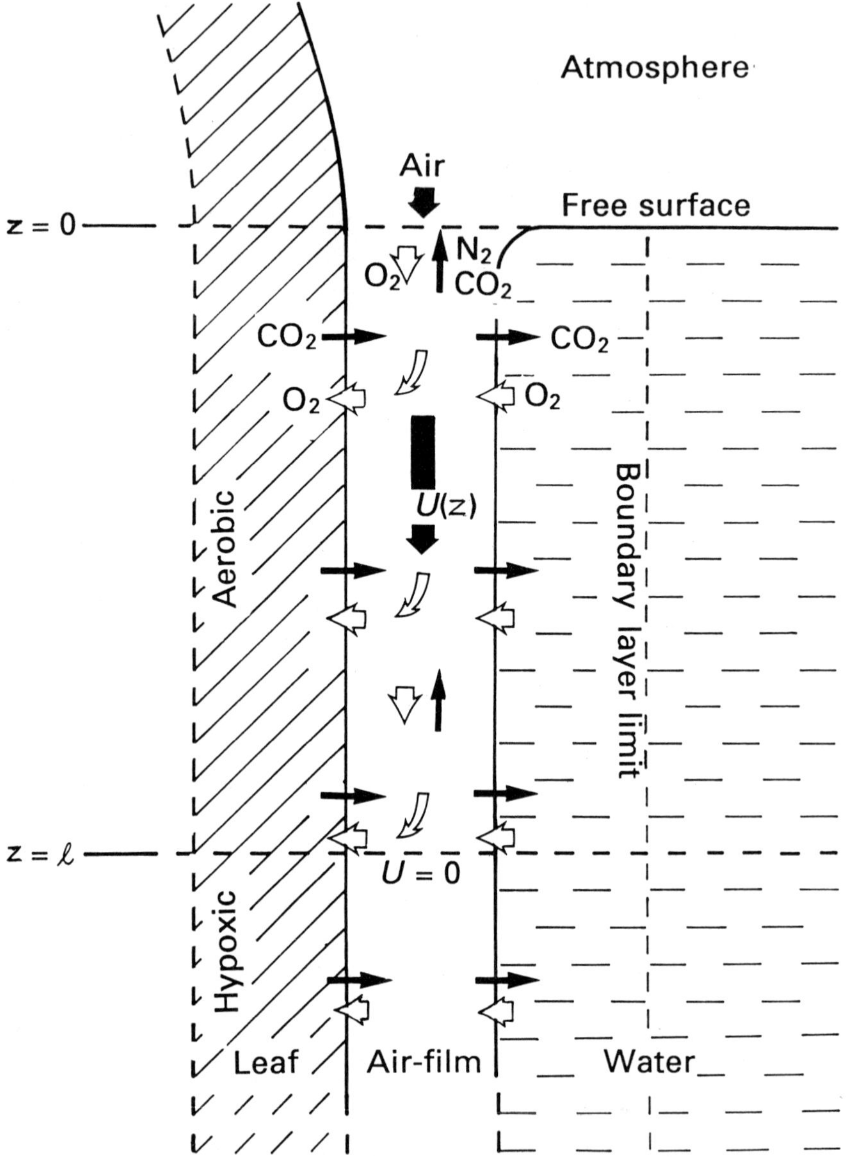

Fig. 4. Diagrammatic representation of longitudinal half-section of a partially submerged rice leaf with its attendant surface gas film. Arrows indicate the normally prevailing diffusive (O_2, N_2 and CO_2) and convective (air) movements.

merged surfaces of the leaves by surface tension interaction with longitudinal surface grooves. It is the loss of respiratory CO_2 from the gas film to the surrounding water (or from roots to the soil: Končalová *et al.*, 1988) which creates the suction for driving the *non-throughflow* convection. The films can play an important role in the oxygen supply to both leaf and the submerged roots, although not necessarily by their support of *non-throughflow* convection.

The gas exchanges are shown schematically in Fig. 4. In the upper portion of the gas film, where there is a longitudinal O_2 gradient, O_2 is supplied to the leaf both from the water and from the atmosphere (via convection and diffusion down the film), while in the lower portion the O_2 concentration is constant down the film and the only passage is across the film from the water to the leaf. This lateral transport will also dominate down the entire length of the leaf during the period immediately after any submergence of the plant because the O_2 concentration is initially uniform, and without a concentration gradient there is no longitudinal diffusion, or convection. Since the vertical convection occurs because of the relative solubilities of O_2 and CO_2 it is necessary to model the transport of both O_2 and CO_2 in this problem.

The method of solution is similar to that which was used to calculate the O_2 concentration down the cortex within a root. Here lateral diffusion (perpendicular to the film) is calculated on the basis of known concentrations in the film, and the associated fluxes to/from the leaf and water are treated as providing sink/source effects within the film when modelling the convection/diffusion down the film. The lateral diffusion between the leaf, film and water is therefore the basis of our modelling of the partly submerged rice plant and is treated as an isolated problem in the following section; the justification for taking the concentration to be constant across the film at any depth when modelling the partially submerged leaf is addressed at the same time. It is also relevant to note that this is the complete solution if plants become fully submerged, and where the film can no longer open to the atmosphere.

The fully submerged leaf

Following the theme of this chapter we will outline the mathematical approach with reference to the fundamental equations presented earlier (p. 258). A detailed description of the method of solution and the range of modifications and switches which are necessary when leaves become partially anoxic (both laterally through the thickness of the leaf and axially along the leaf) will form a separate article. First, it is necessary to describe the simplified structure of the leaf which is used for the modelling. Diffusion within the leaf is presumed to be perpendicular to the leaf surface along channels, the total volume of which is given by the leaf porosity σ, so the diffusion within the leaf will be the same as that of O_2 and CO_2 in air but in a restricted region of the leaf. We will denote the *average* respiration in the leaf by Q_L so the respiration (per unit volume) within the diffusive path is Q_L/σ and seek to find the gas concentrations along the gas pathway, that is, concentrations related to the porous part of the leaf.

The mathematical problem involves finding the O_2 and CO_2 concentration distributions across the leaf/air/water regions as functions of time. We take x as a measure of normal distance so that $x = \text{lth}$ (where lth is the leaf thickness) denotes the interface between the leaf and the film and $x = \text{lth} + \text{ath}$ (where ath is the thickness of the gas film) defines the interface between the film and the water. The concentrations are continuous at $x = \text{lth}$ but discontinuous at $x = \text{lth} + \text{ath}$ because the gases, O_2 in particular, are less soluble in water; here it is assumed that at the film/water interface the concentrations remain in the same proportion as in the equilibrium state, while conservation demands, for example, that O_2 lost from the water is gained by the film. If the leaf is totally aerobic the O_2 concentration at $x = 0$ is positive, but if there is insufficient O_2 to supply the entire respiratory demand the leaf can become partially anoxic and the O_2 concentration is zero for $0 < x < \varkappa(t)$, where $\varkappa(t)$ defines the penetration of anoxia into the leaf. When the leaf is fully aerobic we presume there is no diffusion at $x = 0$ but if the calculations show the concentration becomes negative at $x = 0$ then we conclude the leaf has become partially anoxic and impose simultaneous conditions of zero flux and zero concentration at a point $x = \varkappa(t)$. The extra condition is necessary in order to calculate $\varkappa(t)$ in much the same way as we found the radius of the anoxic cores in roots.

The actual concentrations of O_2 in the porous part of the leaf, in air and in water are denoted by $[O_L](x,t)$, $[O_A](x,t)$ and $[O_W](x,t)$, respectively, and those of CO_2 by $[C_L](x,t)$, $[C_A](x,t)$ and $[C_W](x,t)$. However, in order to accommodate the discontinuities in concentration between the air and water we introduce modified functions $[\hat{O}_W]$ and $[\hat{C}_W]$ defined by $[\hat{O}_W] = \lambda_O[O_W]$ and $[\hat{C}_W] = \lambda_C[C_W]$, where λ_O and λ_C are the relative solubilities in water compared with air. Since diffusion is the sole mode of mass transfer the mathematical equations are extremely easy to formulate and are as follows for oxygen:

in the leaf channels; $0 < x < \text{lth}$,

$$\frac{\partial[O_L]}{\partial t} = DO_A \frac{\partial^2[O_L]}{\partial x^2} - \frac{Q_L}{\sigma} \tag{27}$$

in the air; $\text{lth} < x < \text{lth} + \text{ath}$,

$$\frac{\partial[O_A]}{\partial t} = DO_A \frac{\partial^2[O_A]}{\partial x^2} \tag{28}$$

in the water; lth + ath $< x < B(t)$,

$$\frac{\partial[\hat{O}_W]}{\partial t} = DO_W \frac{\partial^2[\hat{O}_W]}{\partial x^2} \tag{29}$$

where DO_A and DO_W are the diffusivities of oxygen in air and water, respectively, and $B(t)$ is the thickness of the water layer in which the O_2 concentration differs from the equilibrium concentration of O_2 in the surrounding water. For rapidly stirred water or for a confined experiment there is a maximum boundary layer thickness and $B(t)$ needs to be calculated until it attains this maximum value, while for a stagnant environment the boundary layer will continue to grow for all time and $B(t)$ is always unknown.

There are similar equations for the carbon dioxide and these include a source term in the leaf given by $(\varrho_C/\varrho_O)(Q_L/\sigma)$; the factor ϱ_C/ϱ_O accommodates the relative *mass* strengths of the CO_2 source and O_2 sink terms (because equal *volumes* of O_2 and CO_2 are exchanged during respiration).

These equations are solved subject to the conditions that initially, at time $t = 0$, the O_2 and CO_2 concentrations are equilibrium values. While the leaf remains fully aerobic the spatial boundary conditions which are imposed on Eqns 27–29 are those of zero flux at $x = 0$, zero flux and equilibrium concentration at $x =$ lth + ath + $B(t)$ together with continuity in concentration and continuity in flux at the interfaces $x =$ lth and $x =$ lth + ath. The extent of any anoxia is found by imposing the condition that O_2 concentration and O_2 flux are zero at $x = \varkappa(t)$. Two particular methods of solution have been employed to solve the system of equations: first, a fully numerical algorithm which accommodates the detailed variations in concentration across the leaf/film/water, and secondly, an approximate method in which the concentration is assumed to be constant across the leaf and film and that in the water is calculated using an averaged form of the diffusion equation. The motivation for constructing the latter lies in the need to have a very rapid scheme for the lateral diffusion for use in the iterative solution of the partially submerged case discussed in the following section.

A comprehensive description of the mass transfer characteristics is beyond the scope of this chapter; here we highlight only the main conclusion concerning the differences between the flow of O_2 from the water to the film and CO_2 in the opposite direction. Figure 5 shows a typical situation for well-stirred water in which the volume transfer of rate of O_2

from the water to the film and of CO_2 in the opposite direction each increase to a maximum before decaying to give equal (volume) exchange across the film/water interface. During the period before the rates balance there is a markedly higher outflow (of CO_2) than inflow (of O_2) and in consequence the tendency for the film width to decrease; for films open to the atmosphere the effect is to draw air down into the film and create the *non-throughflow* convection mentioned earlier. If the waters are not agitated, the oxygen reserves of the plant and its gas-film reservoir are rapidly depleted, and the modelling shows that partial anoxia can be expected along the full length of the leaves. Since root aeration depends on oxygen transport via the leaf it can be deduced that much of the root system will become anoxic, and this can be confirmed experimentally. On the other hand, the model predicts that agitation of the water around the leaves will help sustain oxygen levels in the films and leaves and, we may conclude, also in the roots; the degree to which this aeration will be

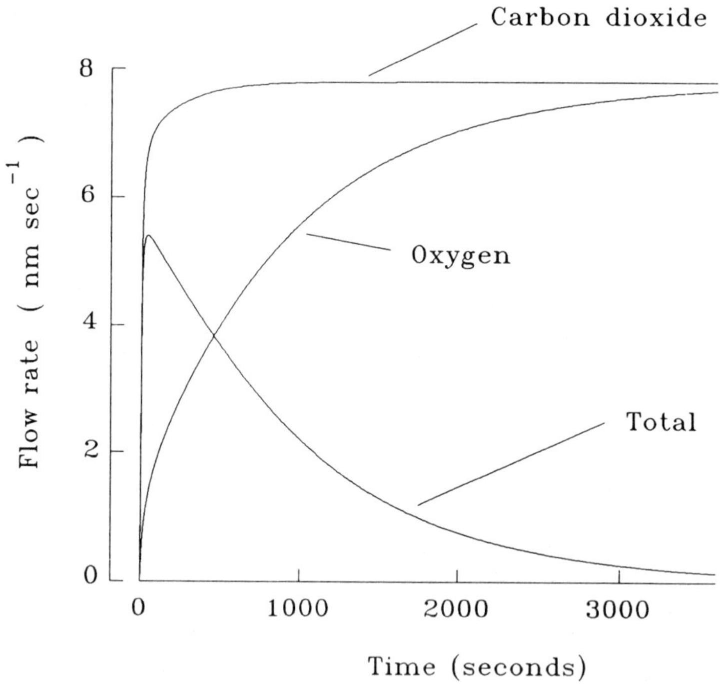

Fig. 5. Typical flow rates of O_2 from the water into the gas film, of CO_2 from the film into the water and the net flow from the film to the water for a submerged rice leaf which remains totally aerobic.

effective depends upon the degree of agitation of the waters and their oxygen content. Since such agitation will often, if not invariably occur in the field, the major advantage to be gained by the presence of surface gas films is probably the facilitation of this lateral oxygen exchange with the flood waters.

The partially submerged leaf and *non-throughflow* convection

We turn now to the modelling of diffusion and convection vertically down the film adjacent to the rice leaf. The concentration of O_2 is presumed constant across the film so the only spatial variation is longitudinal; thus we denote the concentrations of O_2 and CO_2 at a distance z below the water surface at time t by $[O_2](z,t)$ and $[CO_2](z,t)$ together with the mean velocity of gas flow down the gap by $U(z,t)$. The equations governing the O_2 and CO_2 distributions are

$$\left(1+\sigma\frac{\mathrm{lth}}{\mathrm{ath}}\right)\frac{\partial[O_2]}{\partial t}+\frac{\partial(U[O_2])}{\partial z} = D_{\mathrm{O}}\frac{\partial^2[O_2]}{\partial z^2}-Q_{\mathrm{O}}\frac{\mathrm{lth}}{\mathrm{ath}}+\frac{F_{\mathrm{O}}}{\mathrm{ath}} \tag{30}$$

$$\left(1+\sigma\frac{\mathrm{lth}}{\mathrm{ath}}\right)\frac{\partial^2[CO_2]}{\partial t}+\frac{\partial(U[CO_2])}{\partial z} = D_{\mathrm{C}}\frac{\partial^2[CO_2]}{\partial z^2}+Q_{\mathrm{C}}\frac{\mathrm{lth}}{\mathrm{ath}}-\frac{F_{\mathrm{C}}}{\mathrm{ath}} \tag{31}$$

where D_{O} and D_{C} are the diffusivities of O_2 and CO_2 in air; here we take $D_{\mathrm{O}} = 0.201\,\mathrm{cm}^2\,\mathrm{s}^{-1}$ and $D_{\mathrm{C}} = 0.159\,\mathrm{cm}^2\,\mathrm{s}^{-1}$. Strictly speaking these diffusivities are not constant or independent, and we should consider the diffusion of a three-phase gas, namely O_2, CO_2 and N_2. This is a complication which makes the mathematical solution several orders of magnitude more difficult, but since the O_2 and CO_2 concentrations together remain below 25%, each component can be regarded as diffusing independently (with constant diffusivity) relative to the bulk nitrogen; see also Beckett *et al.* (1988). The presence of σ in the first term of each equation allows for the internal (in the leaf) O_2 reservoir to be exploited, the second terms model the convection down the film, the first term on the right-hand sides accounts for diffusion, while the final terms account for the O_2 consumed in the leaf as a sink term with reciprocal CO_2 appearing as a source term, and with O_2 released from the water as a source, and CO_2 solubilised in the water as a sink effect. The quantities F_{O} and F_{C} are the fluxes of O_2 and CO_2 which are calculated in the fully submerged model with prescribed film concentrations.

The mathematical formulation is complete when we impose appropriate boundary conditions. If the O_2 concentration is positive down the film, the length of which is L, then we impose the conditions that $[O_2](0,t)$ and $[CO_2](0,t)$ have their atmospheric values and there is no diffusion at $z = L$. However, if the O_2 concentration is not positive down the entire length of the film it is necessary to find that part in which $[O_2](0,t)$ remains positive; if we denote this length by ℓ the condition at $z = L$ is replaced by the two conditions $[O_2](z,t) = 0$ and $\partial/\partial z[O_2](z,t) = 0$ when $z = \ell$.

At this point it is relevant to notice that although there is no longitudinal transport in the section $\ell < z \leqslant L$ there can still be some respiratory activity owing to O_2 diffusion from the water; only if the water is stagnant or if the O_2 boundary layer in the water is very large will the respiration be zero. The liberation rate of CO_2 in this case is not totally clear: on the one hand, it may cease in the anaerobic section $z > \ell$ or, on the other hand, because of anaerobic metabolism, CO_2 may continue to be produced at a constant rate (albeit perhaps at a lower rate than when the leaf is fully aerobic across a section). Consequently, the differential equation for CO_2 transport down the film is modified to

$$\left(1 + \sigma\frac{\text{lth}}{\text{ath}}\right)\frac{\partial[CO_2]}{\partial t} + \frac{\partial(U[CO_2])}{\partial z} = D_C\frac{\partial^2[CO_2]}{\partial z^2} + \beta Q_C\frac{\text{lth}}{\text{ath}} - \frac{F_C}{\text{ath}} \tag{32}$$

where β is the ratio of CO_2 output from the leaf when it is anoxic compared with when it is aerobic; a constant value of β corresponds to the second situation described in the previous paragraph (i.e. constant CO_2 production when $O_2 = 0$) while the former is modelled by taking β as the fraction of the maximum respiration which can be satisfied by lateral diffusion directly from the water.

Using the submerged solution to find F_O and F_C the final element of the differential equation which needs to be quantified is the convection velocity $U(z,t)$. Since there is no net volume flux into the leaf if it remains aerobic, any flow down the air gap alongside an entirely aerobic leaf is generated by the difference between O_2 diffusing from the water and CO_2 diffusing into the water. The mass of CO_2 entering the water has been denoted by $F_C(z,t)$ which corresponds to a mass sink $F_C(z,t)/\text{ath}$ within the air gap, and similarly the O_2 influx corresponds to a mass source $F_O(z,t)/\text{ath}$; thus the effective *volumetric* loss within the air gap is

$$(F_C(z,t)/\varrho_C - F_O(z,t)/\varrho_O)/\text{ath}$$

and hence the velocity $U(z,t)$ down the gap is

$$U(z,t) = \int_{z}^{L} \frac{(F_C(z,t)/\varrho_C - F_O(z,t)/\varrho_O)}{\text{ath}} \, dz \tag{33}$$

and the velocity at $z = 0$ is $U(0,t)$. If the width of the leaf is 1 cm and the diffusion is from one side only, the volume flowing into the air gap at $z = 0$ is therefore

$$\text{Total volume} = \int_{0}^{L} (F_C(z,t)/\varrho_C - F_O(z,t)/\varrho_O) dz \tag{34}$$

If the leaf is partially anoxic the CO_2 output from the leaf is not precisely equalised by O_2 take-up and a volume flux of magnitude $(L - \ell)Q_O$lth/ath is added to the air gap. Thus, the flow rate down the gap is reduced and the volume flux into the gap at $x = 0$ becomes

$$\frac{(F_C(z,t)/\varrho_C - F_O(z,t)/\varrho_O)}{\text{ath}} - \frac{(L - \ell)Q_O}{\text{ath}} \tag{35}$$

Again, a detailed description of the flow characteristics is beyond the scope of this chapter, the purpose of which is to outline the methods by which plant respiration can be modelled mathematically. It is, however, relevant to include some observations.

The temporal pattern and magnitude of *non-throughflow* convection has now been established experimentally for a number of wetland species (including rice: Beckett *et al.*, 1988), and by modelling, and it has become clear that the convection cannot sustain root aeration or root extension. In stagnant conditions the convective flow pattern is a function of plant respiratory demand and/or degree of submergence. Whatever the demand, however, the typical pattern following the partial submergence shows first an initial peak. The magnitude of this is a function of (i) the plant's maximum potential respiratory demand, (ii) the balance between the lateral loss of CO_2 and lateral gain of oxygen from the surrounding water, and (iii) the potential for diffusive O_2 and CO_2 transport down the leaf gas films. The modelling data confirm that the magnitude of the peak is approximately 0.7 times that which would be driven by the full respiratory demand potential. The peak in convection is followed by a gradual decline towards a quasi-equilibrium at *c.* 1–3 h, and if respiratory demand is low or the level of submergence such that the plant can remain fully aerobic, convection will remain at approximately this level indefinitely. If respiration is high and the degree of submergence such that aerobiosis cannot be maintained by diffusive and convective inflow down the films and lateral O_2 inflow from the waters, then the convective flow

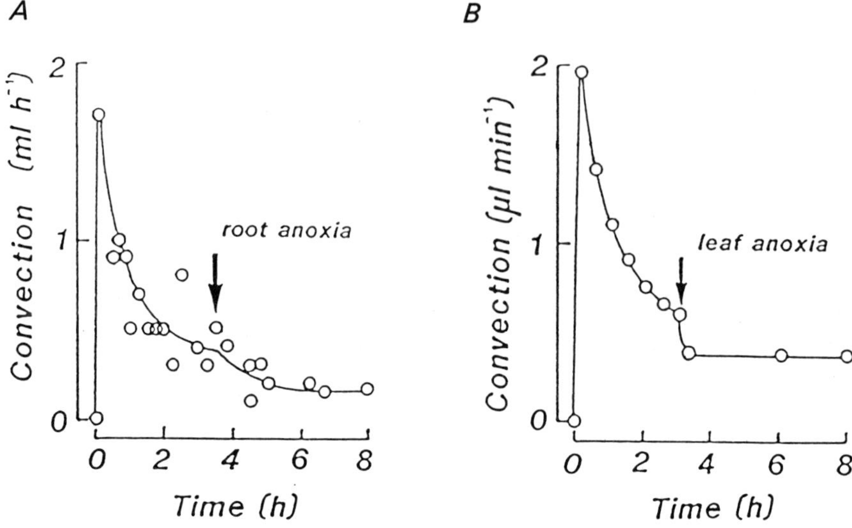

Fig. 6. *A*, Experimental example of *non-throughflow* convection into a whole rice plant after partial submergence to 500 mm compared with *B*, a modelling example of the convection in a single rice leaf submerged to 500 mm.

undergoes another rapid decline to a new plateau; the start of this decline coincides with the onset of anoxia in the distal parts of the system. The magnitude of the convection at the asymptote represents activities occurring in those parts of the shoot system which remain aerobic, their aeration supported chiefly by axial diffusion of oxygen down the gas films. Experimental and modelling examples are shown in Fig. 6; the likely changes in O_2 and CO_2 patterns along and across the leaf, film and boundary layers which give rise to this type of flow pattern can also be found (P. M. Beckett & W. Armstrong, unpublished data). Also, since the convective inflow is of air (21% O_2), it has been possible to deduce that convection even at its peak could provide only 0.147 of the plant's total oxygen requirements, while later at the first and second plateaux, its potential for providing aeration may be much less: approximately 0.088 and 0.052, respectively. It should be noted, however, that the effects of diffusion and *non-throughflow* convection down the gas films are not simply additive (see Table 2), and it can be shown that in the absence of convection, diffusion can, in theory, aerate a path almost as long.

If, on the other hand, the waters are constantly streamed past the shoot system, and boundary layer resistances to gas exchange with the water thus reduced, then even at relatively low flow rates, and even with full submergence, root aerobiosis and extension may be rapidly restored (S. Justin, W. Armstrong & P. M. Beckett, unpublished data).

Phragmites australis

The aerial shoots (culms) of the common reed *Phragmites australis* arise at regular intervals from an underground rhizome which is often several metres in length and up to 20 mm in diameter. The rhizome usually lies buried in anoxic mud at anything up to 1 m below the surface, while the aerial shoots themselves may also be partially submerged. Adventitious roots are borne at the rhizome nodes and on the buried or submerged basal nodes of the aerial shoots. The culms and rhizome are permeated by an extensive interconnecting gas-space system in which porosity may be as high as 60%. It has recently become evident that, in addition to diffusion, aeration in *Phragmites* will be influenced by several types of convective gas flow (Armstrong & Armstrong, 1989, 1990, 1991; Armstrong *et al.*, 1991). These are (i) **humidity-induced convection** (which can be thermally enhanced), (ii) **thermal transpiration** (Reynolds, 1879), (iii) **Venturi-induced convection** and (iv) the ever present *non-throughflow* convection. Of these, the humidity- and Venturi-induced convections are almost certainly the most important.

The *humidity-induced* convection results from a diffusive influx of atmospheric gases via stomata on living leaf sheaths and culm nodes. The constant humidification of the plant atmosphere (2–3% by volume) helps to create and maintain water vapour levels, and at the same time produces a concentration gradient for the inward diffusion of the atmospheric gases from the drier external atmosphere. Since there is also a significant stomatal resistance to Poiseuille flow from leaf sheath to atmosphere, there is a tendency for the plant to pressurise as the atmospheric gases diffuse in. The incoming gases are thus convectively forced along the path of least resistance, i.e. from the shoots into the underground rhizome, and are vented back to the atmosphere via the persistent dead and broken culms. Plug-flow velocities of up to 800 mm min^{-1} for single shoots have been measured under field conditions. *Venturi-induced* flows arise because of the suction pressures developed when winds blow across the ends of previous seasons' persistent tall but snapped flowering shoots. The suction draws air into the rhizomes via other old shoots broken close to ground level.

Knowing the velocity through the rhizome we can assess the likely effectiveness of these convective flows by using the mathematical analysis presented above for the case of *throughflow* convection. By assuming that the concentration is maintained as C_0 (e.g. atmospheric) at $x = 0$ (the inflow point to the rhizome), and that there is no back diffusion into the section at $x = L$ (the outflow point), i.e. $dC/dx = 0$ at $x = L$, the concentration at any distance x along the rhizome from the inflow point is given by Eqn 23 and hence we can deduce that the concentration at $x = L$ is given by

$$C(L) = C_0 - QL/U + QD/U^2\{1 - \exp(-UL/D)\} \qquad (36)$$

It should be noted that this is only applicable if $C(L) > 0$. It is possible to prescribe a combination of parameters which yields a negative value for $C(L)$, but in such cases the rhizome would not be fully aerobic and the basis on which the formula has been constructed is no longer valid. It should also be noted that D is the effective diffusion coefficient (i.e. $D_0\varepsilon$, where ε is the fractional porosity and it is assumed that there is no tortuosity in the diffusion path), and Q is the respiratory demand. It is assumed for convenience that Q and D are effectively uniform along the length of the rhizome.

Several types of prediction concerning the aeration of *Phragmites* can be made by the appropriate use of this equation. For example, the relative importance of diffusion and convection in terms of supportable aerobic rhizome length can be estimated by solving for L for a range of Q, D and U. Alternatively, for a fixed length of rhizome, the relationship between convection rate and the degree of rhizome aeration in terms of achievable oxygen concentrations can be computed. Also, by appropriately varying the value of Q, it is possible to predict, in a simplified way, the likely effects of root and rhizosphere oxygen demands.

We have dealt elsewhere with the influence of diffusion and convection on aerated path lengths (Table 1), and our primary concern here is with the effects of convection on rhizome oxygen concentrations, and in particular to show to what extent an increased sink strength in the form of root and rhizosphere oxygen demand may influence the relationship between convective flow rate and the rhizome oxygen concentration $C(L)$. To facilitate comparison between the mathematical modelling and laboratory experiments, the values assigned to the constants C_0 and D were those which pertain at room temperature (20 °C).

The extent to which convection can be expected to raise oxygen concentrations in a rhizome of fixed length is shown in Fig. 7, where oxygen concentrations in the gas venting from the rhizome (i.e. concentration $C(L)$: Eqn 36), are plotted against the rate of convection. The length of active rhizome was fixed at 500 mm, the respiratory oxygen demand of the rhizome itself was set at 7.5 ng cm^{-3} s^{-1} in accordance with experimental measurements of rhizome oxygen demand made in the laboratory, porosity was taken as 60% ($\varepsilon = 0.6$) (Armstrong *et al.*, 1988) and rhizome diameter was assumed to be 6.45 mm. Curve A represents the condition in which the only oxygen demand is that of rhizome respiration; curve B includes, in addition, the respiratory demand of 10 adventitious roots each bearing 600 laterals; for curve C a respiratory demand of 50 ng cm^{-3} s^{-1} was imposed in the rhizosphere of these roots. Curve D represents the case in which the oxygen demand of the rhizosphere is 500 ng cm^{-3} s^{-1}. The oxygen demands of roots and rhizosphere were determined by substituting appropriate values for porosity, oxygen demand and root geometry into the multicylindrical model of Armstrong & Beckett (1987).

The modelling predictions in Fig. 7 are in close agreement with the experimental observations (Armstrong *et al.*, 1990*a*, 1991): in all cases rhizome aeration was greatly improved by relatively low rates of convection. In Fig. 7 (curve A), where the oxygen demand is from rhizome respiration only, a convective flow of as little as 5 mm min^{-1} was sufficient to raise the rhizome venting oxygen concentration $C(L)$ to 90% of its potential maximum. With the added respiratory demand of 10 roots, the same degree of aeration is achieved with a convective flow of *c.* 20 mm min^{-1} (curve B). To satisfy in addition the lower of the two rhizosphere oxygen demands requires a rather larger increase in flow velocity, to 47 mm min^{-1} (curve C). To accommodate the further 10-fold increase in rhizosphere oxygen demands appears to require a relatively small further increase in flow to 60 mm min^{-1}. Although the rhizome could cope exactly as shown with an internal oxygen demand of 56 ng cm^{-3} s^{-1}, other modelling data have revealed that if the hypothetical root system were unchanged in its geometry and potential activity, the 10-fold increase in rhizosphere demand would probably cause some degree of root anoxia, particularly in the apices of the lateral roots (Armstrong *et al.*, 1990*a*). In other words, no matter how rapid the convection might be through the rhizome, it might not be sufficient to support in full the potential oxygen demands of root and rhizosphere. However, it would be possible to support the same total demand if this were to be spread over more and shorter roots.

In comparison with the above it is encouraging to note that in experiments a humidity-induced convection of *c.* 40 mm min^{-1} was sufficient to raise the venting oxygen concentration to 90% of its presumed maximum, while for a Venturi-experiment the 90% figure was achieved at a convective velocity of 20 mm min^{-1}.

The results in Fig. 7 also reveal some of the limitations of diffusion for effecting aeration in *Phragmites*. For example, diffusion *per se* would not support fully the aeration of rhizome and roots in the presence of even the lower of the two soil oxygen sink strengths (curve C at convective flow = 0). This further emphasises the potential value of convection.

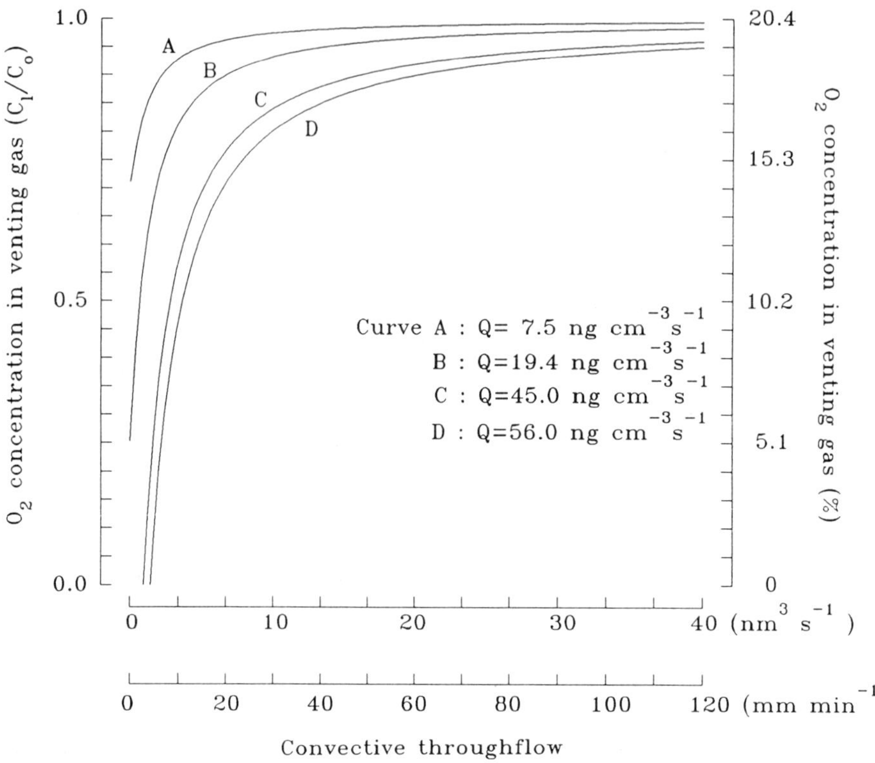

Fig. 7. Modelling predictions which assess the effectiveness of *throughflow* convection rate and respiratory demands on the O_2 concentrations in the gases vented from a rhizome.

Curves C and D also reveal the need for a threshold rate of convection to support fully rhizome aeration when the roots are subjected to moderate or high soil oxygen demand. However, it should be noted that, whereas rhizome oxygen demands *per se* are fully met above these threshold convection rates (*c.* 2.5–5 mm min^{-1}, curves C–D), adequate root aeration may still require faster rates than this, while oxygen losses to the rhizosphere will always continue to increase as a function of rhizome oxygen concentration.

Conclusions

We have established that mathematical modelling can play an important role in developing the conceptual framework for the understanding of root aeration. In particular we have seen how the fundamental equation, which governs simultaneous convection and diffusion, can accommodate the non-homogeneities that characterise roots, can be used to assess anoxic development in both simple and complex geometries and also lead to a quantitative assessment of the roles played by convection and diffusion. The usefulness of mathematical modelling cannot be overstressed; having constructed a reliable solution which accommodates the essential physics, the freedom to introduce a wide range of parameter states into the theoretical model enables the researcher to gain a wide qualitative insight into the behaviour of the system without the need for time-consuming and expensive additional experimentation.

Clearly, there is enormous potential for further model development; the subject is perhaps still only in its infancy. So far as future development of the multicylindrical models is concerned, there is some scope to investigate the declining sub-apical oxygen permeability to radial oxygen entry and loss in roots, and for exploring ways of incorporating simultaneous diffusion or convection of any chemical reductants into the rhizosphere (Ahmad & Nye, 1990; Kirk *et al.*, 1990). Furthermore, we anticipate that extensions of the time-dependent studies could lead to a better appreciation of the role of haemoglobin in roots and will be helpful in exploring the diurnally fluctuating dimensions of the rhizosphere which must be a feature of photosynthetic effects in partially or wholly submerged plants.

References

Ahmad, A.R. & Nye, P.H. (1990). Coupled diffusion and oxidation of ferrous iron in soils. I. Kinetics of oxygenation of ferrous iron in soil suspension. *Journal of Soil Science* **41**, 395–410.

Appleby, C.A., Bogusz, D., Dennis, E.S. & Peacock, W.J. (1988). A role for haemoglobin in all plant roots. *Plant, Cell and Environment* **11**, 359–67.

Armstrong, J. & Armstrong, W. (1989). Light-enhanced convective throughflow increases oxygenation in rhizomes and rhizosphere of *Phragmites australis* (Cav.) Trin. ex Steud. *New Phytologist* **114**, 121–8.

Armstrong, J. & Armstrong, W. (1990). Pathways and mechanisms of oxygen transport in *Phragmites australis*. In *The Use of Constructed Wetlands in Water Pollution Control*, ed. P.F. Cooper & B.C. Findlater, pp. 529–43. Oxford: Pergamon Press.

Armstrong, J. & Armstrong, W. (1991). A convective throughflow of gases in *Phragmites australis*. *Aquatic Botany* **39**, 75–88.

Armstrong, J., Armstrong, W. & Beckett, P.M. (1988). *Phragmites australis*: a critical appraisal of the ventilating pressure concept and an analysis of resistance to pressurized gas-flow and gaseous diffusion in the horizontal rhizomes. *New Phytologist* **110**, 383–9.

Armstrong, J., Armstrong, W. & Beckett, P.M. (1992). *Phragmites australis*: Venturi- and humidity-induced convections enhance rhizome aeration and rhizosphere oxidation. *New Phytologist* **120**, 197–207.

Armstrong, W. (1970). Rhizosphere oxidation in rice and other species; a mathematical model based on the oxygen flux component. *Physiologia Plantarum* **23**, 623–30.

Armstrong, W. (1979). Aeration in higher plants. In *Advances in Botanical Research*, Vol. 7, ed. H.W. Woolhouse, pp. 226–332. London: Academic Press.

Armstrong, W. (1982). Waterlogged soils. In *Environment and Plant Ecology*, ed. J.R. Etherington, pp. 290–330. Chichester: John Wiley.

Armstrong, W. & Beckett, P.M. (1985). Root aeration in unsaturated soil: a multi-shelled model of oxygen distribution and diffusion with and without sectoral blocking of the diffusion path. *New Phytologist* **100**, 293–311.

Armstrong, W. & Beckett, P.M. (1987). Internal aeration and the development of stelar anoxia in submerged roots. A multishelled mathematical model combining axial diffusion of oxygen in the cortex with radial losses to the stele, the wall layer and the rhizosphere. *New Phytologist* **105**, 221–45.

Armstrong, W., Armstrong, J. & Beckett, P.M. (1990*a*). Measurement and modelling of oxygen release from roots of *Phragmites australis*. In *The Use of Constructed Wetlands in Water Pollution Control*, ed. P.F. Cooper & B.C. Findlater, pp. 41–52. Oxford: Pergamon Press.

Armstrong, W., Armstrong, J., Beckett, P.M. & Justin, S.H.F.W. (1990*b*). Convective gas-flows in wetland plant aeration. In *Plant Life under Oxygen Deprivation*, ed. M.B. Jackson, D.D. Davies & H. Lambers, pp. 283–302. The Hague: SPB Academic Publishing hv.

Armstrong, W., Beckett, P.M., Justin, S.H.F.W. & Lythe, S. (1990*c*). Modelling and other aspects of root aeration. In *Plant Life under Oxygen Deprivation*, ed. M.B. Jackson, D.D. Davies & H. Lambers, pp. 267–82. The Hague: SPB Academic Publishing hv.

Armstrong, W., Justin, S.H.F.W., Beckett, P.M. & Lythe, S. (1991). Root adaptation to soil waterlogging. *Aquatic Botany* **39**, 57–74.

Atwell, B.J., Drew, M.C. & Jackson, M.B. (1988). The influence of oxygen deficiency on ethylene synthesis, 1-aminocyclopropane-1-carboxylic acid levels and aerenchyma formation in *Zea mays*. *Physiologia Plantarum* **72**, 15–22.

Beckett, P.M., Armstrong, W., Justin, S.H.F.W. & Armstrong, J. (1988). On the relative importance of convective and diffusive gas flows in plant aeration. *New Phytologist* **110**, 463–8.

Bouldin, D.R. (1968). Model for describing the diffusion of oxygen and other mobile constituents across the mud–water interface. *Journal of Ecology* **56**, 77–87.

Bowes, G. (1987). Aquatic plant photosynthesis: strategies that enhance carbon gain. In *Amphibious and Intertidal Plants.* British Ecological Society Special Symposium 5, ed. R.M.M. Crawford, pp. 79–98. Oxford: Blackwells.

Brandle, R. (1990). Flooding resistance of rhizomatous plants. In *Plant Life under Oxygen Deprivation*, ed. M.B. Jackson, D.D. Davies & H. Lambers, pp. 35–46. The Hague: SPB Academic Publishing hv.

Brandle, R. & Crawford, R.M.M. (1987). Rhizome anoxia tolerance and habitat specialization in wetland plants. In *Plant Life in Aquatic and Amphibious Habitats*, Special Publication Series of the British Ecological Society, No. 5, ed. R.M.M. Crawford, pp. 397–410. Oxford: Blackwell Scientific Publications.

Currie, J.A. (1961). Diffusion within soil microstructure: a structural parameter for soils. *Journal of Soil Science* **16**, 279–89.

Currie, J.A. (1962). The importance of aeration in providing the right conditions for plant growth. *Journal of the Science of Food and Agriculture* **13**, 380–5.

Dacey, J.W.H. (1981). Pressurised ventilation in the yellow water-lily. *Ecology* **62**, 1137–47.

Dakora, F.D. & Atkins, C.A. (1989). Diffusion of oxygen in relation to structure and function in legume root nodules. *Australian Journal of Plant Physiology* **16**, 131–40.

De Willegan, P. & Van Noordwijk, M. (1984). Mathematical models on diffusion of oxygen to and within plant roots, with special emphasis of effects of soil–root contact. I. Derivation of the models. *Plant and Soil* **77**, 215–31.

De Willegan, P. & Van Noordwijk, M. (1989). Model calculations on the relative importance of internal longitudinal diffusion for aeration of roots of non-wetland plants. *Plant and Soil* **113**, 111–19.

Drew, M.C. (1991). Sensing soil oxygen. *Plant, Cell and Environment* **13**, 681–93.

Glinski, J. & Stepniewski, W. (1985). *Soil Aeration and its Role for Plants.* Boca Raton, FL: CRC Press.

Greenwood, D.J. (1967). Studies on oxygen transport through mustard seedlings. *New Phytologist* **66**, 597–606.

Greenwood, D.J. & Goodman, D. (1967). Direct measurement of the distribution of oxygen in soil aggregates and in columns of fine soil crumbs. *Journal of Soil Science* **18**, 182–96.

Große, W., Büchel, H.B. & Tiebel, H. (1991). Pressurized ventilation in wetland plants. *Aquatic Botany* **39**, 89–98.

Hansen, J.I. & Andersen, F.O. (1981). The effects of *Phragmites australis* roots on redox potentials, nitrification and bacterial numbers in the rhizosphere. In *9th Nordic Symposium on Sediments*, ed. A. Broberg & T. Tiren, pp. 72–88.

Hofmann, K. (1990). Use of *Phragmites australis* in sewage sludge treatment. In *The Use of Constructed Wetlands in Water Pollution Control*, ed. P.F. Cooper & B.C. Findlater, pp. 269–78. Oxford: Pergamon Press.

Hunt, S, Gaito, S.T. & Layzell, D.B. (1988). Model of gas-exchange and diffusion in legume nodule. II. Characterization of the diffusion barrier and estimation of the concentrations of CO_2, H_2 and N_2 in the infected cells. *Planta* **173**, 128–41.

Jackson, M.B., Fenning, T.M., Drew, M.C. & Saker, L.R. (1985). Stimulation of ethylene production and gas-space (aerenchyma) formation in adventitious roots of *Zea mays* L. by small partial pressures of oxygen. *Planta* **165**, 486–92.

Justin, S.H.F.W. & Armstrong, W. (1987). The anatomical characteristics of roots, and plant response to soil flooding. *New Phytologist* **106**, 465–95.

Justin, S.H.F.W. & Armstrong, W. (1991*a*). Evidence for the involvement of ethene in aerenthyma formation in adventitious roots of rice (*Oryza sativa* L.). *New Phytologist* **118**, 49–62.

Justin, S.H.F.W. & Armstrong, W. (1991*b*). A reassessment of the influence of NAA on aerenchyma formation in maize roots. *New Phytologist* **117**, 607–18.

Kirk, G.J.D., Ahmad, A.R. & Nye, P.H. (1990). Coupled diffusion and oxidation of ferrous iron in soils. II. A model of the diffusion and reaction of O_2, Fe^{2+}, H^+ and HCO_3^- in soils and a sensitivity analysis of the model. *Journal of Soil Science* **41**, 411–32.

Končalová, H., Pokorny, J. & Květ, J. (1988). Root ventilation in *Carex gracilis* Curt.: diffusion or mass flow? *Aquatic Botany* **30**, 149–55.

Kristensen, K.J. & Lemon, E.R. (1961). Soil aeration and plant root relations. III. Physical aspects of oxygen diffusion in the liquid phase of the soil. *Agronomy Journal* **56**, 295–301.

Kreuzer, F. (1982). Oxygen supply to tissues: the Krogh model and its assumptions. *Experientia* **38**, 1415–26.

Laan, P., Smolders, A. & Blom, C.W.P.M. (1991). The relative importance of anaerobiosis and high iron levels in the flood tolerance of *Rumex* species. *Plant and Soil* **136**, 153–6.

Lemon, E.R. (1962*a*). Soil aeration and plant root relations. I. Theory. *Agronomy Journal* **54**, 167–70.

Lemon, E.R. (1962*b*). Soil aeration and plant root relations. II. Root respiration. *Agronomy Journal* **54**, 171–5.

Leuning, R. (1983). Transport of gases into leaves. *Plant, Cell and Environment* **6**, 181–94.

Luxmoore, R.J., Stolzy, L.H. & Letey, J. (1970). Oxygen diffusion in the soil–plant system. *Agronomy Journal* **62**, 317–22.

Ponnamperuma, F.N. (1984). The effects of flooding on soils. In *Flooding and Plant Growth*, ed. T.T. Kozlowski, pp. 10–46. New York: Academic Press.

Raskin, I. & Kende, H. (1983). How does deep-water rice solve its aeration problem? *Plant Physiology* **72**, 447–54.

Raskin, I. & Kende, H. (1985). Mechanism of aeration in rice. *Science* **228**, 327–9.

Reynolds, O. (1879). On certain dimensional properties of matter in the gaseous state. *Philosophical Transactions* **170**, 727–845.

Robe, W.E. & Griffiths, H. (1990). Photosynthesis of *Littorella uniflora* grown under two PAR regimes: C3 and CAM gas exchange and the regulation of internal CO_2 and O_2 concentrations. *Oecologia* **85**, 128–36.

Sheehy, J.E., Minchin, F.R. & Witty, J.F. (1985). Control of nitrogen fixation in a legume nodule: an analysis of the role of oxygen diffusion in relation to nodule structure. *Annals of Botany* **55**, 549–62.

Smith, K.A. (1980). A model of the extent of the anaerobic zones in aggregated soils and its potential application to estimates of denitrification. *Journal of Soil Science* **31**, 263–77.

Sorrel, B.K. (1991). Transient pressure gradients in the lacunar system of the submerged macrophyte *Egeria densa* Planch. *Aquatic Botany* **39**, 99–108.

Thomson, C.J. & Greenway, H. (1991). Metabolic evidence for stelar anoxia in maize roots exposed to low oxygen concentrations. *Plant Physiology* **96**, 1294–1301.

Van Noordwijk, M. & De Willegan, P. (1984). Mathematical models on diffusion of oxygen to and within plant roots, with special emphasis on effects of soil–root contact. II Applications. *Plant and Soil* **77**, 233–41.

Index